ADOLESCENT LIVES 1

A series edited by
Jeanne Brooks-Gunn

THE RELATIONSHIP CODE

Deciphering Genetic and

Social Influences

on Adolescent Development

DAVID REISS

with

JENAE M. NEIDERHISER

E. MAVIS HETHERINGTON

ROBERT PLOMIN

HARVARD UNIVERSITY PRESS

Cambridge, Massachusetts, and London, England

2000

Library of Congress Cataloging-in-Publication Data

Reiss, David, 1937–
    The relationship code : deciphering genetic and social influences
on adolescent development / David Reiss, with Jenae M. Neiderhiser,
E. Mavis Hetherington, Robert Plomin.
        p.   cm. — (Adolescent lives ; 1)
    Includes bibliographical references and index.
    ISBN 0-674-00054-4 (alk. paper)
        1. Adolescent psychology.   I. Neiderhiser, Jenae M.
II. Hetherington, E. Mavis (Eileen Mavis), 1926–   .   III. Plomin,
Robert, 1948–   .   IV. Title.   V. Series.
BF724.R39   1999
155.5—dc21   99-38680

For Joy Schulterbrandt,

a knowing and supportive friend of family research

Contents

Foreword   ix

Preface   xi

Reader's Guide   xvii

1   Introduction: Reconciling Social and Genetic Influences
on Adolescent Development   1

PART 1
LOGICAL TOOLS FOR ANALYZING
ADOLESCENT DEVELOPMENT   11

2   Relationships and Adolescent Development   13

3   Genetic Influences on Development   44

4   Genetic Analysis of Adolescent Development   69

5   Studying Adolescent Siblings and Their Families   103

PART 2
GENES AND RELATIONSHIPS:
THESIS, ANTITHESIS, SYNTHESIS   145

6   Thesis I: A Theory of Adolescents' Shared and Nonshared
Family Relationships   147

7   Thesis II: Major Findings on Adolescents' Family
Relationships   168

8   Antithesis I: Influences on Stability and Change
in Adolescent Adjustment   206

9   Antithesis II: Influences on Stability and Change
in Adolescents' Families   243

10 Antithesis III: Linking Family Relationships and Adolescent
Development 269

11 Synthesis I: Genetic Influences on Change in Family
Relationships and Adolescent Development 309

12 Synthesis II: The Relationship Code 343

13 Synthesis III: Genetically Informed Portrayals of
Adolescents and Their Families 385

14 Epilogue: The Family 417

Appendix A: Explanation of Methods for Data Presented
in Chapters 8 through 13 429

Appendix B: Explanation of Results 451

Appendix C: Additional Genetic Analyses 481

Glossary 483

References 489

Index of Tables and Figures 511

General Index 519

It is a pleasure to introduce Harvard University Press's new series, entitled Adolescent Lives, with *The Relationship Code* by David Reiss and his colleagues Jenae Neiderhiser, Mavis Hetherington, and Robert Plomin. I could not imagine a better beginning for the series. This inaugural book captures the spirit of scholarship on the development of lives through time, in this case a unique and rather lengthy time, the second decade of life.

The Relationship Code also provides a blueprint for what the next two decades of developmental study might bring—more nuanced portrayals of how biological, psychological, and social processes contribute to the unfolding of lives. I would add economic, anthropologic, and political-scientific processes as well, even though this book does not address these fields. The goal of this volume, and the series more generally, is to bring together the separate worlds that make up adolescent lives, as David Reiss says, rather than relying on somewhat disparate and often immixable views of development.

My belief is that path-breaking studies of the next two decades, like *The Relationship Code,* will systematically break down the barriers between disciplinary views of development. In their place will be offered a more synthetic framework, one that recognizes and celebrates the interconnections between biology and psychology, and in addition situates this intersection within a social, economic, and historical context. *The Relationship Code* is a prototype of such integrative foci. At the same time, this book is not atheoretical; nor is it so integrative that the differences between behavioral geneticists and family sociologists and psychologists are ignored. David Reiss and his colleagues sagely articulate the areas of convergence and divergence among scholars with varying perspectives. They also present their arguments as a thesis, an antithesis, and a synthesis—a brilliant method of crystallizing and testing their premises about the ways in which behavioral genetics influences family interactions and families are experienced. It is my hope that other scholars will adopt the convention of specifying their hypotheses as well as the counterfactuals to their arguments, rather than presenting all-encompassing, difficult-to-test theories of human development. The Adolescent Lives series is based on the premise that lively debate and careful scholarship go hand-in-hand.

The focus on adolescence is deliberate. The second decade of life is characterized by changes in physiology, appearance, internal mood states, identity, relationships, and ultimately residence, means of financial support, and life-style choices. The ways in which adolescents are treated, perceived, and provided with opportunities offer a window on a particular society in a specific historical period. The authors of this volume chose to focus on adolescence because it is a time when the contribution of genetic factors as well as nonshared environmental factors may increase or at least be altered. Additionally, during adolescence sibling relationships may be studied not only by observation but also by direct report.

Aspects of the adolescent experience perplex us. Even reconstructing our own youths and recalling our relationships with parents, peers, and siblings does not seem to lessen adult ambivalence and confusion about this extraordinary period of development. A recent national survey has indicated that about two-thirds of all adults in the United States perceive teenagers very negatively and vastly overestimate the percentage who have difficulties in school and at home. Many scholars wish to temper these notions with more realistic (and sometimes highly complex) snapshots of youth. Ultimately, when we look long and hard and carefully enough, we find what we should have known all along: that teens are as diverse and multifaceted as everybody else.

Jeanne Brooks-Gunn
Virginia and Leonard Marx Professor
Teachers College, Columbia University
November 1999

I wrote this book and prepared its graphics to describe the results of an extraordinary collaborative project. Thirteen years ago Mavis Hethering-ton, Robert Plomin, and I set out to merge two perspectives on psychological development: behavioral genetics and family process. We sought to explore a phenomenon that had great relevance for both perspectives: the distinctive differences between siblings in the same family. Robert was just then preparing his very influential review of startling data from behavioral genetics studies. These findings confirmed that siblings in the same family are quite different in personality, cognitive abilities, and psychopathology. Further, environmental factors played a major role in these differences. Even more significantly, these environmental factors shaping differences in siblings in the same family were much more influential than those environmental factors responsible for similarities between siblings. These distinctive environmental factors are now referred to collectively as the "nonshared environment." Behavioral genetics had provided an estimate of the importance of these environmental factors but few clues as to what they might be. They might be physical or social, prenatal or postnatal, random or systematic.

Mavis, Robert, and I thought they might be siblings' differential experience of their current social environment, particularly in their families. That is, even though siblings grow up in the same family and social community, they may experience them differently. We planned a study to explore that idea. We thought of adolescence as a cauldron of social experience that probably left an enduring mark on each youngster's psychological development. Thus, discovering important nonshared environmental influences during this period might provide clues to the origins of important differences among adolescents and adults. We concentrated on the adolescents' household families but also assessed their experience of their extended family, peers, and teachers. In this work we drew heavily on Mavis's long experience in studying the family as a social system. She and her team had developed many measures of the important family subsystems—marital, parent-child, and sibling—as well as approaches to assessing both the successful and the problematic aspects of psychological development in children and adolescents.

The most unique feature of our work was our use of siblings who varied in genetic relatedness. We used not only identical and fraternal twins but also, in response to an innovative idea of Robert's, full sibs who had not experienced a divorce, as well as full, half-, and genetically unrelated siblings in step-families. Mavis's extensive experience in studying step-families was essential to fleshing out Robert's idea. Our study design provided information on genetic as well as social factors in adolescent development. We also hoped that by extending our genetic design beyond the traditional use of twins, our findings might be more generalizable, although, as many parts of the book point out, we are still not certain about this.

Thirteen years after beginning this unusual venture, we have concluded that the household family is not an important source of nonshared environment for adolescents. Further, our preliminary assessments of other social worlds of the adolescent have provided few, if any, clues as to what the main source might be. Instead, we have encountered a set of striking findings relating to genetic influences. Indeed, the strongest clues emerging from our work may reflect mechanisms of gene expression and a possible role for family process in these mechanisms that we could not have dreamed of when we began our study.

This book describes our major findings and the significance they may have for psychological development. Robert and Mavis graciously supported my role in writing up these findings. They carefully read and reviewed each chapter and made countless comments and corrections that have strengthened every one. When I began this book more than four years ago, Jenae Neiderhiser had just joined our research group at George Washington University after completing graduate studies under Robert's direction at Penn State and serving as a central member of the Penn State project team, a major component of our multisite collaboration. We invited Jenae to join our efforts to write this book.

When Jenae and I reviewed the scores of publications from our project, we decided against a simple tactic: depending heavily on reusing already-published data to tell our story. Rather, we chose a more difficult strategy, opting to redo all the analyses using the simplest and smallest variety of analytic models that would fit the complexity of all the data. Jenae's leadership of the analytic team supporting this book provided a uniform framework for all the analyses readers will encounter here. Jenae supervised this entire program of data analysis and prepared the appendixes.

The data we analyzed for this book reflect the work of four closely linked research units at four separate university sites. George Washington University, under my direction, was the main site. There, George Howe served as project coordinator for our work in earlier adolescence, and Danielle Bussell organized our work in later adolescence. Each, in turn, had the task of harmonizing the work among all other units. Other critical members of the George Washington unit were Sam Simmens, Victoria Wegener, Katherine Matsey, Jeannette Nearing Steward, David Leidner, Erica Spotts, Mignon Murray, Judy Piemme, Melody Millando, and Elizabeth Carroll.

The University of Virginia unit, under Mavis's direction, developed many of the measures used in this study and performed all the videotape coding, perhaps the most ambitious effort at coding of observed family interaction ever undertaken. All the work at the Virginia site was coordinated by Sandra Henderson. Tracey Law headed the coding team. Other central members of the Virginia team were Ed Anderson and Tom O'Connor. In addition, fourteen graduate students and twenty-eight undergraduates helped to refine the coding procedure and perform the coding.

The Penn State unit, under Robert's direction, played the central role in data preparation and data analysis. Rarely has any data-analysis group faced the challenges of this project: filing, cleaning, and preparing data from more than seven hundred four-member families studied on two occasions, and utilizing data drawn from children's self-reports, parents' self-reports, teacher reports, and coded video data. Central members of the Penn State unit included Alison Pike, Shirley McGuire, Beth Manke,

Richard Rende, Heather Chipuer, Michael Rovine, Elana Pyle, and Sylvia Vignetti. Many other graduate students and postdoctoral fellows also provided crucial assistance in the project.

The National Opinion Research Center (NORC) of the University of Chicago collected all the data, using more than fifty interview teams in forty-seven states. NORC achieved excellent levels of cooperation from the families, and the data they collected through interviews, questionnaires, and videotapes were of unusually high quality. This study reflects the first effort to collect and code videotaped data from a nationally distributed sample of families. NORC's experience with collecting data nationwide, using novel data-collection techniques, and their experience in enlisting people for research and interviewing them were essential. The NORC team was headed by Alisu Schoua-Glusberg and Ann Cederlund. The success of NORC's work reflected, as well, the dedication of its interview teams.

Over the four and a half years of its evolution, the manuscript for this book received helpful readings and critiques from Lyman Wynne, George Howe, David Mrazek, Virginia Colin, Elizabeth Knoll, Jo Ann Reiss, and thoughtful and constructive anonymous reviewers for Harvard University Press. Jerry Weiner, the former chair of George Washington's Department of Psychiatry, provided me with valuable leave time to write the book.

The project was supported by two large grants from the National Institute of Mental Health (MH 43373 and 48825); Della Hann, as project officer during much of the period of NIMH support, provided important help at many junctures. Our work was also supported by the William T. Grant Foundation, allowing us to add twins to our sample.

Joy Schulterbrandt was the initial NIMH project officer. Indeed, long before this work was funded, she recognized the potential of this project. She arranged a conference on the nonshared environment at the Center for Advanced Studies in the Behavioral Sciences in Palo Alto; the conference convinced Robert, Mavis, and me that this study was essential. Joy encouraged us to plan this project and came to George Washington for the very first meeting of our fledgling research team. This book is dedicated, with gratitude, to her, not only for her assistance to this project, but also for her visionary support of family research during her distinguished career at NIMH.

The completion of this book is also a tribute to an extraordinary synergy of talent, enthusiasm, and dedication, along with grueling and meticulous work by our large team. It has been the privilege of a lifetime for me to be a part of this effort.

David Reiss
Washington, D.C.
January 1, 1999

Reader's Guide

This book was written for a wide audience of people interested in children and families. At the same time, it is the only full summary of the findings of a major, complex study of psychological development. I wanted the book to provide enough information to researchers about the study to serve as a useful reference for further research on psychological development. More important, I wanted all readers, whatever their professional background, to feel entirely comfortable exploring this book: the rationale of our study, its observational and statistical methods, its findings, and its hypotheses. Toward these ends I provide at the outset a quick guide to the book's major features, which were designed to make our findings fully accessible.

First, the book tells a story in a familiar format. It states a thesis, then an antithesis, and finally a synthesis. This format is designed to emphasize the paradoxes not only in the data from our study but also in the field of psychological development more generally at the close of the millennium. It is also designed to emphasize the importance of hypotheses and speculation as major tools for resolving paradoxes in science.

Second, the first four chapters introduce some of the central assumptions in both psychosocial and genetic research on psychological development. These are intended, in part, to make sure all readers are fully prepared for the description of our study and for the analyses of its results. Indeed, experienced behavioral scientists may be tempted to skip these chapters altogether. But they are not primers; they take a distinctive stand. They show the parallels in the logic of psychosocial and genetic analyses and carefully prepare each reader for our synthesis, which combines both perspectives.

Third, from Chapter 6 onward, the summaries that begin each chapter emphasize the chapter's role in the thesis, antithesis, or synthesis. Fourth, all statistical procedures are fully explained in plain English with many examples of how we go from observations to results of quantitative analyses. This book does not presume the reader has any training in statistics. Further, all the quantitative findings are presented in graphs, not in tables. The visual format is used to emphasize patterns of findings and to help the reader see important similarities and differences of findings across the entire book. Statistically minded readers will find appendixes that provide a technical account of our analyses and tables that provide a more detailed report of our findings. An index of all tables and figures precedes the General Index.

Finally, at the end of the book readers will find a glossary that defines terms commonly used throughout the book.

D. R.

Chapter 1

INTRODUCTION: RECONCILING SOCIAL
AND GENETIC INFLUENCES
ON ADOLESCENT DEVELOPMENT

Until now, psychosocial and biological studies of human psychological development have been conducted in separate worlds. The major research findings from each of these domains present a paradox to all who wish to understand development. On the one hand, persuasive studies argue for the primacy of the social environment of the developing child. On the other hand, there is strong evidence for the role of genetic factors that contribute not only to differences among children and adolescents, but also to the developing child's relationships with family and peers.

Researchers advocating the social viewpoint do not neglect biology, including genetics; but they do not pay explicit attention to biological and genetic factors in development nor systematically include them in their concepts and theories. Biological and genetic researchers acknowledge the importance of the social environment but rarely measure it effectively and comprehensively. With rare exceptions, theories of psychological development built around an understanding of genetics and brain function do not contain a sophisticated or nuanced grasp of social relationships or how they might influence development.

It's time to bring these two lines of thinking together. With this goal in mind, we have assembled an interdisciplinary team of researchers drawn from the fields of genetics, social development, and psychoanalysis. Using a design whose methods and theory give equal weight to genetics and social relationships, we have chosen to study adolescence, a period of singular importance in psychological development. Adolescents who can establish strong social and academic skills, as well as overall psychological maturity, are well prepared to take on the major challenges of adult life.

Additionally, we think of adolescence as a remarkable window on developmental processes more generally. Both the history of work in this field and current research constitute powerful testimony to the yawning chasm between the social and the genetic perspectives in research. We believe that a careful study of adolescence, incorporating both genetic and psychosocial designs, is an important tool in the effort toward reconciliation of these two viewpoints.

We begin our account with the most fundamental convergence between the social and the genetic viewpoints. Both are preoccupied with individual differences among adolescents and with understanding the factors that account for these differences. Indeed, adolescents and their parents are well aware that young people are very different from one another. Some children negotiate the transitions of adolescence with ease; they emerge as competent young adults fully prepared to face more challenges as their lives unfold. In contrast, other children experience difficulties throughout this period: the social, academic, and psychological problems they face are a precarious platform from which to deal with the challenges of both adolescence and young adulthood. Somewhere in between are children who show more transient difficulties in adolescence. Although their problems may at times seem quite severe, they gradually diminish and the adolescents enter young adulthood with reasonably good prospects.

Several important trends in the twentieth century have heightened an interest in these differences in adolescent development. One such trend has been a shift toward child-centered families, accompanied by a preoccupation with children's psychological development. Consequently, parents are now seen as responsible not only for the basic protection of their children, but also for what their offspring become as they reach maturity and establish families of their own. Recent years have also seen a trend toward adolescence as an increasingly important period of choice for young people. Now more than ever they are expected to understand their own wishes and abilities, as well as how to apply these to the demands and opportunities of the adult world. With the growing emphasis on parental responsibility and the widening of choices for young adults, the issue of adolescent failure has come to the fore, particularly in the last two decades. From an economic and occupational standpoint, some adolescents are seen as at great risk of failure: failure to complete the schooling and train-ing necessary for complex and demanding work, and failure to find and sustain satisfactory employment. From a clinical perspective, there is wid-ening recognition that severe psychological difficulties and psychiatric syndromes often appear in adolescence, which may place young peo-ple at risk for drug use, criminality, and suicide, as well as for psychiat-ric disorders and impaired personal relationships throughout their lives (Furstenburg et al., 1987; Kandel and Davies, 1986; Kandel et al., 1986; Moffitt, 1993).

Researchers seeking to understand these differences in adolescent de-velopment have turned, with few exceptions, to the influence of social re-lationships and social institutions in which adolescents are embedded. Al-though most studies have focused on the family, researchers are also looking carefully at the impact of adolescents' friendships, schools, and neighborhoods, as well as at the economic circumstances of their families (Baumrind, 1991; Bell and Bell, 1983; Conger et al., 1991; Conger et al., 1992; Conger et al., 1994; Conger et al., 1995; Dornbusch et al., 1987; Ensminger et al., 1982; Gjerde et al., 1991; Grotevant and Cooper, 1985; Snyder et al., 1986; Steinberg et al., 1992; Whitbeck et al., 1991). In general, all these components of the adolescent's social world have been associated with individual differences in psychological development. Re-searchers have understandably concluded that these associations indicate the important causal role of the social environment in producing the striking differences among adolescents.

New research studies of the effects of genetic factors on adolescent be-havior suggest, however, that the conclusions of psychosocial research on adolescent differences may have to be revised for two reasons. First, data indicate that genetic influences are much more important in adolescent development than previously thought, substantially affecting many as-pects of adjustment, such as self-esteem, cognitive ability, personality, and psychopathology (Cadoret et al., 1983; McGuire et al., 1994; Rende et al., 1993; Rende, Plomin, and Reiss, 1992; Rose and Ditto, 1983; Scarr et al., 1981; Scarr and Weinberg, 1978; Scarr and Weinberg, 1983). More important, different studies suggest that adolescents' genes influence how they are treated by others in their social world. Factors such as parenting, the quality of sibling relationships, and characteristics of peer groups are all affected by young people's genetic profiles (Menke et al., submitted; Pike at al., 1996; Plomin, Reiss, Hetherington, and Howe, 1994; Rowe, 1981).

It is doubtful that single genes are directly responsible for variations in these aspects of adjustment. Indeed, most genetic influences on complex behavior are probably due to several genes acting in concert. Further, the effects of these genes on behavior are mediated by a long chain of intracellular, extracellular, and behavioral events. Making matters even more complex, genetic influences can change rapidly over time. For example, in the short period of time between ages fourteen and twenty months, there is a substantial change in the influence of genes on inhibited or withdrawn behavior in toddlers; 18 percent of such influences at twenty months of age were not operating at fourteen months (Plomin et al., 1993). We now know that some of the genetic effects at fourteen months are replaced by new genetic influences that are activated by twenty months. We report similar phenomena for many behaviors in adolescence in Chapter 8. Thus it is probably incorrect to think of specific genes as determining, in particular and unchanging ways, specific behaviors. This misperception of genetic influence is often referred to as "genetic hard wiring." In all likelihood, genes do not have fixed and irreversible influences on complex behaviors.

In the coming decades we will almost certainly learn more about specific genes that influence both simple and complex behaviors, as well as more about the factors that mediate the influence of these genes. For current purposes, though, we refer throughout this book not to specific genes but to "genetic factors." This term indicates that we have detected the end results in adolescent adjustment of a sequence of steps initiated by a particular set of genes. We know that without these genes the adjustment problem would not occur, but we have not as yet identified the individual genes responsible.

In this book we give serious attention to these genetic influences on adolescent adjustment. Indeed, our work considers a greater range of measures of adjustment than all genetic studies of adolescence up until this point. Perhaps even more important, we report research which suggests that the genetic factors influencing many outcomes in adolescent development are, in large measure, the same genetic factors that influence the family social environments thought to cause those outcomes. For example, the same genetic factors that influence antisocial behavior in adolescents are those that influence the amount of harsh parenting they receive. This is true if we measure harsh parenting and antisocial behavior at the same point in time or three years apart. Chapters 10 and 11 report many of these findings, which imply that the associations reported by environmental researchers between measures of adolescents' social relationships and their development may not be an indication that these relationships cause or influence development. This is true even of the vaunted longitudinal associations in which researchers predict developmental outcome on the basis of prior measurements of the environment. At the center of these new findings is a phenomenon that behavioral geneticists call *gene-environment (GE) correlation,* which refers to the regular association of specific genetic factors with certain environments.

Behavioral genetics provides a second set of findings that questions conclusions drawn from psychosocial research. Genetic studies provide challenging information about environmental influences. These genetic studies confirm, in substantial measure, that environmental influences remain important in shaping differences among adolescents. As will be shown, genetic designs permit an estimate not only of the *heritability* of a particular developmental outcome, but also of the magnitude of environmental effects. Adopting a term from behavioral genetics, we refer to the sum total of these environmental effects as *environmentality.* Behavioral genetics research suggests, however, that almost all these environmental effects must be circumstances that are different for siblings in the same family. The sum total of these sibling-unique effects on psychological development is now known by the term *nonshared environment.* An example of a sibling-unique effect contributing to the nonshared environment might be a family in which a depressed mother withdraws from one of her children but not another, or in which one child is exposed to a terrifying or disorganized neighborhood while a sibling is protected from that neighborhood by parents or other adults.

By contrast, according to genetic data, circumstances that equally affect two or more sibs in a family, known as the *shared environment,* have very little influence on individual differences in development. For example, the level of a mother's depression (neglecting its impact on unique relationships with her children) or the level of neighborhood decay (neglecting differential exposure to the neighborhood among siblings) do not by themselves account for differences among adolescents.

Gene-environment correlations and evidence of the importance of the nonshared environment represent significant challenges to psychosocial theories of development of children, adolescents, and adults. These data do not necessarily invalidate the older theories, but they may require us to revise those theories substantially. We have written this book to clarify what this new genetic evidence means for our understanding of adolescent development. Our fundamental strategy is to examine the data from our longitudinal study to see how they widen and deepen existing psychosocial theories, not how they refute them.

In this book we report the main findings from a twelve-year study of a large and scientifically precious national sample of 720 pairs of adolescent siblings and their parents. We recruited siblings of the same gender who were as close in age as possible. We obtained data on their relationships with other family members as well as with peers, teachers, and best friends in order to gather as much information as possible on the separate, or nonshared, social worlds of our teenage subjects. We sampled six different sibling types in order to explore the role of genetic factors in adolescents' psychological development. These included sibs who were *monozygotic (MZ) twins,* and therefore genetically identical. It also included *dizygotic (DZ) twins;* ordinary siblings in nondivorced families; and siblings whose mothers had divorced and remarried. Siblings in these last three groups share approximately 50 percent of *segregating genes,* or the genetic factors that make individuals different from one another. We also included step-families consisting of both a child whom the mother had brought into the family from a previous marriage, and a second child she had with her current husband in the step-family. These children are *half-sibs* and share 25 percent of their genes. The sixth group consisted of blended step-siblings: one sib was the child of the mother by her former spouse, and the other was the child of the father by his former spouse. These *blended sibs* did not share any segregating genes. All the parents in these step-families had been married for at least five years.

The measures used in our study were as unique as the sample. From the outset we wished not to disprove previous psychosocial theories of adolescent development, but rather to reconcile their findings with the newer challenges from behavioral genetics. Thus we tried as faithfully as possible to measure the same aspects of the environment that had been the focus of many previous psychosocial studies. As with previous studies, we gave particular emphasis to the family. Moreover, in keeping with the most current research on the influences of family on adolescent development, we measured not only parent-child relationships but also marital and sibling relationships. We not only relied on parent and child reports of these relationships, but also trained thirty-seven interview teams to collect videotaped records of interactions among the marital partners, the parent-child dyads, and the siblings as they occurred in the family's home. In accordance with a great deal of current research, we carefully assessed major stressful events in the lives of these adolescents, including the loss of a family member's job, the adolescent's getting injured, the death of a close friend, a girlfriend's pregnancy, the death of a pet, and getting rejected by a college.

In analyzing the data, we had to develop new and often complex statistical procedures that had to accomplish many things at once. First, they had to combine data from many different measurement instruments (we used, for example, four different measures of an adolescent's social behavior) and data from many different sources (for example, the adolescent rated his or her antisocial behavior, but so did teachers and both parents). Second, the designs had to estimate the influence of the shared and nonshared environments, as well as that of genetic factors, on the variables we measured. Third, they had to estimate the extent to which the same genetic factors might be responsible for influencing both our measures of relationships and developmental outcome. These models were difficult for statisticians to develop, and only a relatively small number of specialists, mostly in behavioral genetics, can properly evaluate and understand them. Thus where they are crucial for reporting our results we include them in an appendix. Most of our major findings can be summarized using simple correlations, and these are the mainstay of the data sections in most of the chapters. For readers unfamiliar with statistics, Chapter 3 provides an introduction to the statistics of correlations, as well as a brief primer on how to interpret correlations in order to estimate genetic and environmental influences on measures of adolescent development. We also summarize how more complex models are constructed and tested, but we report the results of these tests very schematically in the body of the text.

Genetic research provides us with new and important data to help us understand individual differences among adolescents. More important, it provides additional logical tools, in the form of research designs and data analysis rarely used in developmental studies, for more precise estimates of influences on adolescent development. Thus a major step in presenting our study is to delineate the underlying logic on which its design and analyses are based. The reader will want to know, from the outset, whether we are describing a whole new set of logical principles or whether we are building on existing ones. Will views and perceptions of human development, with our special focus on adolescence, have to be substantially or completely rewritten in the light of new genetic information?

In subtler (Scarr, 1992) and more challenging terms (Rowe, 1994), major scholars have argued just that. They have advanced recent genetic findings to downplay the importance of nongenetic factors in development. We reach the opposite conclusion. Genetic research should heighten an interest in environmental contributions to development. In adolescence, the most important aspect of the environment may be the relationships in which the adolescent is involved. But to understand fully the potential of genetic studies in this area, we must make more explicit the rules of inference that have been used in the best psychosocial studies. We must inquire how the leading scientists in this field have asked questions about adolescent development, how these questions have been addressed in research designs, how the data have been analyzed, and which principles have been used for drawing inferences. We refer to a sequence of this kind as a "logical principle of research." Such principles distinguish psychosocial and genetic research. But are they incompatible? Do the logical principles of genetics supersede those of psychosocial research?

The answers to these questions are not simple. Indeed, we have spent twelve years planning and executing a single major study to address them. It is now clear that in order to reach comprehensible answers we must lay out the basic logical principles of psychosocial research and do an equally careful job of identifying comparable logical principles for genetic research. An analysis of this kind reveals important consonance between the two approaches and sets the stage for a reconciliation between current psychosocial knowledge and the rapidly emerging data of genetic research. More specifically, the data from our study confirm the importance of social relationships in adolescent development and open up several new possibilities for thinking about why these relationships are important.

First, the data suggest that parent-child relationships are, as psychosocial researchers have concluded, still central to adolescent development. But our findings, along with those from other genetic studies, suggest a very different reason that this may be the case: adolescents share exactly 50 percent of their genes with their parents. Much of what psychosocial researchers interpret as evidence for the social influence of parents on children may be ascribed to this genetic relationship.

Second, the role of genetic factors in shaping links between parental and child behavior may help us understand the influence of social processes in a new way. The data suggest—but do not prove—that these social processes may be part of a mechanism by which genetic factors influence adolescent behavior. It now seems entirely possible that particular genetic differences among adolescents cause their parents, as well as their siblings and friends, to respond to them in a certain way. It also seems possible that these evoked responses play an additional role in adolescent development. That is, genetic factors initiate a sequence of influences on development, but certain social processes are critical for the expression of these genetic influences. Indeed, we present preliminary evidence that specific genetic factors may be linked to specific relationships within the family.

Third, the data from our study mostly confirm previous genetic findings that suggest we must pay special attention to the social relationships that are unique for each sibling in the family if we are to understand the impact of social relationships on adolescent development, above and beyond genetic influences. The data strongly suggest that these sibling-specific, or nonshared, experiences are not straightforward. In some cases they may be experiences that are not only special for siblings but unique for each family as well. In other cases complex situations within families may cause the social experiences of one sibling to undermine or protect the other. For example, if a parent regularly singles out one child for derision and ridicule, the other child may come to feel relatively safe and protected (Reiss et al., 1995).

Our data provide less in the way of firm conclusions than open a new door of inquiry. They sustain a long-held belief that adolescents' relationships are critical for understanding differences among them. But we must decipher the genetic data as well as the social information inherent in these relationships, for they are both crucial for unlocking this *relationship code*. Our ideas about a relationship code constitute a hypothesis, not a conclusion that is rigorously drawn from the data we have collected. We have named this book after a plausible but unproved idea for one simple purpose: to open up new ways of thinking about human development.

Part 1

LOGICAL TOOLS FOR ANALYZING
ADOLESCENT DEVELOPMENT

Chapter 2

RELATIONSHIPS AND ADOLESCENT
DEVELOPMENT

By organizing this book around an idea rather than a proven "fact," we hope to encourage dialogue between genetic and psychosocial perspectives on development. This chapter begins that dialogue by summarizing some of the logical principles that underlie how researchers draw inferences about social and genetic influences. We provide examples of how psychosocial findings are crucial to a better understanding of genetic effects on adolescents' psychology, as well as examples of how genetic findings can clarify questions that have been raised in the study of social influences.

After three decades of investigation, researchers can now make a very strong case for the role of adolescents' relationships in differences among them. In this chapter we outline the logic of this research and summarize some of the main lines of evidence. Space considerations prevent us from reviewing the hundreds of good studies in this field, and so we often refer readers to more detailed reviews. Our aim here is to show that inferences about the role of family relationships, friendships, and social settings in adolescence are based on clear scientific principles. Reviewing these principles will make it easier to recognize that the major contribution of behavioral genetics is to provide additional logical tools for making inferences on development more precise. As will be shown, behavioral genetics research does not introduce "biological processes" into an equation that has, until now, centered solely on psychosocial influences on differences in adolescents. Indeed, it is time that we discarded the old debate of nature versus nurture, or biological versus psychosocial factors, as predominant influences on individual development at any point in the life span, especially adolescence. For several reasons, the logical principles of genetic research lead to a rejection of this dichotomy even more forcefully than do logical principles developed in more traditional psychosocial research.

First, psychosocial research designs have already pointed to the likelihood that biologically influenced dispositions may be important in the development of differences among adolescents. For example, psychosocial research suggests the importance of temperamental differences among children in their response to divorce (Hetherington, 1991). Second, behavioral genetics designs are richly informative about psychosocial influences on development. For example, these designs have pointed to the importance of family experiences that are unique to each child in the family.

Third, though at this stage in behavioral genetics research we learn very little about the biological details of how genes influence behavior, new advances may soon yield a more precise biological portrait of genetic mechanisms. Current behavioral genetics research can tell us that genes are important in individual differences in intelligence, for example, but it has taken only the most elementary steps in identifying which genes are important, and it has not yet detailed how these genes express themselves in intelligent behavior.

Fourth, good behavioral genetics designs must draw heavily on the findings of more traditional psychosocial research in order to maximize their research results. As will be shown when we present data from our own study, behavioral genetics analyses are particularly useful when psychosocial research has established strong associations between certain types of social relationships, on the one hand, and individual, psychological adjustment, on the other. One example of strong and repeated findings is the association between adverse parenting and antisocial behavior in the adolescent.

Fifth, the finding that some differences among individuals are attributable, in large measure, to genetic variations among them does not exclude the importance of psychosocial factors in the expression of these genetic differences. Indeed, data indicate that genetic differences among individuals may require psychosocial processes in order to be expressed as differences in adjustment in adolescence as well as at other points in the life span.

Parallels between the Social and Genetic Analysis of Development

Scientific psychology's concern with individual differences among adolescents dates to the dawn of the twentieth century, when the psychologist G. Stanley Hall published his influential book *Adolescence: Its Psychology and Its Relations to Physiology, Anthropology, Sociology, Sex, Crime, Religion and Education* (Hall, 1904). This book and other contributions by Hall provided a major connection between academic psychology and other fields concerned with teenagers, chiefly education (Modell and Goodman, 1990), and it launched a sustained scientific interest in adolescent development. Because Hall's work emphasized young people's vulnerability to influences outside themselves, it became a blueprint for wide-ranging studies of social influences on development at this phase. Interestingly, it was Hall—as the president of Clark University—who invited Freud to America to meet the leading American psychologists of the day, including James, Titchener, and Cattell.

In subsequent years, psychoanalysis significantly influenced how professionals and lay people alike thought about adolescence. Analysts such as Blos (Blos, 1962) called attention to how young people rework inner conflicts about sex, whereas others such as Erikson (Erikson, 1963) delineated the importance of the social context—culture, peers, and family—for successfully completing what many regarded as adolescents' most important task: to develop a distinct sense of themselves and their future. This psychological status came to be known as "ego identity." As will be shown in Chapter 13, our own genetically informed research gives good reason to return to some of the central ideas of psychoanalysis.

On the surface, the roots of genetic analyses of individual differences seem vastly different from those of social analyses. But perhaps the most important difference is that genetic research has never given much emphasis to the adolescent period. The field of behavior genetics received its impetus from three sources: Darwin, Mendel, and Galton. Darwin sought to explain the origin of separate species of animals, Mendel focused on similarities and differences between "parent" and "offspring" peas, and Galton focused on the aggregation of genius in families. All three posited intrinsic features of individual entities—animals, vegetables, or humans—that influenced their physical appearance, adaptation to their environments, and the characteristics of their offspring. Darwin, however, clearly recognized the role of environmental challenges in the emergence of species. Galton, too, appreciated the role of social forces. Indeed, in the final chapter of *Hereditary Genius* he provides an extended discussion of how social and cultural factors alter the distribution of hereditary factors in a population through age of marriage.

Because early researchers emphasized either external or internal forces in the shaping of individual differences, the search for the social and genetic roots of such differences proceeded along two different tracks for nearly a century. Each perspective developed its own logic for drawing inferences from data. Even more important, each developed its own community of researchers whose terminology and technology often seemed designed to prevent rather than promote the exchange of ideas between the two intellectual communities. The two perspectives were even further alienated when genetic data were misused in the service of anti-Semitism, the persecution of the Gypsies, and racism in America and elsewhere.

Genetic influence on intelligence, self-control, sexual behavior, and personality remains a very hot topic, mainly because social forces outside the research community have tended to seize on genetic data to prove or disprove primitive racial and sexual theories. But even researchers themselves, most of whom have tried to stay above the fray, have wittingly or unwittingly exacerbated the split between the two lines of analysis. From Galton's time forward, genetic researchers have displayed an almost studied naiveté about social processes. Indeed, even allowing for the times in which he worked, it is difficult to read Galton's magnum opus, *Hereditary Genius,* without being appalled at its racist overtones.

A more telling example of this naiveté can be found in the history of the twin method in behavioral genetics. As the next chapter shows, this method, which was first introduced in 1924, was built around the *equal-environment assumption* that parents and others treat identical twins more or less as equally as they treat fraternal twins. As almost any parent of twins could attest, this assumption is absurd. It was not critically examined by geneticists until the family therapist Don Jackson wrote a withering critique of genetic research and the equal-environment assumption in particular (Jackson, 1960). This sophisticated review was part of a trend whereby geneticists started to improve their methods, but it is more important as an indicator of just how easy geneticists made it for researchers in other fields to dismiss their entire corpus of work.

If geneticists were myopic, most researchers of the social environment—with notable exceptions—were densely blind to the emerging fields of quantitative, population, cytological, and molecular genetics. A toxic mixture of ignorance, obliviousness, and myth-making kept almost all research on psychological development free of genetic inquiry. Many social researchers were, of course, keenly sensitive to the serious misuse of genetic data and ideas by racists and demagogues. Others, however, could not extricate themselves from a myth of biological determinism. If a human trait is heritable, they believed, it is fixed at birth and cannot be influenced by upbringing, education, or social policy. As will be shown, dramatic evidence to the contrary has been readily available to researchers and lay persons alike for at least forty years. One of the most conspicuous examples of the effect of the environment on genetic influence is the very successful dietary treatment for the highly heritable mental deficiency caused by phenylketonuria.

Today, thankfully, this climate is changing. Various efforts at reconciling these two important research traditions are under way. One such example is the project that inspired this book. Funded by the federal government, our study received enthusiastic support from many review panels consisting of geneticists as well as psychosocial researchers. This climate has encouraged more wishes than accomplishments, however. Indeed, there is still no clearly established way to integrate genetic and social data into a common or integrated understanding of development. This book is an effort to define one approach to this reconciliation. The arguments we advance rest on the following tenets:

- Hard facts are difficult or even impossible to come by in the behavioral and psychological sciences. Thus we draw reasonable inferences from data that are often collected under less-than-ideal circumstances.
- These inferences can be understood only if we understand the scientific reasoning behind their formulation, including the assumptions behind the design of a research project, the way its data are analyzed, and how those data are used to make generalizations.
- Both the social and the genetic analyses of development are becoming

mature sciences, and their logical research principles can now be described clearly. They have produced impressive results that often generalize across many studies and have been converted into practical applications of great benefit to children, adolescents, and adults.

- Bringing together genetic and social analyses for understanding adolescent development is an enterprise not in comparing "facts" but rather in comparing the logical principles of the underlying research.

- This comparative analysis reveals very important overlaps between the two scientific endeavors. Indeed, social researchers anticipated in their own work findings that emerged with much greater clarity in genetic research. For example, social researchers have detected many instances in which individual attributes of teenagers seem to shape the social relationships in which they become engaged. This phenomenon was pursued and illuminated quite dramatically by genetic researchers in subsequent studies. The impact that teenagers have on how they are treated by parents, sibs, peers, and teachers forms a cornerstone of a new developmental theory of adolescent development to which both social and genetic researchers can now make a truly collaborative contribution. Alternately, genetic research opened new avenues for social researchers and enhanced the importance of the logical principles undergirding their central efforts. For example, genetic research pointed to differences in how siblings in the same family were treated by parents, sibs, and others in their social world. The specific nature of these differences, however, can only be determined by social research.

These tenets lead us to propose a reconciliation between social and genetic analyses that starts from a clear understanding of the logical principles of research in each area. Significantly, almost all the logical principles of genetic analyses have analogues among the logical principles of social analyses. We present these principles in two ways. In this chapter and the next two we summarize some of the central logical principles of social and genetic analyses of development and provide illustrations of their use.

We then review most of these logical principles a second time in Chapters 7 through 13. We also report the findings of our study, which are based on those principles. In the course of discussing these data we continue our efforts to reconcile them with comparable findings from more

Table 2.1 Classification of logical principles of social and genetic analyses of development

Level of analysis	Social	Genetic
Cross-sectional	Synchronous analysis of relationship characteristics and adjustment	Associations of genetic similarity in pairs of related and unrelated individuals and their similarity on (a) developmental outcomes or (b) social behavior
Longitudinal	Observations of change and stability of relationship process over time in association with change and stability in developmental outcomes	Estimates of genetic change and stability and their association with change and stability of (a) adjustment or (b) social behavior
Experimental or randomized intervention	(a) Systematic reallocation of families or peers into unfamiliar groups (b) Controlled trials of theory-specific therapeutic interventions (c) Controlled trials of theory-specific preventive interventions	Little used in genetic research on human development, but a few examples include (a) dietary interventions (b) pharmacology (c) stress exposure

traditional social analyses of adolescent development that do not attend to genetic influences.

We introduce our review with a simple scheme that is known to all researchers in both psychosocial and genetic areas. This scheme has three basic concepts and clarifies some of the analogies between psychosocial and genetic analyses. These three categories are summarized in Table 2.1. The fundamental distinction here is among *cross-sectional, longitudinal,* and *experimental* (or *randomized intervention*) studies.

We begin with the last of these three, *experimental* studies. In a broad range of sciences, strong inferences can be drawn from experiments. Some of these are conducted in laboratories, whereas others are carried out in more natural, or "field," settings. The basic logic behind an experiment comes from the ability of researchers to alter or control one or more circumstances and produce the exact outcome predicted by their theories. For example, the early geneticist Gregor Mendel purposely crossed wrinkled and smooth peas to test a genetic theory. He predicted that the results of the first cross would be plants bearing all smooth peas, but that in the second cross three-quarters of the plants would bear smooth peas and one-quarter would bear wrinkled peas. Mendel had three related hypotheses in mind. The first was that any inherited characteristic is transmitted from one generation to the next by two "elements," now known as "genes." Thus in pure-bred peas smoothness is caused by the two elements SS and wrinkledness by two others, ww. The second idea was that in the process of fertilization the two elements are separated or segregated such that each "offspring plant" receives one element from each "parent" plant. All the offspring of this first cross would then have the elements Sw. But what would they look like? Here Mendel formulated a third hypothesis—the idea of dominance. Whenever a plant had both S and w, the S would always dominate the w; as a result, the peas would be smooth. Only when there were two ws would the peas be wrinkled.

Testing these hypotheses required a second cross among the plants produced by the first cross. If Sw was crossed with Sw, one-quarter of the "offspring" would receive an S from each "parent" and thus be SS. By the same process, one-quarter would be Sw, one-quarter wS, and one-quarter ww. According to the theory of segregation and dominance that linked paired elements, the second generation would have one-quarter wrinkled peas because only the ww type would be wrinkled.

In this experiment the specific crosses between pea plants are known as the *independent variable* and the appearance of peas in subsequent generations is the *dependent*, or *outcome, variable*. Mendel observed that the dependent variable, the proportion of wrinkled peas, was exactly as he had predicted from his theory. Thus this two-step experiment provided strong experimental evidence for a theory that, with important modifications, has held up for more than 130 years.

Many sciences such as astronomy and archaeology do not lend themselves so nicely to the decisive method of experiments. Researchers studying human psychological development have debated whether their research designs should come closer to those of the peas or the stars. A common consensus is that experiments reflect a very effective logical principle for designing research projects on psychological development but that their use is limited by practical and ethical factors. As a result, students of human development must often depend on observed associations between independent and dependent variables. For example, researchers on children and adolescents have noted an association between divorce and behavioral and academic difficulties in children. More specifically, children of divorced parents show more difficulties than children whose parents are not divorced. A typical study in this genre will identify one group of children whose parents are currently divorced and a second whose parents are together. The researcher will then compare children in these two groups using measures of academic performance, social competence, and psychiatric symptoms. Because their status—as children of divorced parents or not—and their psychological competence are measured at the same cross-section in time, a study of this kind is called *cross-sectional.*

Many researchers conclude from such data that divorce causes or influences the psychological status of the children of the divorcing couple. But the reverse may actually be the case. Very difficult, misbehaving children may add a substantial burden to marriage and contribute to its collapse. In order to examine this possibility, researchers have recently reported on *longitudinal* studies, in which children from nondivorced homes were followed over many years; in some cases the parents got divorced. The researchers were startled to observe that many of the behavior problems (particularly in boys) thought to be a *consequence* of divorce actually *preceded* it (Block et al., 1986; Cherlin et al., 1991)! This certainly rules out divorce as the sole cause of a child's difficulty. Perhaps the child's behavior did contribute to the divorce, or perhaps sustained, pre-divorce conflict in the family was the real culprit. As will be shown, behavioral genetics offers a different way of thinking about a finding of this kind.

Longitudinal studies are becoming increasingly popular in the social analyses of adolescent development. Indeed, behavioral geneticists have recently used longitudinal designs with some startling results. In both genetic and social analyses, longitudinal studies are an effort to disentangle cause and effect, particularly when experiments are not feasible. Behavioral genetics studies have thrown new light on psychosocial longitudinal studies and have suggested that inferences drawn from them might need to be reconsidered.

Nevertheless, social researchers are not comfortable relying on longitudinal studies for strong inferences, and thus they have tried to adapt the power of experimental studies to the challenges of research on psychological development. For example, to test the role of marital conflict in child development, researchers have worked with couples to improve spousal relationships during the early years of a child's life to see if behavioral difficulties in the child can be prevented in subsequent years (Cowan and Cowan, 1992). As couples are assigned randomly to a treatment or a no-treatment control, the family researcher can gain some control over the independent variable, approximating Mendel's control over the crossing of smooth and wrinkled peas. Mendel, however, could cross his peas without influencing them in other ways, such as controlling the amount of fertilizer or sunlight they received, either of which might have influenced the surface of his peas. Social researchers, by contrast, can treat one group with a planned intervention and not treat those in the control group, but things don't always turn out as they had hoped. Sometimes the treatment has unrecognized components that influence the research subjects, and sometimes the careful attention paid to members of the control group has an unintended therapeutic impact, known as the "placebo effect."

Studies Using Cross-Sectional Data

All studies using social analyses of individual development can be grouped into two broad categories. The distinction rests on which chapter of the developmental story the investigator uses as a starting point. It is now widely acknowledged that psychological development begins *in utero,* that a child comes into the world with certain features of temperament and responsiveness to the social environment, but that, along the way, social relationships continue to have their own impact on the child. Many social analyses begin the story in the middle, focusing on the effects of these social relationships. In this group of studies researchers concentrate on the unique role of social processes in shaping individual development. They search for characteristics of relationships that are influential across a broad range of children or adolescents. Studies in this domain usually require a measure of social process (or more than one), as well as its association with some important measures of adjustment, such as depression or antisocial behavior in the child or adolescent. Although virtually all researchers today know that children's behavior can influence the behavior of their parents, the typical aim of these cross-sectional studies is to explore the reverse: the extent to which children and adolescents are influenced by the relationships in which they are involved.

A second group of studies focuses on earlier chapters of the developmental story. These studies ask which characteristics children display prior to entering or being influenced by a relationship, and how these individual characteristics influence children's choices of relationships. An even more important question follows: How do these characteristics of individuals and characteristics of their relationships codetermine adjustment? For example, adolescent temperament seems to play a major role in whether peers accept or reject the adolescent as an active group member. The adolescent's temperament and the rejection by peers, in turn, are strongly associated with psychological problems in the young adult (Lerner et al., 1989).

Studies in the first group are the simplest. As will be shown, they have their analogues in genetic studies and can be called *association studies.* That is, they study the simple association between relationship characteristics and development. The second group of studies focuses on the impact of enduring, perhaps congenital, characteristics of individuals and their development. They are part of a broader class called *moderating and mediating studies,* emphasizing the notion that social relationships are evaluated for how they might influence a developmental process that is initiated *in utero* and continues through adulthood.

Association Studies

There are two basic strategies within association studies. The first strategy is stimulated by a wish to understand more fully the impact of major social trends, such as the rise during the last three decades of the divorce rate or the increasing number of unwed teen mothers who are raising their infants. The second strategy is to focus on more subtle influences within the family. Research in the first genre typically studies the effects of major trends on child and adolescent development. Divorce (versus marriage) and teen motherhood (versus young-adult motherhood) are then the independent variables in these simple association studies. There is a strong social motivation for conducting these studies in order to better inform policy that seeks to minimize the negative impact of these trends on child development. There are also substantial scientific motives for this genre of work. The most distinctive advance comes from the straightforward means of assessing the independent variable in the study.

Many studies have been conducted to answer three related questions about divorce and teenage parenthood: Do these major disruptions have significant impacts on the offspring of divorced or young parents? If so, how do these disruptions influence children? And, finally, what factors might mitigate these negative impacts?

Researchers conducting association studies are also concerned with the impact on children and adolescents of other serious parental problems. For example, large amounts of data have been collected both on the impact of mental and physical illness in parents on the adjustment of their children, and on the short- and long-term impact of parental death. Typically, researchers who focus on divorce, teen motherhood, or parental disability concern themselves primarily with the effects of a specific disruption on family life and on children. As a consequence, they examine a very wide range of possible consequences for children and adolescents, including school performance, social adjustment, and psychopathology.

This line of investigation makes an important contribution to the overall logic linking social relationships and differences among adolescents. For example, because of the widespread social concerns about divorce and unwed teen pregnancy, many studies on the link between these disruptions in family life and child adjustment have been completed and reported in detail. Further, the definition of family disruption is so obvious and straightforward that we can readily compare the results of one study with those from another. For example, a recent review systematically summarized ninety-two published studies of the impact of divorce on children and adolescents (Amato and Keith, 1991). These studies provide a confident statement on the association between a major family disruption and child or adolescent adjustment. The accumulated data can be reviewed for their generality across samples, across recent historical periods, and across countries and races. Further, the impact of the disruption can be ascertained for children who differ in age and gender.

Because these disruptions appear so conspicuous and so stressful, many researchers have assumed that they must be causing the problems with which they have been associated in research. We have already mentioned two longitudinal studies that cite the appearance of problems in children prior to divorce, casting doubt on this straightforward assumption. New genetic data also shed more light on the effects of teenage motherhood and divorce on child development. Evidence now suggests that genetic influence on the liability for divorce (McGue and Lykken, 1992) and many of the risk factors for teen pregnancy, such as impulsiveness (O'Connor et al., 1998) and early menarche (Fischbein, 1977), also show genetic influence. We will return to the implications of these new findings for interpreting the results of these studies.

The second type of cross-sectional association study focuses less on obvious disruptions and more on subtle variations in family life, such as the quality of parental discipline. Some of the studies of major disruptions take this approach as well in order to understand how, for example, divorce might affect young children. Does divorce disrupt competent child development because it disrupts parental discipline? Researchers focusing on the quality of parent-child relationships, in contrast to those studying conspicuous family disruption, select independent variables—subtle family processes—because of their theoretical link to the dependent variable, the measures of adjustment under study. For example, the quality with which a parent monitors the behavior of a teenager is a reasonable independent variable if one is studying a teen's ability to master self-control and stay out of trouble.

This genre of research clarifies more clearly than others the properties of a social relationship that are associated with adolescents' psychology and with their observable patterns of behavior. Clarifying this linkage is as crucial for progress in genetic studies as it is for advancement in psychosocial studies. For example, a number of well-designed psychosocial studies have established an association between aversive parenting and antisocial behavior (Patterson, 1982; Patterson and Capaldi, 1991; Patterson,

Crosby, and Vuchinich, 1992; Patterson, Reid, and Dishion, 1992; Reiss et al., 1995). Moreover, a particular form of aversive parenting has been central: escalating coercive behavior whereby both child and parent attempt to influence each other through threats, nagging, and occasional violence. Indeed, the behavior that successfully gets the parent to *stop* the fight seems to become an indelible feature of the victor's repertoire. The crucial observation in this work is that the child who is victorious in these confrontations is often on the way to becoming antisocial; parental surrender is associated with increasingly coercive behavior on the part of the child in the home and then in crucial settings outside the home. Specific and carefully observed links like these, between social transactions and persistent patterns of behavior, are a preparatory step for more detailed and subsequent psychosocial research; they are even more important for genetic analyses that seek to weight the relative importance of genetic and environmental mechanisms that might account for these striking observations.

Researchers who study positive adolescent development have sought to link the quality of social relationships not just to specific behaviors of adolescents, but also to evidence that they have successfully completed one or more of the broad "tasks" of adolescent development. Three psychological tasks of adolescence in particular have been a focus of much of this work. First, adolescents must develop increasingly clear appraisals of their own abilities and weaknesses while also maintaining adequate levels of overall self-acceptance and self-esteem (Harter, 1990b). Assessment of success in this task has a central role in our study. Second, adolescents must develop increasing capacities for taking the perspective of others, tolerating ambiguity, and developing long-range plans: tasks that are subsumed under the concept of "ego development" (Loevinger, 1979). In our study we assess a component of success for this task as well. We focus on adolescents' capacity to initiate and sustain their own activities while keeping others' viewpoints clearly in mind. Finally, adolescents must remain flexible enough to continue to explore new avenues for their own energy and development, a task that has been termed "identity exploration" (Erikson, 1968; Marcia, 1966).

Researchers have taken each of these three areas of adolescent development and determined the types of relationships that would support or retard them. To see how this works, let us take the area of identity exploration. We can learn about variations in adolescents' flexibility and curiosity through a careful, systematic, and standardized interview of adolescents themselves, inquiring, in particular, about their experience exploring different options and values in domains such as occupational choice, politics, and friendships. Researchers have set out to determine, for example, which types of families would promote this important exploration. They have reasoned that families should show a balance between comfort in expressing their own views, on the one hand, and providing respect and support to people with different views, on the other. According to this hypothesis, balance is important not only in the parents' relationship with the adolescent, but also in the siblings' relationships with the adolescent and the parents' relationship, with each other. Results from research confirmed these expectations, but with some interesting particulars. For girls, such a balanced relationship, whether between a parent and child, the adolescent and a sibling, or the mother and father, was associated with high levels of identity exploration. For boys, the relationship with their fathers was more important than all the others. Further, fathers who showed willingness to express their own views had daughters with broader and deeper identity exploration, whereas supportive fathers had sons with greater exploration (Grotevant and Cooper, 1985; Grotevant and Cooper, 1986). The level of a girl's identity exploration was also related to the quality of her ties with her siblings and with her mother, as well as to the quality of her parents' marriage. For example, a girl showed more of her own identity exploration when her mother respected her father's views and when her father made his own positions clear. These qualities of marital interaction were not associated with levels of identity exploration in boys.

Observations like these, as indicated, can be interpreted in two very different ways. One approach is often used by clinicians who work with entire families. They are sensitive to patterns in the transactions of all members of a family unit and often suggest that the psychological functioning of any single member is an index or manifestation of these family-wide patterns; that is, certain family dynamics cannot be said to cause or influence individual development, but individual development itself, normal or pathological, is an indirect measure of the characteristics of the family as a system. For example, the thought disorder characteristic of schizophrenic patients has been described by some family therapists as an index of a family process governed by covert rules. These rules are elaborated implicitly by the whole family in order to maintain ties among members while at the same time concealing, from themselves and outsiders, competition among family members and the successes and failures of each of them (Palazolli et al., 1978).

A sharply contrasting approach sees the relationships among family members as highly responsive to stable and distinctive characteristics of individual members. That is, instead of individual psychological processes being embedded in a powerful web of family relationships, social patterns are a barometer of individual moods, personality, or skills. Among the most notable examples have been studies of the effect of children's and adolescents' temperament on the quality of their parents' behavior toward them. Of particular interest have been reaction patterns that many researchers have labeled "difficult temperament," which has included irritability, distractibility, excessive activity, and lack of adaptability to new situations. Some studies suggest that in younger children and in adolescents, a difficult temperament may elicit negative and aversive responses from parents, particularly when parents are under stress and have few social supports (Hetherington, 1991). Temperamental features of parents can also influence their behavior toward their children (Elder, 1986). In addition, many studies support the strong influence of psychiatric disorders in parents on their interaction with their children.

This second perspective raises doubts about the influence on development of the adolescent's current relationship with his or her family. Perhaps it is the other way around: features of the adolescent's personality or levels of adjustment influence the family. Or perhaps the adolescent's adjustment is influenced by comparable characteristics in the parent, but the mechanisms for this influence do not involve the current relationship between parent and child. For example, it is well established that antisocial children tend to have antisocial parents (and, of course, vice versa). Perhaps this parent-child similarity is due to some mechanism other than current adverse family relationships. Longitudinal studies, with observations of families and adolescents at several different time periods, have been designed to unravel and weight these possibilities. As will be shown, in situations where individual attributes of either parents or children are at least partly heritable, genetic research provides additional insights into this fundamental question.

Most researchers adopting this second approach in their study of the role of social relationships in adolescent development have begun with particular outcomes in children and adolescents, focusing on a broad range of maladjustments termed "antisocial behavior." The amount of work in this area is almost certainly a reflection of the widespread concern about serious misbehavior in adolescents. The components of antisocial behavior that have been studied include serious disobedience, oppositional behavior, fighting, lying, stealing, and cheating in younger children, and all of these plus drug and alcohol use, sexual misconduct, and criminality in adolescents. More recently, researchers have also focused on the role of social relationships in depression and suicidal behavior in teenagers.

The same basic task confronts researchers in this area as confronts those studying major disruptions in relationships. When family or other social relationships are thought to be important influences on antisocial behavior or other psychopathology, the researcher must clearly define which social processes are likely to be related to psychopathology in the adolescent and why. Studies have shown that very different kinds of family processes are correlates of antisocial behavior. For example, antisocial behavior is much less frequent, particularly in pre-adolescent boys, when parents monitor their activities closely and provide firm discipline when their sons' behavior is out of line (Patterson, 1982). Fewer studies have focused on antisocial girls.

Mediating and Moderating Studies

We have made an important distinction in our discussion of cross-sectional studies. Some studies choose to link major manifestations of disrupted relationships—such as divorce or teen pregnancy—to levels of adjustment in children or adolescents. These studies establish an association between the disruptions and adjustment but do not clarify any mechanisms that explain that association. A second set of studies takes a closer look at qualities and processes within specific relationships, including, for example, mutual coercion between parents and sons or openness in expressing ideas in fathers and daughters. These relationship qualities are theoretically "closer" to the levels of adjustment because there are fewer links in the presumed causal chain between the two. For example, it is easy to conceive how coercive cycles in a parent-son relationship can lead to impairment in self-control in sons. As noted, coercive cycles reinforce children's bad behavior. An ineffective parent first responds to an aggressive remark by a coercive child, for example, with a countercoercive behavior such as yelling. In the third move in this sequence the child escalates the verbal aggression until finally, in the fourth move, the parent caves in and retreats. The difference between the second move, parental yelling in response to aggression on the part of the child, and the fourth move, parental withdrawal, is an important one. The third move, the child's continued aggression, is reinforced because the parent becomes less aversive from move two to move four as a response to the child's persistent aggression.

Current research combines measures of conspicuous disruptions in relationships with a more fine-tuned examination of patterns of relationships. The leading idea is that disruptions in a relationship do not have much effect on children directly. Rather, these large-scale disruptions exert their influence by first reshaping the more detailed sequences in particular relationships. For example, one recent study compared the psychological adjustment of early-adolescent boys and girls living with their mothers after a recent divorce with those living with nondivorced parents. As predicted, the psychological adjustment of the first group of youngsters was impaired. The researchers also studied the quality of mother-child relationships in both families, focusing particularly on conflict between the two, problem solving, and positive, supportive communication (Forehand et al., 1990). These mother-child characteristics account, in a statistical sense, for at least half of the negative effects of divorce on child adjustment: the divorced mothers had more conflict with their children and were less positive and less frequently engaged in talk addressed to resolving disputes between them than were the nondivorced mothers. The investigators reasoned that the effect of divorce on mother-child patterns of relationships was responsible for much of its impact on children's adjustment. In other words, parent-child relationships may be said to mediate the impact of divorce on the psychological adjustment of children.

The concept of mediation becomes crucial when we introduce genetic data into studies of family relationships and adolescent adjustment. For example, researchers are no longer interested simply in whether genetic factors influence adjustment. For both solid theoretical and practical reasons, we want to know which particular processes mediate between the presence or absence of certain genetic factors and psychological adjustment. Genetic data can provide explanations very different from those of psychosocial research on the mediating role of relationships. For example, it may appear as if parent-child relationships mediate the effect of divorce on child adjustment. But the same genetic factors might influence divorce liability, parent-child relationships, *and* adjustment, and may account partially or completely for these mediated associations.

The concept of moderation is related to mediation and is equally important. Moderation refers to factors that influence the association between a measure of relationships and a measure of adjustment. The literature on divorce again provides some interesting examples. For instance, divorce has a much stronger impact on social adjustment in boys than in girls. We can say, therefore, that the child's gender *moderates* the relationship between divorce and the social adjustment of the child (Amato and Keith, 1991). The phenomenon of moderation is sometimes referred to as a "statistical interaction." Thus gender of the child and divorce are said to interact in their influence on social adjustment. For reasons that are not entirely clear, frequently replicable moderating or interaction effects have been difficult to demonstrate except in a few areas of research (Wachs, 1992). Nonetheless, when they occur they are of extreme interest. When we add genetic information to a study of psychological development, moderating effects or interactions play a central role in understanding how genetic and social processes fit together in development. We consider this topic briefly in Chapter 4 and return to it in Chapter 14.

Studies Using Longitudinal Data

As shown, cross-sectional studies are an important first step in psychosocial studies as well as in genetic studies. But researchers usually need more information to resolve dilemmas raised by contrasting explanations of the findings. Thus their next step, typically, is to use longitudinal studies when measurements of adolescents' adjustment and of their social environment are made on at least two different occasions. Such studies can help resolve some of the interpretive dilemmas of cross-sectional studies, but they also open up new issues. There are many different strategies for designing longitudinal studies, but we consider only two here. First, we look at *temporal sequencing studies,* which make an initial attempt to disentangle cause and effect. Typically these studies measure changes in the quality of individuals' social relationships and compare these with changes in their competency or their psychopathology.

Second, we consider *assortative affiliation studies,* which, like many of the temporal sequencing studies, focus on disentangling individual and social influences. This approach studies the hypothesized influence of peer groups on psychological development in adolescents. Researchers adopting this method ask if the characteristic of an adolescent's peer group influences the adolescent's behavior, or vice versa. That is, do adolescents choose their peers because they are similar to begin with, or do they change as a result of spending time with their peers?

Temporal Sequencing Studies

Longitudinal studies linking social relationships to adolescent psychopathology offer good illustrations of the temporal sequencing approach to understanding development. In the area of adolescent psychopathology, three closely related strategies help to clarify temporal sequencing of changes in social relationships and changes in adolescent behavior. First are studies that focus on unfolding patterns in adolescent psychopathology; these studies are often a prelude to those that explore the role of social relationships. Second are studies that observe the temporal sequence of changes in social relationships and changes in adolescent behavior. The third type of study capitalizes on natural experiments in communities such as the assignment of adolescents to different schools or the assignment of different punishments for stealing. Such studies allow us to observe how variations in these practices influence adolescents.

Temporal sequencing studies have led to a growing body of knowledge from longitudinal studies on the development of psychopathology itself. For example, in pre-adolescents, antisocial behavior in the home and at school is regularly associated with deficient academic skills. As children become early adolescents, however, two new factors generally make their appearance: substance abuse and trouble with the police (Patterson, 1993). Longitudinal studies have also shown that once serious antisocial behavior begins, for many children it is remarkably constant across time. But these same longitudinal studies have shown different patterns of onset. For example, these studies show that some children develop aggressive behavior in the preschool years and continue antisocial behavior through childhood and early adolescence; these children often show serious criminal behavior in late adolescence and early adulthood (Farrington, 1988; Patterson, 1993). Examination of a broader band of psychiatric disorders in adolescents has revealed a significant difference between disorders that developed before the age of ten and those that began later; the former were often associated with reading disorders whereas the latter were not (Rutter, 1979).

With these changes in the dependent variables in mind, longitudinal studies can proceed directly to the main task: clarifying the temporal sequencing of changes in relationships with changes in psychopathology. For example, longitudinal studies can tell us which family processes precede the onset of psychopathology and, equally important, note what proportion of children exposed to that process develop difficulties. If longitudinal studies show that family processes that were hypothesized causes of psychopathology do not occur until the psychopathology is well established, the family causal hypothesis is severely weakened. As will be shown, the reverse is not the case: establishing the temporal priority of the family process is far from proof that the process caused the psychopathology. The same reasoning could apply to the problem of temperament and family process raised in the previous section. Careful longitudinal studies could establish whether certain features of children's temperament preceded or followed adverse parental treatment. Again, if the development of a difficult temperament follows rather than precedes adverse parenting, the hypothesis of its causal role in parental relationships is severely weakened.

There are two main reasons that temporal priority does not prove cause. First, an unmeasured factor, often called the *third variable,* may be acting to produce one circumstance in an observed temporal sequence and then may also, at a later time, produce the subsequent circumstance. This is likely to be the case when the same genes influence both parenting responses and childhood psychopathology. A genetic factor might first influence parental behavior and then later influence a child's antisocial behavior. Data from our study will support this sequence.

A second, related limitation is the problem of understanding and measuring the developmental precursors of critical variables. We may find in longitudinal studies, for example, that adverse parenting does indeed precede antisocial behavior and falsely conclude that parenting causes antisocial behavior. Antisocial behavior, however, may be preceded by more subtle behavioral and temperamental features in the child (Lyons-Ruth et al., 1993; Tremblay et al., 1994) that researchers may have failed to measure or that may be unknown but are very disruptive to parents. In this sense, the developmental precursor of antisocial behavior is acting as a third variable; initially it provokes adverse parenting and subsequently it evolves into observable antisocial behavior. If this precursor is partially heritable and the disorder itself is partially heritable, behavioral genetics research can be very helpful in identifying not only the precursor itself but the extent to which genetic and environmental factors account for its developing into the clinically significant disorder.

Despite these limitations, research on child and adolescent psychopathology has produced some of the most persuasive longitudinal studies linking social relationships to adolescent development. These studies are compelling because they either search for different relationship factors associated with disorders showing different patterns of onset, or identify both social factors that precede early manifestations of the disorder and different factors occurring later that precede the subsequent manifestations of that disorder. For example, recent research suggests that ineffective parental monitoring and discipline of younger children initiate the early-onset variety of antisocial behavior, but that deviant and delinquent peer groups as well as more general influences of street culture initiate the later-starting variety (Patterson, 1993).

A third use of longitudinal studies, in the area of psychopathology, capitalizes on "natural experiments." These are typical practices, procedures, or circumstances in a community to which some adolescents are exposed and others are not. For example, a recent research project in England sought to understand whether incarceration of adolescents convicted of crimes increased or decreased the likelihood of subsequent criminal behavior after their release. The larger question of this investigation concerned factors that maintain antisocial behavior once it is well under way. Two contrasting theories were at stake. One, a theory of deterrence, predicted that failure to catch and jail offending adolescents would reduce their fear of consequences for their behavior and thus increase the chances that they would commit more criminal acts. A contrasting theory posited that incarceration "labeled" an adolescent as a socially recognized criminal and that this label would serve unintentionally to spur more criminal acts. A sample of adolescents was assessed for criminal behavior before any were incarcerated and then was reassessed several years later to compare the criminal behavior rates of those who were jailed with those who escaped detection and incarceration. Criminal offenses rose in the former and declined in the latter. Again, researchers suspected that many third variables came into play. They questioned, for example, whether the incarcerated boys were more criminal to begin with, or whether they differed in some other measurable way from those not incarcerated. The answer was "no" in both cases, supporting the view that incarceration is a cause of maintaining or accelerating criminal behavior in adolescents (Farrington, 1988).

Assortative Affiliation Studies

We return here to the potential importance of peers in the development of adolescents.

Researchers now agree that adolescents' peer experiences can be divided into three domains. The first of these is their relationships with crowds, the large social groupings characterized by young people themselves with simple and stereotypic titles such as "jocks," "brains," or "druggies" (Brown, 1990). Crowds reflect the way adolescents sort themselves by reputations that they engender or that are hung on them willy-nilly. Crowds may be distinguished from cliques, which are much smaller groups of adolescents who spend time together, know one another fairly well, and often engage in joint activities. Friendships, of course, are more intimate dyadic ties.

A distinctive feature of all three groups is that there is more similarity in values, long-range plans, activities, and behavior patterns among adolescents within them than among adolescents in different social groups. This similarity among people who are in relationships with one another is referred to as *homophily*. Homophily of adolescent social groups is one of two central characteristics of adolescent social arrangements that provide researchers with a new window on the importance of peer relationships in adolescent development. The second feature is the relatively temporary nature of adolescent relationships; indeed, young people develop and then break off relationships with cliques and friends often enough that research can ascertain the characteristics of the adolescents before they enter the clique or friendship. They can observe more completely changes in the adolescents during the period of relationship and then observe the adolescents at the conclusion of the relationship. These two features permit researchers to explore the extent to which a young person's relationships influence his or her development, as well as the extent to which a pre-existing characteristic of the adolescent leads him or her to select a particular relationship. New genetic data have heightened interest in this issue. Recent findings indicate that heritable characteristics of adolescents appear to influence the characteristics of the peer groups they join. Indeed, the genetic influences on these choices are among the strongest in our study (Manke et al., 1995).

Over the last two decades, social researchers in this area have anticipated these genetic analyses. The most complete study of this kind was conducted among New York State high school students almost two decades ago. Researchers focused on several similarities between adolescents in friendship dyads, including drug use, educational aspirations, and political orientations. The study noted substantial similarity among friends-to-be before they linked up, but also increasing similarity as the friendship developed. The researchers concluded that pre-friendship characteristics of the adolescents influenced their choice of one another but that, once connected, the friends continued to influence one another toward greater homophily (Kandel, 1978). A similar balance of influence between the effects of adolescents' characteristics on their selection of friendships and the subsequent influence of friendship on the teens has been observed in a study of adolescent sexual behavior (Billy and Udry, 1985). A third study reported results suggesting that friendship groups served to reinforce tendencies toward adolescent criminal behavior that had already been observed in early childhood (Fergusson and Horwood, 1996). Other studies of cliques and friendships have given greater weight to the importance of pre-existing characteristics of adolescents in selecting friends who are similar than to influences of the relationship once it begins.

Studies of assortative affiliation represent a powerful approach to disentangling the role of an adolescent's individual characteristics from the role of social relationships in psychological development. But they are subject to an interpretive dilemma similar to that faced by other purely psychosocial designs. We can measure assortative affiliation only for those characteristics of adolescents that can be clearly measured before they join friendships and cliques. As will be shown, behavioral genetics designs allow us to assess the influence of all heritable characteristics of adolescents, whether we can measure them or not, on their choice of particular friendships and cliques. Behavioral genetics widens quite dramatically the scope of individual characteristics we can take into account in delineating the balance of their importance in the adolescent's initiating, maintaining, and terminating peer relationships.

Studies Using Experiments

As longitudinal studies are always limited by the possibility that there is some unmeasured third variable confounding results as well as our interpretations of them. Thus it is fortunate that studies of psychopathology in adolescents provide three other tools for measuring influences on psychological development, all using planned rather than adventitious interventions into the lives of adolescents. The first is brief reassembly of the family. Researchers using this method ask parents to interact with children whom they meet for the first time in a study. Meanwhile, their own offspring are asked to interact with the parents of those children. The second tool, therapeutic interventions designed very specifically to test a developmental theory, are applicable only in investigations of psychopathology. Such interventions are used on a randomly selected sample of parents and their offspring; the untreated subsample serves as a control. The third approach, a preventive intervention, is applicable to all children and adolescents, and is designed either to promote competent development or to prevent psychopathology.

Reassembly of Families

In both psychosocial studies and genetically informed studies, research designs become much more difficult, time-consuming, and expensive the more definitively they address causal mechanisms. Longitudinal studies, for example, require researchers to revisit their sample at least twice and, in many instances, more often. Moreover, some research subjects may not wish to participate in the same study for a second or third time. These dropouts pose enormous difficulties for any longitudinal analysis. Similarly, therapeutic and prevention studies, which we will describe shortly, are expensive and cumbersome. Here too, dropouts pose major headaches for the researcher.

A particularly clever strategy for short-circuiting some of the difficulties of intervention designs is to observe parents and children interacting in biologically related pairs and then in unrelated pairs. In the second condition the child and parent are not only unrelated but are meeting each other for the first time. In one study boys with a history of misbehavior were compared with those without behavior problems. The results were quite striking. Boys with conduct problems elicited negative reactions and control attempts from both their own mothers and the stranger-mothers. By contrast, when interacting with nonproblem children, mothers of children with conduct disorders and mothers of normal children reacted similarly (Anderson et al., 1986).

A closely related approach is to study the impact of treating a child for a pharmacologically responsive behavior problem and then observe the subsequent effect of that treatment on the family. Although there is no effective pharmacological treatment for disobedience and misbehavior, stimulants such as Ritalin are effective for attention problems in children that are often associated with secondary problems such as refusal to obey parental instructions. In a number of studies, pharmacological treatment of the child has been shown to improve not only parents' relationships with their children, but also teachers' relationships with their affected students (Barkley, 1989; Cunningham and Siegel, 1987; Cunningham et al.,

1985). In one study of this kind it was clear that these improvements in parent-child relationships, including increased warmth and decreased criticism, occurred only when the child responded to the drug (Schachar et al., 1987).

These two closely related approaches—family reassembly and effects of controlled trials of child treatment—provide some of the most powerful evidence that children's characteristics can evoke particular responses from their parents. They also underscore the important influences of enduring characteristics of children on their relationships with their parents and others. These evocative effects are important in understanding how genetic factors in a child can have a substantial influence on his or her social relationships.

Therapeutic Interventions

Therapeutic interventions can be aimed specifically at relationship difficulties that are thought to influence the onset or maintenance of psychopathology. A successful therapeutic outcome of such an intervention can be strong evidence supporting a theory of the causes or etiology of a disorder if (a) the intervention can be shown to have changed the offending relationship pattern; and (b) this measured change in relationship pattern precedes and predicts changes, in response to treatment, in the psychopathology itself.

Therapeutic interventions can be thought of as controlled, rather than natural, experiments. Subjects are assigned at random to the treatment condition or to no-treatment contrast groups. The larger the total sample of subjects, the more likely it is that the group randomly assigned to the treatment will be similar to that assigned to the control, reducing the chances that a third, unmeasured variable might account for an apparent advantage of the treatment over the comparison condition.

As a logical principle for verifying hypotheses, however, even well-conducted treatment studies have two major limitations. The first concerns secondary, or nonspecific, complications of an established disorder. By the time an adolescent has a diagnosable disorder, three separate sets of factors combine to influence the seriousness of his or her condition and its amenability to treatment.

The first are the original etiologic factors: the risk factors, the absence of protective factors, and the immediate provocative factors that initiated the disorder. The second are maintenance factors, or circumstances that may maintain a disorder once it starts but had no important role in its initiation. We have already cited one example of a maintenance factor: incarceration of misbehaving young people appears to exacerbate their antisocial patterns. Incarceration may have this effect because it involves teens in an even harder, more delinquent peer group, interrupting what little chance they have left to acquire positive skills and labeling them for life with a prison record. Incarceration, then, perpetuates or intensifies a serious adjustment difficulty but cannot be counted among its causes. A treatment for incarcerated youngsters may fail not because it does not address the core etiologic mechanisms of antisocial behavior, but because the secondary consequences of a jail experience may make the youngsters unusually resistant to treatment.

A study done some years ago resulted in a similar finding for schizophrenic patients and individuals with other chronic psychotic disorders. As a consequence of prolonged hospitalization, now almost unheard of in the managed-care era, as well as nonsupportive relationships with hospital staff, psychotic patients experienced "social breakdown syndrome," characterized by the loss of the most elementary social skills and motivations (Gruenberg, 1967). This state made these patients very difficult to treat, even though the same treatment given to them was effective in more recently diagnosed patients. Indeed, there is accumulating evidence that a secondary breakdown in functioning in schizophrenia may start very soon after the first major clinical episode.

The third set of factors relates to the treatment milieu itself: the impact on the adolescent of being labeled a "patient"; the willingness and ability of the adolescents and their families to participate in treatment; the influence of the referral source on the treatment process (a court or the family may have brought the adolescent for treatment); and many specific features of the treatment setting itself, including the characteristics of the therapist and how he or she is perceived by the patient.

These three sets of factors are not easy to disentangle, but as a consequence of any of them, a well-designed treatment may fail even if it addresses the underlying core mechanisms of the disorder. Thus the absence of a good therapeutic response doesn't necessarily invalidate an etiologic theory.

A second limitation of treatment studies as hypothesis testing is that even well-focused therapies may affect more behaviors in the patient or more social patterns in the family than the targeted behaviors proposed by an etiologic theory. Thus therapeutic interventions that elicit unintended but benign effects may produce a good therapeutic response, not because they directly address the proposed etiologic factors, but because they offset them. In Chapter 11 we present genetic evidence that illuminates this problem. In our study, evidence suggests that mothers, fathers, and siblings each engage in separate but important adaptive strategies that suppress antisocial behavior. For example, mothers appear to respond with warmth and support to early forms of sociability in their children. This may have the effect not only of enhancing the child's sociability, but also of suppressing the child's antisocial behavior. An effective therapeutic intervention, aimed at reducing antisocial behavior through improved disciplinary practices, might inadvertently activate a warm and supportive parenting pattern of this kind. The investigator may not be aware of the importance of this supportive behavioral system and, as a result, may fail to measure its activation by the intervention. As a consequence, the effect of the intervention on the target behaviors of poor discipline, and the consequence for changes in these targeted behaviors, may be too narrowly interpreted. This is another version of the third-variable problem. In this instance a positive outcome of a treatment does not necessarily prove an etiologic theory. A variety of strategies, reviewed below, have been developed to deal with this difficulty.

Preventive Intervention

Preventive intervention trials offer important advantages over therapeutic trials as a logical principle for testing psychosocial theories. There are three component strategies in this approach (Gordon, 1983). *Universal preventive interventions* are designed to be administered to an entire community or population. The most obvious example is fluoride in the drinking water. *Selective preventive interventions* are designed for populations identified as at-risk. The risk is estimated from the status of the child, not from the child's current psychological functioning. Examples of children whose "status" is "at-risk" include all children of alcoholics, or all children who have had a parent die or who have witnessed a homicide. This strategy requires that the psychological functioning of all individuals who are entered into the study be within normal limits at the time of the intervention. *Indicated preventive interventions* are designed for individuals whose behavior or experience is showing some signs of abnormality or a characteristic predictive of disorder, but who have not developed a recognizable disorder. In our brief review we treat only universal interventions, since they are most distinct from the logic of theory-testing therapeutic trials.

Universal preventive interventions have two distinct advantages over treatment studies. First, like other efforts at prevention, they can address the early risk, protective, and provocative factors that initiate the disorders before the complications of diagnosed illness become established. Second, unlike any other intervention, the universal preventive intervention does not label the children and families participating in the study. The confounding factors involved in treatment settings, those reviewed in the previous section, are completely bypassed. Further, some prevention designs seriously address the third-variable problem.

A good example of a universal intervention that accomplishes these important objectives is a major study in Baltimore designed to test two parallel theories concerning the early precursors of adolescent depression and antisocial behavior (Kellam and Rebok, 1992). Low school achievement is posited as the antecedent of the former and aggressive behavior, mixed with shyness, as an antecedent of the latter. The Baltimore study designed two interventions suitable for first-grade classrooms. Shy but aggressive behavior was addressed by a classroom exercise that helped children respond to different demands of the teacher and setting. In one such exercise, cooperative games rewarded groups of children for the good behavior of all of them; this encouraged children to be responsible for one another's cooperation and self-control. A second intervention also focused on groups but encouraged each child to be responsible for enhancing the learning capacities and speed of the others. Classrooms and schools were assigned to one or another of these interventions or to control groups in which children received ordinary classroom treatment typical of Baltimore public schools.

The use of two different interventions is one logical strategy for assessing the specificity of each of them. Researchers in the Baltimore study, which is ongoing, expect the cooperative-behavior game to reduce shy and aggressive behavior in those children who have high levels of both, and then to show that this intervention reduces antisocial behavior later in childhood and in early adolescence. They have comparable expectations for the cooperative-learning curriculum and its long-term effect in reducing depression. Crossover effects, for example, if the cooperative-learning curriculum reduced shy and aggressive behavior or the cooperative game reduced depression, would weaken the hypothesis. In early reports from this project the good behavior game did have a specific, short-term impact on aggressive behavior, and the mastery-learning task an equally specific effect on achievement (Dolan et al., 1993). Moreover, the use of the good-behavior game in both first- and second-grade children did have an impact on aggressive behavior when children were observed in sixth grade. The impact of the mastery-learning task on subsequent depressive symptoms was less clear, thus raising useful questions about the initial etiologic theory (Kellam, Rebok, Ialongo, and Mayer, 1994; Kellam, Rebok, Mayer, Ialongo, and Kalodner, 1994).

It is widely believed that if a disorder can be successfully treated or prevented by psychosocial interventions, it is unlikely that genes play a major role in the development of that disorder. Genetically influenced disorders will someday, this implicit argument goes, be treatable by biological methods. But as will be shown, successful psychosocial interventions may actually support *genetic* theories of the causes of some psychiatric disorders and may, in fact, be the best prevention or treatment for genetically influenced disorders of behavior and psychological adjustment.

Chapter 3

GENETIC INFLUENCES
ON DEVELOPMENT

This chapter continues to lay out the logic of psychosocial and genetic studies. It centers on three fundamental logical principles in drawing inferences from genetically informed designs. First, it clarifies that samples in behavioral genetics always consist of twosomes: twins, siblings, or parent-child pairs. Samples in psychosocial studies, by contrast, are typically constructed of single individuals. Second, it summarizes how genetic inferences can be drawn from twin and sibling designs and from adoption designs. Just as with psychosocial designs, however, there are problems in drawing unambiguous inferences from these methods, and we review some of these in this chapter. Third, the chapter describes how genetically informed designs are a rich source of data on the nongenetic influence of the environment.

The study of the heritability of behavior is as old as history. For example, it has been known for thousands of years that dogs can be bred for temperamental and behavioral traits, and to this day selective breeding of animals remains an important part of studies of the genetic influence on behavior. During the last century and a half, many new approaches have added to our understanding of the role of genes in both animal and human behavior. These include evolutionary theory, population genetics (the study of factors that alter the frequencies of genes in populations), and—most recently—the direct study of DNA and RNA using the tools of molecular genetics. But the branch of genetics that is currently most relevant for studying adolescent development has been termed for some years "quantitative genetics." This field of inquiry asks a simple but fundamental question: What role do genes play in accounting for differences among individuals in areas such as personality, cognitive abilities, or psychopathology? More recently, researchers in this field have expanded their inquiry to include understanding not just individual differences at a single point in time, but also differences in how individuals develop both skills and deficits across time. This new and highly relevant approach may thus be called *quantitative, developmental behavioral genetics* (Plomin, 1986).

Genetic Differences among Individuals with Behavioral Differences

Quantitative, developmental behavioral genetics is concerned with differences among individuals within large groups or populations. It asks what proportion of these differences are due to genetic differences among individuals. For example, suppose we are interested in differences among adolescents in antisocial behavior in the population of the United States. To begin with, we need a sample of adolescent research subjects that is reasonably representative of this population. Then quantitative behavioral genetics can estimate the proportion of the differences in antisocial behavior in this adolescent sample that can be attributed to genetic differences among them. This proportion is the heritability and can vary from 0 to 100 percent, with 100 percent implying that individual differences in behavioral patterns among members of a particular population are attributable entirely to genetics, a circumstance that is unheard of in this area of research. Conversely, most behavior patterns that have been studied—personality, cognitive abilities, psychopathology, and others—show at least moderate heritabilities.

Quantitative, developmental behavioral genetics deals with the sum total of genetic influences, from one gene operating alone to produce a particular developmental outcome, to many genes operating sequentially or simultaneously to do the same. But quantitative behavioral genetics cannot identify individuals who have particular genes; nor can it estimate the genetic liability for behavioral disorders or for positive developmental outcomes for any single individual. There is a precise analogue to these logical principles in psychosocial studies. They, too, focus on populations and estimate the proportion of differences among individuals that can be attributed to particular social influences such as family type or quality of peer relationships. In most cases, psychosocial studies cannot precisely estimate the liability for behavioral disorders for any single individual on the basis of psychosocial risk factors.

In order to estimate the importance of genetic factors for a developmental outcome, quantitative behavioral genetics must measure the association between genetic variation and the outcome. Because molecular genetics has not advanced enough to enable us to locate and identify the specific genes related to most developmental outcomes of importance in adolescence (or other developmental periods, for that matter), quantitative genetics uses an intriguing, indirect strategy: its samples are always composed of pairs of subjects rather than individuals alone. Typically these are twins, other siblings, or parent-child pairs. In measuring the association between genetic factors and developmental outcomes, all quantitative behavioral genetics studies use a two-step process.

First, researchers measure the similarities between individuals within a pair. This can be done in two ways. To illustrate each of these we will focus on studies of depressive disorder. Let us say that we use a system for classifying depression such that there are only two choices: an adolescent meets all the criteria and is rated "definitely depressed," or the adolescent does not meet all the criteria and is rated "not depressed." If we have thirty pairs of identical twins, we can then count the number of pairs where both twins are depressed, where one is depressed and not the other, and where neither is depressed. Let us say that in our small sample we observe ten pairs where both were depressed, ten where neither was depressed, and ten where just one was depressed. Two different measures of association can be used here; both use only pairs where at least one twin is affected. A "pairwise concordance" is the proportion of pairs of individuals where both are affected, here 50 percent. A "probandwise concordance" is the proportion of depressed twins in concordant pairs, here 20/30 = 67 percent.

A second form of measuring association, and one we use exclusively in this book, is a *correlation*. Like many conceptual tools in behavioral genetics, correlations, illustrated in Figure 3.1, owe their origins to Francis Galton. A correlation is used when a difference between individuals can be measured quantitatively, such as height or weight. Moreover, the intervals on the scales used to measure the difference must be approximately

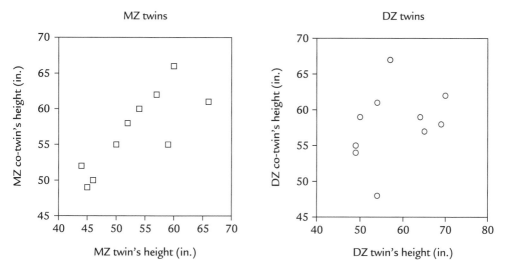

Figure 3.1   Correlations for height in MZ and DZ twins

equal. For example, the difference between an adolescent who is sixty inches tall and another who is seventy inches tall is the same as that between one who is fifty inches tall and one who is sixty inches. The correlation asks how predictable the weight or height (or any other measure) is in one member of a pair if we know that value for the other. Figure 3.1 shows data contrived to illustrate two correlations. In this example the height of a co-twin is predicted from the height of the twin. In the identical (monozygotic, or MZ) twins, in this example, the height is quite predictable from one twin to the other, but for fraternal (dizygotic, or DZ) twins it is less predictable. The more predictable one sibling's measure is from that of the co-sibling, the higher is the *correlation coefficient*. In this case the correlation coefficient, r, is .64 for MZ twins and .29 for DZ twins. These correlations can be represented as bar graphs: the higher the bar graph, the greater the value of r. In genetics we typically use a special form of correlation known as the *intraclass correlation*, which assesses not only the predictability of one pair member's score from the other, but also the extent to which it is the same in value.

The next step in the logic is to compare these correlations across subsamples of pairs that differ in their genetic similarity. For example, we have just compared monozygotic twins who are genetically identical to each other, since they derive from the same sperm cell and egg, and dizygotic twins, who share approximately 50 percent of their genes for reasons we will explain later. If genes are associated with behavior, we would expect greater intraclass correlations for the group of sibling pairs more closely related genetically. That is, if a measure of behavior is influenced by genes, there should be more predictability from one individual to another when those individuals are closely related genetically than when they are not. For even greater contrast, we can introduce a third group of sibling pairs in which one child has been adopted into a family and the second is the biological child of the parents. These siblings share no genes if the biological parents of the adopted child were not blood relatives of the adopting parents and the adoption agency did not successfully match biological and adoptive parents on characteristics related to our study. If genes are important to the developmental outcome we are measuring, we would expect much lower correlations for this group than for the other two.

Figure 3.2 shows data contrived to illustrate the comparison among three expectable correlations of identical twins, fraternal twins, and unrelated or adopted siblings. The x axis, which we usually call the *independent variable,* is the degree of genetic similarity among the pairs, and the y axis, the *dependent variable,* is the correlation in that sample of pairs. The height of the bar graphs equals the size of the correlation. The difference among the heights, then, is the difference among the correlations. We can also draw a line between the bar graphs as shown. The steepness of this line is its slope, and this directly corresponds to the proportion of individual differences accounted for by genetic factors. We refer to an orderly decrement in correlations, one that is associated with declining genetic similarity among pairs, as a *genetic cascade.*

There is a direct correspondence between this form of association and that used in psychosocial inferences. Usually a psychosocial variable, like a genetic variable, is associated with a developmental outcome by correlation. The psychosocial variable compares groups of *individuals,* however. For example, let us say that three groups of adolescents were exposed to three different levels of adverse parenting: low, moderate, and very severe. On an imaginary scale we might assign these three groups the scores of 0,

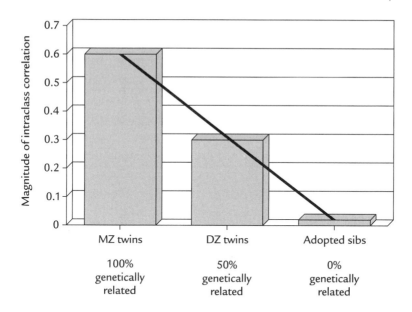

Figure 3.2 Comparison of intraclass correlations across subsamples

50, and 100, respectively, on the *x* axis. If adverse parenting is associated with antisocial behavior, the 0-group should have very low scores, the 50-group moderate scores, and the 100-group very high scores on the *y* axis. Note that here, in contrast to genetic analyses, the *y* axis uses mean scores of groups of individuals rather than correlations of pairs of individuals.

Genetic Mechanisms and Inferences

Although genetic and psychosocial analyses share a preoccupation with individual differences in populations, and reason by associating an independent variable with a dependent variable, there are two important differences in the fundamental logic of their inferences. First, genetic analyses take account of environmental influences, but psychosocial analyses cannot encompass genetic analyses. As shown in the last chapter, when psychosocial studies explore the role of relatively stable individual attributes as *initiators* of social processes that influence measures of adjustment or that interact with social processes to influence adjustment, they may be indirectly estimating the role of genetic factors in psychosocial processes.

Nongenetic factors may contribute to stable differences among individuals, however, and genetic factors may contribute to change in individual characteristics. Thus psychosocial studies can only speculate about genetic influences. In contrast, genetic analyses can provide more secure estimates of the magnitude of genetic influences as well as environmental and nongenetic influences on any measure of adjustment.

Second, quantitative behavioral genetic analyses draw a special logical force from certain well-established features of genetic mechanisms. Here we can only summarize briefly the most relevant mechanisms, but interested readers can consult excellent introductory texts for more information on the subject (see, for example, Plomin et al., 1997).

- When a sperm cell or an egg is formed, all the genes of each parent are divided exactly in half. Thus, an egg contains 50 percent of the mother's genes and a sperm cell exactly 50 percent of the father's. Therefore, each child shares exactly 50 percent of his or her genes with each parent.
- The genes that end up in one sperm cell or in one egg are not the same as those that end up in subsequent sperm cells or eggs. The assortment is, basically, a random process. It is theoretically possible for two sperm cells produced at different times to have the exact same genes, but this is extremely unlikely given that there are about 100,000 genes in the human genome. It is more likely that any egg or sperm cell will have approximately 50 percent of its genes in common with any other sperm or egg. Thus, full-born sibs will have approximately 50 percent total genetic overlap with any given sib, on average.
- MZ twins are the product of a single sperm and a single egg. Within ten days of fertilization, the fertilized egg splits into the germ of two separate individuals who develop *in utero* and subsequently are genetic carbon copies of each other. DZ twins are the result of two nearly simultaneous fertilization events; two different sperm fertilize two different eggs, both of which go on to develop *in utero* and subsequently are as genetically similar as ordinary siblings.
- Genes are, biologically speaking, very conservative. This was Mendel's fundamental discovery, although the term "genes" was not applied un-

til fifty years after his famous experiments with peas. Genes replicate themselves exactly both in the course of ordinary cell division and cell replication and in the production of sperm and eggs. Except for unusual events that destroy, distort, or replace them, such as radiation, virus infection, or gene therapy, they remain chemically invariant across broad stretches of time. Once individuals are endowed, at fertilization, with their complement of 100,000 genes, only unusual or destructive events alter those genes. The new gene therapies are unusual efforts to produce positive alterations of genes in certain cell lines. It is clear that behavioral patterns of individuals or of those in their social world cannot, under usual circumstances, influence their own particular complement of genes: in this sense genes are a primary cause, and all biological and behavioral events in any individual are subsequent to them—a clarity of causal ordering unknown in other branches of behavioral science.

- Gene expression is not foreordained, however, and the degree to which many, if not most, genes express themselves in behavior or other observable manifestations depends on many factors. Some of the molecular mechanisms that regulate gene expression have now been worked out. As will be shown, it is now plausible that psychosocial processes are also part of the chain of events by which genes express themselves in behavior. Because these variable mechanisms of gene expression are so important for so many genes, almost certainly those that regulate behavior, behavioral genetics findings do not support notions of strict biological determinism. Just because a behavioral characteristic is heritable does not mean that if certain genes are present the behavior will always occur. Natural variation in the operation of genetic expression may mean that in many cases the behavior of some individuals will not reflect the genes that put many others at risk for certain behaviors. Similarly, a full understanding of the pathways of genetic expression would allow us to use planned interventions to enhance the effect of positive genes and suppress the effects of negative ones. Again, in the area of behavior these interventions—once perfected—are as likely to be psychosocial interventions as biological ones.

Natural Experiments: Siblings and Adoption

These critical principles of genetics have led quantitative geneticists to use two core strategies to estimate genetic influences on behavior: siblings and adoption designs. The sibling approach has centered on the contrasts between monozygotic and dizygotic twins, but more recently other types of siblings, such as half-siblings, have also been used in genetic studies (Schukit et al., 1972). The twin method is the most important and frequently used sibling design, and the one to which we devote the most attention here.

SIBLING DESIGNS Twins have fascinated people for millennia. The fraternal twins Esau and Jacob are believed by many to be critical in the line of succession from Abraham to the establishment of both Jewish and Moslem civilizations. Castor and Pollux were divine progeny of Zeus and Leda. Two pairs of identical twins are central to the hilarious and poignant plotting of Shakespeare's *Comedy of Errors.* During the nineteenth century, the circus impresario P. T. Barnum played on public fascinations by widely exhibiting Siamese twins. Only in the last seventy years has the natural phenomenon of twinning been used for serious scientific investigation of genetic influences on behavior. But the special attention twins continue to receive has fed lingering concerns about how accurately we can generalize what we learn from twins to other people. Thus it has been essential that researchers supplement the twin method with other approaches, as we do in the current study.

About one in eighty births in the United States is twin; about one-third of these are identical twins, and an additional third are same-sexed fraternal twins. These two groups are critical for the comparisons we have already described. Twins, particularly identical twins, are not only special as objects of public fascination; they also have special circumstances of development *in utero.* As a fetus grows, it develops a sac with two layers to provide protection and nourishment: an outer chorion and an inner amnion. All fraternal twins have their own chorions and also their own placentas. In some cases the inner, amniotic sac is fused. By contrast, in approximately two-thirds of identical twins there is only one chorion. In such cases the twins develop from a fertilized egg that splits late in the ten- to fourteen-day period when identical twinning is possible. In almost all these cases, there is some interconnection between the blood supplies of the two twins. In between 5 and 25 percent of cases the blood of one twin, called the donor, flows to the other twin, the recipient. This is called "twin transfusion syndrome" and can lead to significant differences in hemoglobin level and birth weight between the identical twins and may constitute the first chapter of nonshared environmental experiences, in this case the nonshared intrauterine environment.

The contrast between identical and same-sexed fraternal twins is attractive to researchers because it seems to be a nifty "experiment of nature." As in experiments, certain factors seem to be controlled or similar between the two types of twins: their gender, their age, and the fact that they are reared by the same parents, with the same siblings, in the same family. The issue of differences in age between ordinary siblings is a major one. Not only might many parents treat a first-born differently from a later-born child, but the children might be born into very different family contexts. For example, a grandparent who was very close to the family may have become seriously ill and died between the birth of the first child and the appearance of the second child (Walsh, 1978). Equally important, a major historical event may have occurred during a different phase of development for siblings of different ages. For example, the age of a child at the time of the onset of the Great Depression had a lasting influence on his or her subsequent development (Elder, 1974).

Despite the conspicuous advantages of twin studies, researchers have voiced two reservations about their use. First, they are concerned that parents and others treat identical twins more similarly than they do fraternal twins. Second, they wonder if the special status associated with being a twin makes it difficult to reason from twins to more ordinary singletons.

It is plausible that same-sexed twins would have more or less the same environmental experiences; differences in similarity between identical and fraternal twins could then—in this natural experiment—be attributed entirely to the differences between these two groups of siblings in their genetic similarity. The assumption about equal similarity in environmental experiences between fraternal and identical twins, known as the "equal-environment assumption," is the logical cornerstone around which all twin research has been established. As critics have maintained, however, parents and others may treat identical twins more similarly than they treat other siblings simply because they are twins. In that case, the greater similarity of identical twins would be a trivial result of this special labeling.

In recent years, and after considerable prodding by their nongenetic counterparts, genetics researchers themselves have given much thought to this potential flaw in the twin method, and their research has addressed the problem carefully. They have collected data on differences in the family experience of identical and fraternal twins. To be sure, identical twins are treated more similarly than fraternal twins. They are dressed alike, spend more time together, are more likely to have the same teacher in school, and are more likely to be treated similarly by their parents. Nonetheless, these specific differences between identical and fraternal twins are unlikely sources of differences in similarity on a range of developmental outcomes. For example, twins who are treated similarly, or who share the same room, are no more likely to have similar personalities or cognitive and academic skills than twins who do not share the same room or who are not treated similarly (Loehlin and Nichols, 1976).

Recently, more convincing tests of the equal-environment assumption have been carried out. One approach takes advantage of variations in both similarities in fraternal twins and inaccurate obstetrical diagnoses. Both of these factors lead some parents to believe that twins who are biologically fraternal are identical and vice versa. Two studies have shown, however, that the estimates of heritability are about the same for correctly labeled twins as for incorrectly labeled twins (Goodman and Stevenson, 1991; Scarr and Carter-Saltzman, 1979).

Even more convincing are recent studies comparing identical and fraternal twins who have been reared apart. Let us imagine a twin pair, A and B, separated soon after birth and raised by adopting families, also A and B, who have no contact with each other. The key question concerns the similarity in rearing practices of adopting families A and B. There is no reason to believe families A and B will treat their adopted twin more similarly if their adopted child happens to be an identical twin rather than a fraternal twin. Even if they know they have adopted an identical twin, family A and family B have no way of coordinating or even knowing about each other's rearing practices. Thus, it is crucial that heritability estimates in studies using twins reared apart are comparable to those in which the twins are reared together. It should also be noted that uterine circumstances such as fetal transfusion syndrome are environmental factors that are *less similar* for identical twins than for fraternal twins and would lead to underestimates of genetic influence by the twin method.

The second concern researchers have about using twin studies is that twins might be too special. On the one hand their celebrity status may mean that factors important in their development, such as physical appearance, may be less important for nontwins, or vice versa. This aspect of twinning has not received much systematic or direct attention. On the other hand, in some areas twins are at a distinct disadvantage in comparison with singletons. They are born three to four weeks earlier than singletons and, as a consequence, have lower birth weights and are shorter. These differences begin to disappear in childhood. Similarly, twins show some delay in language acquisition early in development (Plomin et al., 1990). As will be shown, our study revealed a great deal about the comparability of twins and nontwin siblings on many measures in adolescence, and shed new light on parent-child and sibling relationships for twins as well as for their nontwin siblings.

Perhaps genetic researchers' best defense against such problems is to explore other sibling approaches. If the results of studies using other sibling pairs are comparable to those utilizing twins, our confidence in these results is strengthened. Step-families provide another natural experiment for comparing groups of siblings. When a mother (or father) has two children by a former spouse and then remarries, those two children are—genetically speaking—like ordinary siblings and fraternal twins. They share approximately 50 percent of their genes. If the mother brings to her new family a child from a former marriage and then has a child by her current spouse, the two children—because they have a common mother—share 25 percent of their genes. They are "half-sibs." A third type of sibling pairs we call "blended." A step-family of blended siblings consists of children from the mother's previous marriage as well as from the father's. These children share no genes if the prior spouses were not related and the current spouses did not pick each other on the basis of the similarity of the genes under investigation.

The step-family genetic model is not without its own problems. Two in particular stand out. First, remarriage may occur relatively late in a child's life. For example, in our study, all the siblings were ten to eighteen years old, and we included only those families in which the current spouses had been married for at least five years. The marriages or co-parenting arrangements in half-sib families had to be longer than the age of the youngest sib, by definition. Thus the environments of the three groups of siblings may have differed. Blended sibs may have lived in separate households until their birth parents divorced. They may have lived in a common household only for the few years since the step-family had been formed. By contrast, full sibs in step-families have shared the same mother their entire lives and have had the same biological father for their earlier years and the same step-father later. The divorce process itself, in most cases, goes on for many months, if not years, and there is often a gap between the final divorce and remarriage. Thus there may be many months or even years when the full sibs are not living with a common father. Half-sibs have the longest period of uninterrupted similarity of rearing parents, a time extending from the birth of the youngest of two step-sibs. On the basis of differences in uninterrupted similarity of rearing parents, however, the half-sibs score highest, with the full sibs and blended sibs roughly tied. If uninterrupted similarity of current rearing parents was the central factor in sibling similarity, we'd expect the following pattern of sibling correlations on outcome measures: half-sibs *greater than* full sibs *equal to* blended sibs. If the amount of time spent living with the same rearing parents was key, then we would expect the following pattern of sibling similarities: half-sibs *greater than* full sibs *greater than* blended sibs.

Two research findings suggest that these differential percentages of sharing the same household and the same rearing parents are unlikely to influence comparisons of correlations across these groups. First, as noted and as we review in more detail in Chapter 6, there is very little evidence that common family experience accounts for sibling similarity on almost any measure of developmental outcome. Most data suggest that siblings are similar because they share some of the same genes rather than the same environment (Plomin, Chipuer, and Neiderhiser, 1994; Plomin and Daniels, 1987). Second, in our preliminary analyses we have examined the length of time children have shared the same current household parents. While additional analyses are under way, our initial results suggest that this variation in shared parenting had negligible effects on the similarity of step-siblings on a number of measures (Plomin, Reiss, Hetherington, and Howe, 1994). We return to this issue in greater detail in Chapter 5.

A second major difficulty with the step-family approach arises when parents feel different about their biological children versus their step-children. The parental experience has its analogue in twins and can be summarized, from the perspective of the parent, as follows: "*These are identical twins,* so I will treat them similarly." For some step-parents the comparable experience is, "This is *my biological child,* so I will treat her more warmly and give her greater freedom." Cinderella was the victim of this family dynamic. Other step-parents may have a different view about how to treat their biological children: "This is *my biological child,* so I will take responsibility for careful supervision even if it means getting into a fight."

These considerations suggest that parents will treat children differently when one of the children is biological and the other is a step-child. When both children are biological or both are step, the mean differences should be fewer. As shown, however, genetic inferences are based on correlations, not on mean differences within pairs of siblings. It is entirely conceivable that parents' behavior toward their biological child is predictable from their behavior toward their step-child; that is, if a father, say, is very mean to the step, he will also be mean to the biological child. If he is kind to the step, he will also be kind to his biological child. In sum, the phenomenon of biological parentage predicts contrasts between parents of mixed children (step and biological) and parents of children of the same status (both biological or both step) in mean differences of parenting, but does not necessarily influence the predictability of parenting from one child to the other.

We can examine this assumption empirically, asking if parenting practices are more consistent or predictable across two children for same-status children than for mixed-status children. If so, we can construct expectations for sibling similarities in parental treatment on the basis of varying patterns of biological relatedness. For a mother the expected similarity is full sibs equal to half-sibs (in the second case both children are her biological children) greater than blended sibs. For a father we would also expect high correlations for full sibs (since neither is his biological child), but lower correlations for half-sibs and blended sibs, since they consist of one who is his biological child and one who is not. Thus the crucial comparison to determine the effect of biological relatedness is between half-siblings. We examined these comparisons for six measures of parenting. The average correlations across these six measures were identical for mothers and fathers. While our data suggest that there are reliable differences in *mean or average* levels of parenting toward a biological and a step-child, there is no difference in predictability of behavior. For example, a mother who treats her biological child harshly is very likely to treat her step-child harshly even if she treats her biological child somewhat better.

ADOPTION DESIGNS   We have already discussed two ways in which adoptions can be combined with sibling research to make genetic inferences. The study of twins reared apart and of step-families exploits for scientific purposes the willingness of adults to assume the care of children who are not their biological offspring. But the most widely used adoption approaches are simpler and more straightforward. There are two major types of adoption studies, and they are distinguished by how researchers obtain their samples. The distinction is also a logical one, however.

In the first type the researcher studying a particular condition typically begins by locating parents who have given up their children for adoption. Usually the parents themselves are not contacted; rather, the researcher uses various public records documenting whether the biological parents had a serious mental illness or a criminal record. These parents fall into two categories: those who have the condition under study and those who do not. For example, the affected parents might be people with alcohol or drug addictions, or sustained criminal behavior, or schizophrenia. The control group usually consists of people who are relatively normal. The fundamental question is the status of the offspring. The simplest approach is to confine the sample to offspring who are well into the age of risk for a particular disorder. Thus in many studies of this kind the offspring may be late teenagers or adults. If all or almost all the affected offspring are the biological children of affected parents, then this is strong evidence for the importance of genetic influence. This design is a useful companion to twin and sibling studies.

In one important study, this type of adoption design was modified and strengthened (Tienari et al., 1985; Tienari et al., 1987; Tienari et al., 1994). Children were followed up, in many cases, before they passed through the age of risk. Moreover, their adoptive families were also included in the study. Such a design allows for a more direct study of the role of the family's social milieu. More important, it is a powerful design for estimating the interaction between genetic and environmental factors in psychological development. We return to this topic in the next chapter.

In the second type of adoption study researchers typically begin with the children who have been adopted. In some cases the adopted children are divided into those who are affected with a condition and those who are not. Researchers then make an effort to contact, or obtain records on, all the biological parents of those children. In addition, and this is the crucial distinction, the adopting parents of those children can also become subjects of the study. In one important instance a variation of this design studied continuous variation among children in personality and cognitive competence rather than simply dividing children into two categories: ill or healthy. This study directly assessed the adopted-away child, the adoptive parents, the birth mothers, and many of the birth fathers (Plomin, DeFries, and Fulker, 1988). In addition, it included a control group of birth parents who were raising their own children. A study design of this kind allows for a critical comparison that is directly analogous to the sibling methods just discussed. We can compare three correlations: (a) correlations between the characteristics of the biological parents and their adopted-away offspring; (b) correlations on the same measure between the adoptive parents and their adopted offspring; and (c) correlations between the birth parents rearing their own children and their biological offspring. Figure 3.3 shows a comparison of this kind between parents' body weight and that of their four-year-old children. For a characteristic that is substantially influenced by genetic factors, we would expect high correlations between birth parents and their offspring, whether or not they reared their own children, and much lower correlations between adoptive parents and the children they are rearing. This is the pattern shown in Figure 3.3.

This more complete adoption design sometimes contains one more component. Often, adopting parents have a biological child of their own.

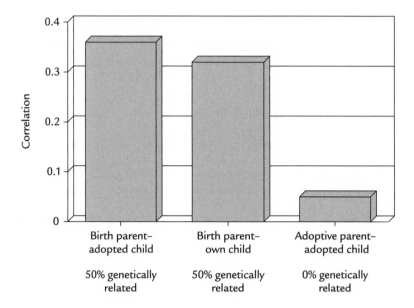

Figure 3.3 Adoption study correlations in body weight between parents and their four-year-old children (Plomin, DeFries, and Fulker, 1988)

This biological child shares no genes, of course, with the adopted child. This sibling pair is identical, in a genetic sense, to the blended siblings in the step-family design.

Critics of adoption studies have raised three major objections to this type of research. The most important cuts to the core of the studies' logic: Do adoption agencies intentionally and successfully match characteristics of the biological parents giving up their children for adoption with those of the adopting parents? For example, would an agency worker be inclined to find an adopting parent who was a doctor or a lawyer if the biological mother was herself at the same occupational level or, if the mother was a teen, if her father was at that level? This practice, known as "selective placement," could account for similarities between biological parents and their adopted-away offspring and masquerade as genetic effects. Fortunately, selective placement can be estimated if the researcher has sufficient information about the biological parents, the grandparents of the biological parents, and the rearing parents. In well-designed studies the effects of selective placement are trivial (Plomin, DeFries, and Fulker, 1988).

The second major objection, similar to concern about twins, is that adopting families may not be comparable to more typical, nonadopting families. Critics fear that adopting families might be less motivated to influence the development of their children than are other families. Perhaps adopting families feel less responsible for the developmental outcomes of their adopted children and hence discipline them less effectively or expend less effort supporting and directing their positive developmental efforts. Or, conversely, it may be that almost all adoptive families are competent and effective. Thus any sample of adoptive families may not include incompetent families. Without this variation in competence, researchers will not be able to show the importance of social influences. In either case—poorly motivated families or a restricted range of families—it could be argued that adoption studies will underestimate the effect of the family environment and, consequently, overestimate the effect of genetics.

Seventy years ago, when adoption studies were first used, this question of whether adoptive families are like "average" families might have had real force. Even thirty years ago, when the first adoption studies in the field of schizophrenia had a monumental impact on psychiatric research, this question still had some meaning. Now the concept of a "typical" family is nearly meaningless. With the number of never-married mothers increasing and divorce and remarriage continuing at a high rate, there is no typical or standard family that is an obvious comparison group. Nonetheless, some systematic comparisons have been made between the parenting practices and family environments of adopting families, on the one hand, and those of two-parent families with biological children, on the other. Differences between these two groups of families tend to be small or nonexistent. Where differences are found they are as likely to indicate greater involvement of adoptive parents in child-rearing efforts as to indicate less involvement (Plomin, DeFries, and Fulker, 1988; Rhea and Corley, 1994). For example, in one study of families with young children, adoptive families showed greater maternal involvement as well as restriction and control than did nonadoptive comparison families (Rhea and Corley, 1994).

The question of variability in the level of competence in adoptive families is more open. It is important to note that recent adoption studies have found a great range of competence in adoptive families (Tienari et al.,

1985; Tienari et al., 1994). Indeed, some adoptive parents show relatively high rates of legal, marital, and psychiatric problems (Cadoret, Winokur, Langbehn, Troughton, et al., 1996; Cadoret et al., 1995), whereas others show very low rates.

The third important objection to adoption studies returns, again, to the womb. Various aspects of maternal health can influence fetal development. The most notable and best-documented are the fetal effects of substance abuse, such as nicotine, alcohol, cocaine, and heroin (Chasnoff et al., 1985; Chasnoff et al., 1989; Chasnoff et al., 1992; Chasnoff et al., 1993; Hatch and Bracken, 1986; MacGregor et al., 1987). Substance abuse during pregnancy is associated with a broad range of problems and adverse circumstances faced by the mother, and the impact of these substances on the fetus may masquerade as a genetic effect. A partial solution to this dilemma is to compare correlations between attributes of fathers and their adopted-away offspring with the same correlations for mothers and children. Indeed, a comparison of this kind is one way of separating and then estimating prenatal effects in contrast to genetic and postnatal effects on development. In some cases of adoption, however, the paternity of the adopted-away child may be uncertain. Further, biological fathers are often difficult to recruit into studies for direct assessment. One of the major adoption studies in the United States recruited only about 20 percent of an eligible sample of biological fathers (Plomin, DeFries, and Fulker, 1988). Often police, welfare, and health records can be used to fill in some gaps, particularly in countries like Sweden, where such records are very comprehensive (Cloninger et al., 1982).

In addition to these objections, which apply to adoption studies conducted over the last four decades, critics point to radical changes in the adoption process itself in recent years. In particular, birth mothers often choose the families that will adopt their children, and, whether or not they influence the choice, they often maintain some kind of contact with the adopting family. Thus, adoptive parents and even the adopted children may gain a clear impression of important psychological characteristics of the birth mothers.

Moreover, it is possible that some birth mothers may, for a while, play some ancillary role in rearing their own children. The adoption design should provide a firewall preventing the transmission of psychological characteristics of birth parents to the adoptive family—except via the child's genes. A more open adoption process may confound this aim.

As part of an extended pilot study in preparation for a new adoption study, David Reiss and Jenae Neiderhiser, along with a number of colleagues at the University of Iowa, Iowa State University, and the Oregon Social Learning Center, examined the consequences of these recent changes in the adoption process. Unpublished data provide substantial reassurance, at least for adoption studies focusing on very young children. Although most birth mothers do play a role in selecting the adoptive family, they tend to pick families unlike themselves and their own family: there is little or no correlation between their own characteristics and those of the families they choose. Further, although many adoptions are open and both birth and adoptive parents wish to keep in touch, there is, in fact, little contact between them, and even this rapidly diminishes over time. Moreover, variations in this contact have little or no effect on the association between birth-mother characteristics and child characteristics, as reported by parents. Finally, although adoptive parents who have contact with birth parents do form an impression of the biological parents, this impression appears to have little effect on the association of birth-parent characteristics and adopted-child characteristics.

Siblings, Adoption, and the Environment

Convention in science can be as heedless as it is in other human affairs. Thus it is more tradition than logic that has led most researchers to see evidence for genetic influence alone in twin, sibling, and adoption studies. In fact, these same research strategies provide an even richer picture of nongenetic environmental factors influencing development. Indeed, it is this feature of genetically informed research designs that prompted us to undertake the social and genetic analyses of adolescent development described in this book. Further, this use of genetically informed designs is crucial to our analyses and conclusions. We describe this approach later in this chapter and in subsequent chapters. Here we present three important concepts about the environment that are derived from twin, sibling, and adoption studies: environmentality, nonshared environment, and shared environment.

ENVIRONMENTALITY The concept of environmentality refers to all factors influencing development that are entirely independent of genetic influence. As shown, this includes all factors in the intrauterine environment. It also includes all the physical, biological, and social influences that impinge on the individual from birth onward. Like its companion concept, heritability, environmentality can be expressed in percentages from 0 to 100. The percentage reflects the proportion of total variation among individuals on a given characteristic, say, height, that can be explained by nongenetic, environmental factors. Environmentality is simply 100 minus the computed heritability for any characteristic. It can be computed from genetically informed designs in many ways. In twin studies environmentality is 100 minus twice the difference between the correlations for identical and fraternal twins. For twins reared apart, it is 100 minus the correlation of the identical twins. For adoption studies, environmentality is 100 minus twice the correlation between biological parents and their adopted-away offspring. In fact, it is a special strength of genetically informed research that heritability and environmentality can be estimated in so many different ways. The more similar the estimate from each approach is to the others, the more confidence we can have in it.

Like heritability, environmentality can vary from population to population depending on the occurrence of risk and protective factors in that population. For example, consider a country with only two social strata: a vast number of poor people who are at or near starvation and a few "nongenetic elite." In a country of this kind the major source of differences in weight will be a social variable: economic and social status. In a country with an adequate food supply for almost all its individuals, however, genetic influences on weight will be very important and environmentality will be much less so.

Even when environmentality, as a computed statistic for a particular population, is low, environmental factors may still be important in accounting for individual differences in development for two reasons. First, significant and sustained changes in the environment can have a major impact on characteristics that are, under many circumstances, highly heritable. For example, in a French study researchers obtained an unusual sample (Capron and Duyme, 1989). They found children who were adopted away from parents whose occupational ratings placed them in a low social class with unskilled laborers and farm workers. Half these children were reared by parents whose occupations placed them in a high social class with, for example, physicians and professors. A comparable group of children given up for adoption by upper-class parents had been adopted into two different groups of rearing parents, lower and upper class. The social class of the biological parents presumably is strongly associated with their intelligence, which many studies show is highly heritable. As expected, no matter what their rearing experience, adopted-away children of lower-class parents had lower IQ scores than children of higher-class parents. But the IQs of adopted children raised in upper-class homes, presumably rich in intellectual stimulation, were also much higher than those of children raised in lower-class homes. In fact, in these unusual circumstances, the effect of rearing was equal to the presumed genetic influence of the biological parents.

Second, the area of gene expression makes it clear that environmental factors are more important than might be implied by environmentality. Environmentality refers only to those environmental effects on development that are entirely independent of genetic differences in individuals. Genetic factors may, however, initiate differences in certain chains of development that require many environmental links before they are complete. Data we present later in this book, for example, suggest that heritable personality traits of adolescents may disrupt their relationships with their parents. These disrupted relationships may be a crucial step in the process whereby the personality traits evolve into serious antisocial behavior.

NONSHARED ENVIRONMENT Genetically informed designs have given special importance to siblings and environmental influences. In most studies genetically unrelated siblings are little more alike than two individuals picked at random from the population. This is true whether researchers measure cognitive skills, personality traits, or many forms of psychopathology. This is a remarkable finding when one considers that these siblings have grown up in the same family for many years. It is reasonable to expect that some general stamp of this common experience should leave its mark on most or all siblings in a family. Similarly, most studies of identical twins—whether they are reared together or reared apart—show that there are substantial differences between them. Such differences could occur only if environmental factors influencing the development of one identical twin were different in some way from factors influencing the other. Again, these differences can include intrauterine and postnatal influences and maybe also intrafamilial as well as extrafamilial influence.

Taken together, these findings from genetic studies argue that each sibling in a family develops in his or her own social world, and that it is the differences in these social worlds that matter most for psychological development. During childhood, differences in how parents treat children may be most salient; during adolescence, differences in peer groups might be important; and during adulthood, siblings may marry spouses who differ on many characteristics. Preliminary evidence supports all these as important components of the nonshared environment (Bennett et al., 1987; Daniels, 1986; Dunn and McGuire, 1993).

SHARED ENVIRONMENT The shared environment is the complement of the nonshared environment. The environmentality for any outcome measure is accounted for entirely by the sum of shared and nonshared environmental influences. Indeed, individual differences on any measure can be accounted for entirely by heritability, the nonshared environment, and the shared environment. The importance of the shared environment can be directly estimated in at least three ways: the correlation between adopting parents and their children, the correlation between any pairs of genetically unrelated siblings, and a comparison of correlations between identical twins reared apart and those reared together. As will be shown, antisocial behavior in adolescents shows the influence of the shared environment in several studies. One hypothesis is that twins may commit antisocial acts together. If twins, or any siblings, equally influence each other in any domain of development, genetic analyses will record this as a component of the shared environment (Rowe, 1983b).

Analyzing Anonymous and Specific Independent Variables

As noted, the use of a genetically informed research design and analysis permits estimates of the relative importance of genetic factors, as well as estimates of shared and nonshared environmental factors and any developmental outcome we can measure. These analyses do not, however, permit us to know which genes are important or which specific shared or nonshared environmental factors are the most important influences on any given developmental outcome. For example, consider data collected in a large Minnesota study of adopting families of older adolescents with comparisons to families in which parents are rearing their biological children of the same age (Scarr et al., 1981). Almost all the adopted children had been adopted shortly after birth and had lived together in the same home for approximately eighteen years. Figure 3.4 shows data for a personality factor called neuroticism. This is usually measured by a questionnaire that assesses how likely an individual is to become anxious in certain social or physical situations. As Figure 3.4 shows, the correlations between pairs of biological sibs, who share 50 percent of their genes, is modest but significant from a statistical point of view. The correlation between the adopted siblings is very low, however, and not significantly different from zero. We can conclude that the heritability of neuroticism is 2 x (.28 − .05) = 46%. This leaves environmentality at 54 percent. The correlation of the unrelated (adopted) sibs is a direct estimate of the

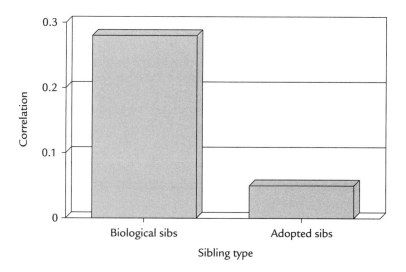

Figure 3.4 Correlations of neuroticism measures between pairs of biological siblings and pairs of adopted siblings (Scarr et al., 1981)

shared environment, 5 percent. This means that 49 percent of adolescents' differences in neuroticism are due to nonshared factors (and to a much smaller extent, to lack of perfect reliability of the measures used in the Minnesota study). Another way of saying this is that .49 − .05/.49, or 90 percent, of the environmental influence on this personality trait is of the nonshared type.

This analysis tells us that the major environmental influences on adolescents' proneness to anxiety must be different for sibs in the same family. This rules out a number of influences, such as the family social class or the level of the parents' anxiety, all of which are *shared* by siblings in the same family. The influential nonshared variables, however, remain anonymous. There are three possibilities for the kinds of nonshared variables that might be influential here: one or two very highly influential factors, not shared by siblings, may influence one sibling to become anxiety-prone and the other to remain calm, even under stress; a number of factors, each with its small contribution, may combine to account for sibling differences; or there may be no systematic nonshared influences—that is, the important factors may vary from family to family. An example of this last type of influence may be the death of a grandmother in one sibling's first year of life, before the other was born. Such an event would constitute a nonshared experience that influenced anxiety-proneness in that particular family and not in others. In another family one sibling might have been afflicted with a serious chronic infection in early childhood that spared the second child, a circumstance that is relatively rare and not likely to have occurred in most other families in a research sample.

Other behavioral genetics approaches allow researchers to examine the association of *specific* independent variables with developmental outcomes. Thus we might measure the extent to which siblings are treated differently by their parents, and then measure the association between that differential parenting and anxiety-proneness. Behavioral genetics analyses can help dissect this specific relationship to determine if it is due to environmental or genetic mechanisms, a distinction to which we return in the next chapter.

Chapter 4

GENETIC ANALYSIS

OF ADOLESCENT DEVELOPMENT

The association between social relationships and individual adjustment explored by psychosocial researchers can also be examined using genetic tools. As shown, psychosocial research, without the help of genetics, has already encountered phenomena that anticipate in important ways findings that will be more firmly established by genetic analyses. But data from genetically informed research can shed new light on the importance of these phenomena. In this chapter we review genetic research using the same format we used for psychosocial research.

First, we review studies that gather data in a single cross-section in time, distinguishing studies in which the independent variable is anonymous from studies in which it is specific. In some ways this distinction in genetic studies corresponds to the distinction between psychosocial studies that explore indexes of general social dysfunction and adjustment and those that measure specific social processes and their relationship to adjustment. Second, we review studies that use longitudinal data collected at two or more points in time. Finally, we consider research that employs planned interventions to test developmental theories.

Studies Using Cross-Sectional Data

We can identify four cross-sectional approaches for genetic studies, just as we did for psychosocial studies in Chapter 2. We have ordered our review of them to show the logical interconnection among the approaches—particularly how, taken together, they illumine the role of social relationships in development. We begin with the most traditional genetic analyses: the association of genetic and environmental factors with individual differences in adjustment. A fundamental feature of this and all subsequent genetic analyses is that they are always simultaneously estimating the influence of both genetic and environmental factors on adjustment. Twin and sibling samples are especially informative about environmental influences because they can distinguish between shared and nonshared environments.

If we can demonstrate genetic influences on measures of adjustment, then we can take a second step in exploring the role of social relationships in adjustment: we can estimate genetic and environmental influences on the social relationships themselves. If we find that there is a substantial genetic influence on these measures, as well as on outcomes of interest, we can proceed to a third and crucial step in genetic analysis: gene-environment (GE) correlation.

The analysis of GE correlation asks whether the genetic factors that influence our assessment of an adolescent's level of adjustment are the *same* genetic factors that influence the characteristics of his or her social relationships. If the answer is "yes," we can estimate the degree to which the observed association between social relationships and adolescent development can be attributed to these common genetic factors. Using analogous reasoning, we can also determine the extent to which an observed association is influenced by shared and nonshared environmental factors.

The final cross-sectional tool, also essential for understanding the impact of social relationships on development, is the analysis of *gene-environment (GxE) interaction.* This strategy asks whether genetic factors express themselves more readily in some environments than in others. As shown, psychosocial researchers address a comparable issue without the benefit of genetic analyses to ask how individual characteristics of adolescents "fit" with characteristics of their social settings, families, peer groups, and schools. Genetic analyses extend considerably the precision of these analyses.

Step One: Analyzing Genetic and Environmental Influences on Adjustment

The first cross-sectional approach is to estimate the association between "anonymous" genetic and environmental influences and measures of adjustment. This step bears an important similarity to an early step in psychosocial research: to estimate the association between a crude measure of the social world—such as whether or not a parental divorce has occurred—and one or more measures of adolescent adjustment. In both cases, we cannot say much about specific mechanisms of influence on development, but they are an important starting point for cross-sectional data.

Because heritability estimates can vary from one population to another, tend to be significant for a very broad range of measures of adjustment, and do not by themselves help us understand developmental processes, they are hardly newsworthy when considered alone. The important news concerns what we can do once we have established notable heritabilities for adjustment: we can deploy an impressive arsenal of tools from behavioral genetics, much more sophisticated than those used to determine heritability, to elucidate developmental processes. Heritability estimates are a springboard for cross-sectional, longitudinal, and intervention strategies that can tell us a great deal about development and its genetic and environmental determinants.

Despite its limits, establishing heritability is a crucial first step in both genetic and environmental analyses of adolescent development. Indeed, it is not one step but three separate phases. The first step is picking the salient areas of individual adjustment for exploration using genetic analyses. The second step is comparing patterns of differential heritability among related areas of adjustment. The third step is determining which adjustment variables have genetic influences in common and which have quite distinct genetic influences. These three steps, taken together, enable us to develop an initial impression of the role of genetic and environmental factors in psychological development.

SALIENT MEASURES OF ADJUSTMENT   In order to reconcile their approaches to adolescent development, both psychosocial researchers and geneticists need to agree on the aspects of adjustment they seek to understand. To a limited extent they have already done so. Both groups, for example, have an interest in antisocial behavior and depression. In the area of adolescent personality development, however, psychosocial investigators have focused on two broad aspects of adjustment that have rarely been investigated by geneticists: (a) the adolescents' appraisal of their own competence in several areas, along with their feelings of general self-worth; and (b) the adolescents' appraisal of their position in their social world, including friendships, sex roles, current and future occupational activities, and religion. The former grouping is usually referred to as "perceived self-competence," and the latter, following Erik Erikson (Erikson, 1963), as "identity formation." Psychosocial researchers have selected these areas because they are intimately related to the major developmental tasks of adolescence.

Major genetic studies of personality in adolescence, by contrast, have focused on domains more closely related to temperament or to simpler, more elementary responses that are stable over time. These include such characteristics as introversion, impulsivity, social anxiety, cheerfulness, and timidity (Loehlin and Nichols, 1976; Scarr et al., 1981). Genetic researchers have touched on areas that have preoccupied psychosocial researchers by studying simple self-perceptions, attitudes, and interests such as self-ratings of popularity, interest in religion, science, drawing, and business (Loehlin and Nichols, 1976). In general, modest to substantial heritabilities are found for all of these, with heritabilities tending to be higher for twin studies (Loehlin and Nichols, 1976; Nichols, 1978) than for adoption studies (Scarr et al., 1981). Since both geneticists and psychosocial researchers have been concerned with psychopathology, both have examined antisocial behavior and depression in adolescents (Capaldi and Patterson, 1994; Faust et al., 1985; Moffitt, 1993; Neiderhiser et al., 1996; O'Connor et al., 1992; O'Connor et al., submitted; Sim and Vuchinich, 1996).

Undoubtedly, the traits explored by behavioral geneticists are pertinent to adolescent adjustment and development. Psychosocial researchers, however, have taken these elemental reaction patterns and attitudes a step further, focusing on how they become organized into patterns of self-perceptions and comparisons with others. Our study embraced both traditions. On the one hand, we included traditional assessments of temperament such as emotionality and sociability (Buss and Plomin, 1984). On the other hand, we explored more complex domains of adjustment such as specific and general self-perceptions. For example, we included a multidimensional measure of adolescents' self-perceptions (Harter, 1988). This assessment comes much closer than traditional assessments both to measuring aspects of personality that capture a core experience of adolescents, and to illuminating differences in their development. This assessment focuses on the adolescents' own ideas about their areas of competence. They rate themselves in seven domains: scholastic, athletic, and social competence as well as physical appearance, morality, friendship, and overall self-worth.

The question, of course, is whether these complex self-images can be influenced by genetic processes. The answer is "yes," but only for some of them (McGuire et al., 1994). As shown in Figure 4.1, we found in our study that the pattern of correlations conformed, for the most part, to genetic expectations. Identical twins show the highest correlations in heritability of scholastic self-confidence. Next come fraternal twins, full sibs in nondivorced families, and full sibs in step-families, all of whom share 50 percent of their genes and show correlations that are roughly the same. The half-sibs and blended sibs clearly show the lowest correlations, although there is no appreciable difference between them. Recall that a

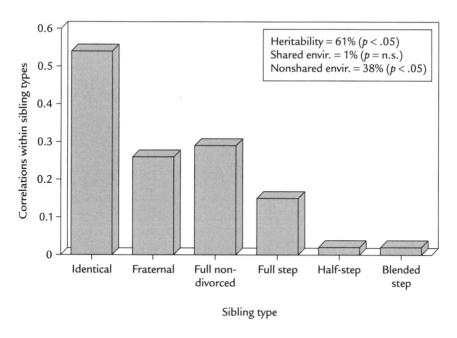

Figure 4.1 Heritability of scholastic self-perception (McGuire et al., 1994)

fairly even slope like this across the six groups is called a genetic cascade. It is possible to estimate the heritability directly from the differences among the correlations, as already indicated. For example, the heritability should equal twice the difference between identical and fraternal twins, in this case (54% − 26%) x 2 = 56%. The heritability is twice the difference because the contrast between identical and fraternal twins covers only half the spectrum of possible genetic relatedness between siblings. For example, the contrast between identical twins and blended sibs provides a direct estimate of heritability, in this case, 54% − 2% = 52%. Heritability should also be twice the difference between full step-sib and blended sib. However, in Figure 4.1 this is only about 26 percent.

Fortunately, geneticists have developed statistical models for computing all possible comparisons here (fifteen in all) and comparing them with those generated by genetic expectations. If the observed pattern closely conforms to that expected by genetic theory, then this computational procedure will yield high heritability estimates. Throughout the rest of this book we refer to these estimates of heritability, derived from all possible comparisons among groups, as *model-testing heritabilities.* The genetic models estimate the percentages of differences among individuals in a population that are due to genetic differences; they also estimate differences among adolescents that can be attributed to nonshared and shared environments. The model also computes the likelihood that such estimates are due to chance. In general we report only estimates that are likely to be due to chance fewer than 5 times in 100. Usually such estimates are noted with one or more asterisks or an indication of chance probabilities ($p < .05$). Thus Figure 4.1 presents the correlations within each group that we actually observed. The box shows the model-testing heritability, shared and nonshared environmentalities, and the probability that any of these estimates reflect simple chance findings.

In order to estimate shared and nonshared environmental influences using this model, we also do some computation directly from the data in Figure 4.1. The shared environment, for example, can be directly estimated by the correlation of genetically unrelated pairs, the blended sibs. Since such siblings share no segregating genes, the only factors that can make them similar are shared environmental. Since the correlation is .02, we can thus estimate the shared environment, from our simple computation, at 2 percent.

An important but indirect approach to estimating nonshared environment is based on the assumption that genetic and environmental effects add up, in a simple, linear way, to influence development. If this is the case, then genetic and shared and nonshared environmental factors must add up to 1. Under this assumption, we can simply subtract the heritability and the shared environment from 1. The remainder is the nonshared influence. Let us say we use the heritability estimate derived from twins, 56 percent. Then the nonshared component is $1 - (56\% + 2\%) = 42\%$. We can estimate the nonshared environment more directly by subtracting the identical twin correlation from 1; this equals 44 percent. This operation reflects our assumption that there are only two reasons that identical twins are not alike, that is, two reasons they fail to correlate at the level of 1. First, there may be an error in our measure; second, the environments of the identical twins may be different.

The assumption of additive environmental and genetic effects is central to the models tested here, as well as to most behavioral genetics designs. It assumes that nonadditive effects, such as gene-environment interactions, are small for the sample and for the measures we use here. We return to this important assumption below.

PATTERNS OF DIFFERENTIAL HERITABILITY   Although finding heritabilities of adjustment measures is nothing new, we can use simple associations of genetic differences and adjustment differences to explore genuinely newsworthy possibilities. The first of these is differential heritability among different measures of adjustment. Thus, in our data about adolescent self-perceptions, we find an intriguing contrast. Where specific self-perceptions, such as self-perceived scholastic abilities, show strong evidence of heritability, the adolescent's overall sense of self-worth shows no significant heritability. The observed correlations, across the six groups we studied, are shown in Figure 4.2. The identical twin correlations are modest, but greater than those of fraternal twins. A twin study would suggest that general self-worth is heritable (at the approximate level of 60 percent). The correlations for full sibs and half-sibs in step-families are, however, much higher than expected and—as a result—the model-testing heritability is 29 percent, which, although substantial, is not statistically significant. This is an instance in which our mixed-sibling design may be self-correcting. In this case a more exuberant estimate of heritability de-

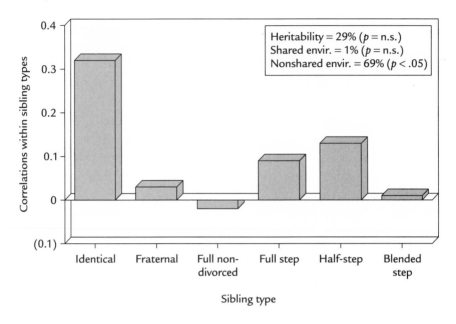

Figure 4.2 Heritability of general self-worth (McGuire et al., 1994)

rived from the twin portion of our design is muted by more modest findings from the step-family portion of the design.

Figure 4.3 shows the patterns of heritability and shared environment and nonshared environments for all seven measures of adolescent self-perceived competence. Several important findings can be noted. First, the heritabilities for self-perceived qualities of friendship, morality, and overall self-worth are small and not significant, whereas those for physical appearance, general social attractiveness, athletic ability, and scholastic ability are significantly heritable. Second, environmental variables—which in this analysis are all anonymous (see Chapter 3)—have an important effect on individual differences on all seven of these self-perceived competence ratings. Third, the nonshared component of these high environmentalities is, by far, the most important type of environmental influence.

What might account for this differential pattern of heritability? Specific components of self-perceived competence may reflect heritable skills. For example, genetic influence on self-perceived scholastic competence may reflect heritable components of intelligence, or the heritable component of self-perceived social abilities may reflect a social and affiliative temperament that is also heritable. Self-perceived social competence reflects the self-perceived achievements of popularity and social standing, whereas sociability reflects a tendency to connect to or withdraw from people and is observable at a very early age.

Nonetheless, given the sizable genetic influence on some of the components of self-perceived competence, it is surprising that adolescents' overall perceptions of self-worth show no significant heritability. This genetic finding fits a conclusion drawn by psychosocial studies (Harter, 1990a). Data from these studies suggest that global self-worth is a product of the adolescent's perception of competence in a particular area of his or her own development and the importance of that area to the adolescent. Thus, a specific self-perceived competence, such as scholastic ability, may partially reflect a heritable skill related to academic success. But adolescents may have the capacity to discount areas in which they are deficient: "That's not important to me." In such cases their overall self-esteem may

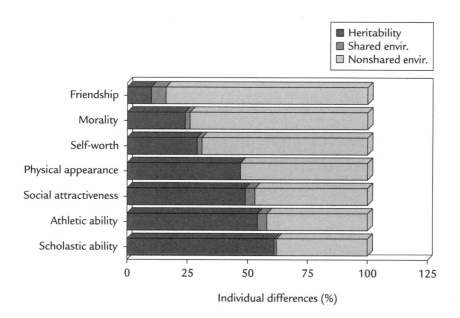

Figure 4.3 Differential heritability of self-perceived competence (McGuire et al., 1994)

escape the influence of heritable skills and aptitudes. In this instance, a pattern of genetic data concerning several aspects of self-perceived competence in adolescents contributes to an understanding of the developmental process of self-perception and self-esteem.

ADJUSTMENT VARIABLES   In our discussion of how genes influence individual differences in complex self-perceptions such as perceived competence, we advanced a hypothesis that inherited, simple skills and temperamental features might lead to these self-perceptions. Behavioral genetics can also address this issue using a set of techniques called "multivariate genetic analyses." Although these techniques are discussed more fully later in this book, we introduce some concepts related to these analyses here to show how they can provide some additional clues to development.

Multivariate analytic techniques permit us to ask, for any two heritable traits, if the genetic factors that influence one trait are the same genetic factors that influence the other. This is a very broad question and can be broken down into more specific questions, each with its own relevance to understanding development. Here we phrase the question in the following way: Does genetic influence on a simple skill or trait lead to the development of self-perception in the general area within which that trait falls? For example, does the effect of genetic influence on self-perceived social competence reflect genetic influence on the temperamental feature of sociability?

As explained in Chapter 3, genetic analyses can be deployed to examine specific independent variables (sociability) and dependent variables (self-perceived social competence). This is because both sociability and the self-perception of social competence are heritable (the heritabilities equal 44 percent and 49 percent, respectively), and sociability and self-perceived social competence are themselves correlated (the correlation equals .35). For these multivariate analyses we do not use correlations within each sibling type, as in conventional or univariate genetic analyses. Rather, we use a *cross-correlation* for each analysis: within each group of siblings we correlate the sociability of one sibling with the self-perceived social competence of the other; we then compare cross-correlations for their adherence to genetic expectations. We provide a more complete

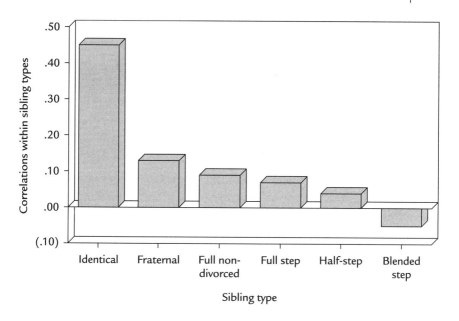

Figure 4.4   Cross-correlations between sociability and social competence (McGuire et al., 1994)

explanation of cross-correlations in Chapter 8 (see, in particular, Figure 8.2).

Figure 4.4 shows bar graphs that convey the cross-correlations we observed. Just as behavioral geneticists have developed statistical models for examining the simple associations between genetic variability and outcome variability, so too have they developed models for multivariate analyses like these. This model estimates the percentage of an association between two variables that is accounted for by genetic and shared and nonshared environmental factors. We can represent this by a stacked bar graph analogous to that in Figure 4.3. The format here is a bit different, however. The total width of the bar reflects the observed correlations between the two variables. As noted, the correlation was .35 for sociability and self-perceived social competence. Within each bar is the proportion of that correlation accounted for by genetic, shared environmental, and nonshared environmental factors.

The data revealed one other elemental or simple attribute of the adolescents we studied that was both heritable and correlated with one of the scales of perceived self-competence: a measure of verbal IQ. According to the data, the heritability of verbal IQ was 31 percent and its correlation with perceived achievement in school was .28. Figure 4.5 displays the two correlations 0.28 and 0.35 and the components accounted for by genetics, shared environment, and nonshared environment.

These stacked bar graphs look quite different from those in Figure 4.3. Here genetic mechanisms account for almost all the observed associations between two variables. We will return repeatedly to the finding that when we examine the simple association between genetic variability and outcome variability in so-called univariate analyses, we find that nonshared environmental factors are very important, but when we look at specific associations between one variable and another, such factors often have almost no role. In this case the preponderance of genetic components tells us, for example, that the association between verbal IQ and self-perceived scholastic competence is not attributable, exclusively, to *any* environmen-

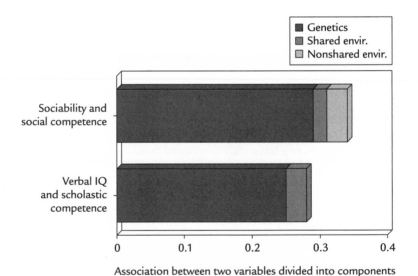

Figure 4.5   Genetic and environmental components of the association between sociability and social competence and verbal IQ and scholastic competence (McGuire et al., 1994)

tal mechanism. This simple point is, by itself, very important because it sets limits on developmental theorizing: these data suggest that no developmental theory accounting for this observed association between verbal IQ and self-perceived scholastic competence can exclude genetic mechanisms.

Within that major constraint there are three developmental theories that could explain these data. Genetic factors may have a direct and relatively early impact on verbal IQ. Differences among children on this measure then influence the quality of their schoolwork and other intellectually challenging activities, which, in turn, influence their own perceptions of scholastic competence. Or it could be the other way around. Genetic factors may influence more directly self-perceptions of abilities, encouraging children to take on intellectually challenging circumstances, which, subsequently, enhance their verbal IQ. Or genetic factors may influence a characteristic of adolescents we did not measure, which, in turn, influences both verbal IQ and self-perception of scholastic competence. This is yet another version of the third-variable problem discussed in Chapter 2. Later in this chapter and in subsequent chapters we present genetic models that help to unravel these possibilities.

Step 2: Clarifying Mechanisms That Link Adolescents' Social Worlds and Their Adjustment

Given that characteristics of youngsters most likely influence the social world in which they live, it is reasonable to expect that some of these characteristics are heritable, such as temperament, personality factors, and social skills. Behavioral geneticists have speculated that there are two types of such influences (Plomin, 1986; Plomin et al., 1977; Plomin et al., 1985; Scarr and McCartney, 1983). The first type involves heritable aspects of physical features and temperament that may elicit warmth or rejection from parents and other important people in the growing child's social environment. Because the children's characteristics are eliciting, without much determined action on their part, a response from others, many geneticists refer to these child influences as "evocative." As noted, however, genetic factors influence the abilities, self-perceptions, interests, attitudes, and values of children and adolescents. These, in turn, may prompt a child to seek out certain kinds of relationships with others. A young child with genetically influenced verbal skills, for example, may actively request a parent to read to her. Geneticists have referred to these child influences as "active."

Psychosocial researchers often neglect a third important reason a child's or adolescent's genes may appear to influence his or her social world. Offspring share exactly 50 percent of their genes with their mother, exactly 50 percent with their father, and approximately 50 percent with full siblings. Apparent influences of the child's genes actually reflect, instead, those of their parents. Geneticists have referred to this form of relationship between genes of an offspring and characteristics of the parents or siblings, including the latter's behavior toward the offspring, as "passive."

Our study provides substantial evidence, detailed in Chapter 9, that genes influence how adolescents perceive their social relationships. Other studies have noted such genetic influences in measures of how adolescents perceive their family environments, including their perceptions of how they are treated by their parents and siblings (Daniels and Plomin, 1985; Rowe, 1981; Rowe, 1983a). Many of these observations have been confirmed by studies in which adults have reported, retrospectively, on their childhood and adolescent years (Baker and Daniels, 1990; Plomin, McClearn, Pedersen, Nesselroade, and Bergeman, 1988).

These findings, although important, are not particularly remarkable. After all, it is not surprising that genetic factors—which influence so many cognitive, perceptual, and attitudinal differences among children and adolescents—would also shape how they perceive important relationships within their families. What is important is that more recent data suggest the influence of children's genetic factors on how their parents report treating them (Braungart, 1994; Goodman and Stevenson, 1991; Rende, Plomin, and Reiss, 1992). These data suggest that genetic factors extend their influence into relationships beyond the representation of those relationships in the mind of the child or adolescent. Perhaps most persuasive to psychosocial researchers is the fact that genetic factors influence directly observed interaction between parents and their offspring, and children and their siblings (Dunn and Plomin, 1986; Dunn et al., 1985; Dunn et al., 1986; Rende, Slomkowski, Stocker, Fulker, and Plomin, 1992). Moreover, recent evidence indicates that adolescents' genetic factors may influence their relationships with their peers (Daniels et al., 1985), and that there are genetic influences on adults' retrospective reports of the emotional climate of their school classrooms (Plomin, 1994). All these findings argue for the role of genetic factors in several components of social relationships that are central in the world of the adolescent.

Figure 4.6 anticipates the findings of our study on this subject, discussed in detail in Chapter 9. We include the figure to illustrate several important features of the genetic analysis of social relationships. Figure 4.6 is a more complicated version of Figure 4.1. It shows correlations within groups of siblings for measures of conflict and negativity in the mother-child and father-child relationships.

Each of the measures is an aggregate of several scales that assess conflict and negativity in parental treatment of children (described more fully in

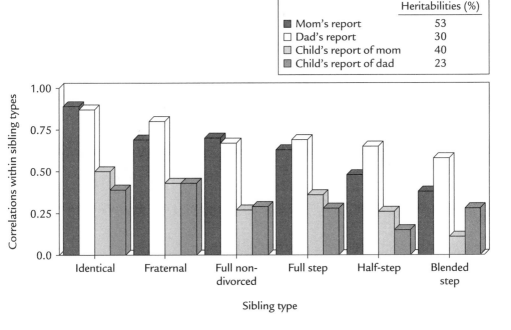

Figure 4.6 Parent and child reports of conflict and negativity (Plomin et al., 1994)

the next chapter). This treatment includes frequent arguments, severe punishment, verbal aggression, and parents' frequent yielding to coercive behavior on the part of the child. The graph depicts an aggregate of these measures as reported by each parent and each child in our study. Data on genetic influence on directly observed interaction among family members are reported in Chapter 9. In a glance we can see that heritabilities are substantial for three of the four measures. Further, parents' reports show as much, if not more, influence of the offspring's genetic factors as do the child's reports. We also note a substantial difference across all groups between parents who tend to see themselves as treating their children similarly (the correlations are all high) and children who report very different parental treatments (the correlations are all low). These differences in bias or sensitivity between parents and children do not seem to influence the role of genetics in individual differences among adolescents in measures of parenting. They do, however, influence our estimates of the role of the shared environment: these estimates would be higher for parental reports than for child reports. We examine this contrast between parents' and children's reports in detail in Chapter 7.

Step 3: Assessing Gene-Environment Correlation

Once we have determined that genetic factors influence many aspects of the adolescent's social world, we must ask if these are the same genetic factors that influence adolescent outcomes such as personality characteristics. If the answer is "yes," then genetic research designs move the study of relationships and development into a whole new domain. As shown, the social analyses of adolescent development rely heavily on observed associations between characteristics of social relationships and characteristics of adolescents, either at a single cross-section in time or as they both change across time. If the same genetic factors are influencing both the social relationships and measures of adjustment, however, we run into the classic third-variable problem. Genetic influences on relationships and outcomes can masquerade as purely social phenomena *if they are not measured directly.* None of the strategies that have been so painstakingly elaborated by social analysis can even guess at these third-variable effects of genetic processes.

There are five ways to understand how genetic factors might operate as the heretofore concealed third variable. Let us take as an example parental conflict and negativity and an adolescent's personality characteristics. Our analysis here follows from previous work on genetic influences on observed associations between environmental and outcome variables (Plomin et al., 1977; Plomin et al., 1985; and Scarr and McCartney, 1983) and is summarized in Figure 4.7. This analysis extends our discussion of ways in which adolescents' genes might influence their social environment. "Child adjustment" includes differences among children on any dimension of interest to developmentalists and might include antisocial behavior, depression, and social responsibility. "Parenting" is more specific and could include any characteristic of parenting relevant to a developmental hypothesis. The same analysis could be made for "sibling behavior" toward the adolescent and even other genetically related family members. Complexities are introduced when we consider other aspects of the adolescent's social world. These will be discussed in Chapter 7. In Figure 4.7 the term **Gc** refers to the gene or set of genes that influences child characteristics, and **Gp** refers to the gene or set of genes that influences parenting.

The first way in which genetic factors might influence an observed association between parenting and child outcome is referred to as the "passive" model. As noted, this effect is due to the fact that children share exactly 50 percent of their genes with each biological parent. For example, children who perceive themselves as being socially successful and popular on a measure of self-esteem such as we used in our study may have parents who treat them with great warmth. Does that mean that parental warmth influences adolescent self-perceived competence, or the reverse, that self-perceived competence elicits parental warmth? Neither may be the case. Parental warmth may just be the adult manifestation of the same genetic factors that lead to self-perceived competence in adolescents. Indeed, several studies suggest a continuity between attributes of children and adolescents on the one hand—attributes that are likely to be heritable—and the nature of their social relationships when they reach early adulthood and mid-life (Brooks, 1981; Elder, 1986; Kandel et al., 1986), on the other. As Figure 4.7 indicates, in a simple case in which only a passive mechanism influences observed associations between parenting and

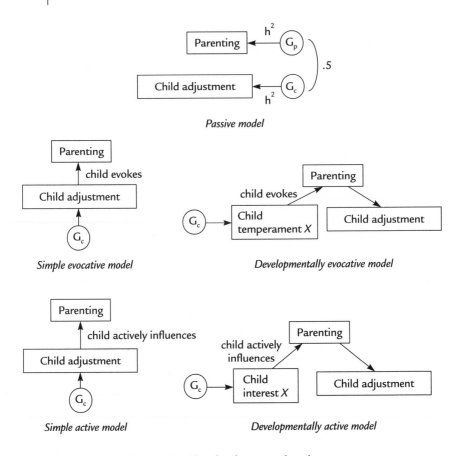

Figure 4.7 Five developmental pathways

adolescent characteristics, the correlation between the parenting characteristic, let us say, negative parenting, and the child outcome, say, antisocial behavior, is attributable to three factors: (1) the 50 percent overlap in genes between parent and child; (2) the heritability of parenting; and (3) the heritability of antisocial behavior.

A second possibility is that the genetic factors influence child outcome (let's continue with self-perceived social competence), and, in turn, this child outcome elicits parental warmth and support. We call this developmental sequence "simple evocative." As shown in Chapter 2, a social analysis of development attempts to detect these effects of children on their parents by using longitudinal designs, which allow us to observe that child self-perceived competence sometimes precedes in time the development of parental warmth. These designs cannot, however, detect the simple evocative path because genetic influence on any child characteristic may itself change over time. Some research suggests that substantial changes in genetic influence may occur in periods as short as six months (Plomin et al., 1993). Thus, genetic effects cannot be ruled out simply by showing that parental effects occur prior to child outcomes. The detection of these genetically influenced evocative effects requires genetically informed research designs.

A third possibility introduces another child characteristic into the development sequence. For the moment we can speculate that this is a heritable temperamental feature of the child that has the capacity to elicit certain types of parental behavior. Let us call this "child temperament *x*." According to this scheme, child temperament *x,* which is a heritable characteristic in the child, elicits parental conflict, which in turn influences the adolescent's perception of his or her own competence. This might be similar to the temperamental feature "sociability" that we used in our study, a characteristic that appears extremely early in child development. This sequence can be called "developmentally evocative."

A fourth possibility is that genetic factors influence the adolescent actively to select certain relationships such as involvement in conflictual and negative peer groups. In our study we observed a substantial genetic influence on parental reports of adolescents' peer-group characteristics (Manke et al., 1995). Thus genetic influences of this kind are quite plausible. Active processes may also account for the influence of a child's genes on parents. In this case the child actively and strategically requests a pattern of responses from parents. For example, a child with strong interests in social activities might engage parents in a variety of leisure and holiday activities involving other adolescents and adults.

As with evocative forms, there are two possibilities here as well. "Developmentally active" sequences imply that a heritable characteristic of the child leads the child to select a particular kind of relationship, and that the relationship, in turn, influences the child's more conspicuous manifestations of adjustment, such as self-perceived competence. We do not term the unknown factor "temperament *x,*" since the concept of temperament does not seem sufficiently developed in the child to influence choices of relationships and relationship patterns. Rather, we term this characteristic "child interest *x*," which is meant to embrace preferences, personality features, abilities, values, and interpersonal needs, all of which have been shown to have genetic influence. "Simple active" associations, to continue our specific example, imply only that genetic factors influence the more fully formed level of adjustment, such as self-perceived competence, which in turn leads the child to become involved in warm and supportive relationships.

This five-part schema can be simplified. Simple evocative and simple active models imply that heritable aspects of child adjustment influence parent-child relationships (and other family subsystems such as the sibling relationship). They can be called, simply, "child effects." Developmental active and evocative models imply that the parent-child relationships (or other family subsystems) have an important effect in amplifying or mediating genetic influences on adjustment. They can be called "parent effects." This distinction is elaborated in Chapter 10.

In part it is the mission of our study to distinguish, using research data, which of these pathways best represents how genetic factors influence an association between parenting and child outcomes. In order to do so, we must determine whether observed associations between relationships and development are attributable in some way to genetic factors. If they are, then it is imperative that these new pathways be explored. Existing social theories of development that do not outline a distinctive role for genetic factors would, in the face of these data on gene-environment correlation, no longer be tenable. In other words, if genetic factors are central in these associations, we must chart a new course for understanding development. As will be shown, this does not mean discarding older theories but rather using them to improve genetic studies and then using genetic studies to revise substantially the older theories.

Researchers have not as yet reported data to distinguish among these five forms of genetic influence on covariance, and so it is not possible to determine if genetic factors account for these observed associations. Our study of nonshared environments has collected longitudinal data that may be helpful in making some of these distinctions. Indeed, our study is among the first to take the initial crucial step here: to examine the genetic component of covariance between family process and any aspect of adjustment. Our analyses also employ correlations and compare them across the six groups of siblings; the correlations are, however, cross-correlations. That is, we seek to predict the dependent variable as measured in one sib in the pair from the independent variable measured for the other sib. If genetic factors account for the observed correlations between independent and dependent variables, then these cross-correlations should be high for MZ twins, lower and approximately equal for DZ twins and full sibs in nondivorced and step-families, even lower for half-sibs, and lowest for unrelated sibs. A pattern of this kind would be difficult to explain by any mechanism other than genetic.

Figure 4.8 shows examples of two covariances analyzed in this way: the relationship between the mother's conflict negativity (MNEG) and the adolescent's antisocial behavior, in one analysis (Pike, McGuire, Hether-

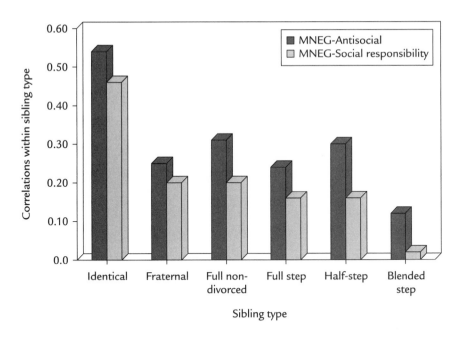

Figure 4.8 Genetic and environmental components of the association between mother's negativity and adolescent's antisocial behavior and social responsibility (Pike et al., 1996)

ington, Reiss, and Plomin, 1996), and the adolescent's social responsibility in another. We measured antisocial behavior by a combination of child report, parent report, and direct observation. We measured social responsibility by a combination of self-report from the child and reports of each parent.

Figure 4.8 shows the cross-correlations for each of the six groups of siblings we studied. Neither of these patterns of cross-correlations across the six groups precisely fits genetic expectations. The pattern expected by genetic theory would be: MZ > DZ = nondivorced full sibs = step–full sibs > step–half-sibs > step–unrelated sibs. For antisocial behavior, the correlations between half-sibs are somewhat higher than genetic expectations. For social responsibility, the comparisons within nondivorced and within step-families meet genetic expectations, but step-families, in comparison with nondivorced families, are higher than a simple genetic model would predict. As with the analysis of variance on a single measure, multivariate-model testing procedures examine simultaneously all fifteen comparisons here. Indeed, almost all these separate comparisons do conform to genetic expectations. Thus very high proportions of both covariances are explained by common genetic factors. For antisocial behavior, the proportion is 72 percent, and for social responsibility, 92 percent.

Using the same techniques, we have also examined the genetic components of mother's conflict negativity with adolescent's depressive symptoms and father's conflict negativity with both adolescent depressive symptoms and antisocial behavior (Pike, McGuire, Hetherington, Reiss, and Plomin, 1996). In all cases, common genetic factors account for well over 50 percent of the association between environment and outcome. A similar pattern holds for father's negativity and adolescent social responsibility, as well as for mother's warmth and support and social responsibility. In this series of analyses, only father's warmth and support and social responsibility showed common genetic factors accounting for less than 50 percent of the covariance (46 percent). We present the complete set of findings on this topic in Chapter 10.

Let us compare these findings with those usually obtained by family researchers. We can begin by computing the correlations between environmental and outcome variables for the whole sample. This, of course, ignores the genetic information available in the data. For example, the correlations between mother's conflict/negativity, warmth/support, and father's conflict/negativity, warmth/support, on the one hand, and adolescent's social responsibility, on the other hand, are −.45, .46, −.40, and .39, respectively. The corresponding correlations for depressive symptoms are .33, −.15, .37, and −.12, and for antisocial behavior, .60, −.28, .57, and −.23. Because of the large sample size in our study, all of these are significant at the .05 level or beyond. Armed with respectable correlations of this kind, psychosocial researchers would develop psychosocial theories either to account for the influence of the family on both positive and negative aspects of adolescent development, or, allowing for child effects, to explore ideas about reciprocal psychosocial influences where the child is also understood to influence the parents through psychosocial means. Our genetic analyses of these correlations indicate that these purely psychosocial theories must be revised.

Step 4: Studying Gene-Environment Interactions

The definition of gene-environment interaction is analogous to that given for person-environment interaction in Chapter 2. Recall that environmental moderation refers to the influence of distinct differences in the environment on the continuity between early-child characteristics and later adjustment. Under some environmental circumstances these characteristics can lead to good adjustment, whereas in others they may lead to poor adjustment. Genetic influences on adjustment vary depending on the characteristics of the environment. Interactions can be detected using twin designs. For example, we can examine the heritability of a measure of adjustment by comparing twins living under one environmental condition with twins living under another. One recent study showed that the heritability of alcohol consumption in women was much higher in unmarried twin pairs than in married pairs, suggesting that being married may buffer some of the genetic influences on drinking behavior (Heath et al., 1989).

Studies using identical twins reared apart or adoption studies are more sensitive to gene-environment interactions than are studies examining siblings raised by their own parents. For example, one study used a large sample of identical twins reared apart to show the interaction between qualities of early family life, as remembered from adulthood, and genetic influences in the development of personality characteristics (Bergeman et al., 1988). The genotype of a co-twin can be roughly estimated from the phenotype of the twin sibling if we know that the phenotype is highly heritable. For example, if the twin sibling scores very high on a heritable trait such as extroversion, we would expect the genetically identical co-twin to have a high genetic liability for extroversion. If the twin sibling scores very low on extroversion, we would expect the co-twin to have a low genetic liability. Using the phenotypic level of extroversion in the twin sibling, we can divide the co-twins into low- and high-liability groups and see whether the difference in personality between these two groups is greater under certain environmental circumstances than under others. For example, one such circumstance is the level of family control. In the study of twins reared separately, when the family was remembered as very controlling, few twins with low genetic liability became extroverted. But co-twins with high liability (as determined by high levels of extroversion in the twin sibling) became extroverted despite their controlling family. Under conditions of low family control, most twins, with and without liability, become extroverted. In other words, the genetic influence on extroversion is enhanced in families in which there is a high level of control. This study of twins raised separately found that these interactions explained only 2 to 6 percent of personality differences among the twins.

Adoption studies, particularly those that examine extreme differences among birth parents, have also demonstrated gene-environment interactions. For example, in one study the adopted-away offspring of birth parents with severe antisocial personality disorders or alcohol or substance abuse were compared with adopted-away offspring of birth parents with no record of serious psychopathology (Cadoret et al., 1995). The adoptive parents were assessed for the presence or absence of serious legal problems, psychopathology, or marital difficulties. When adoptive parents had few or no problems of this kind, there was little difference between the two groups of adopted-away offspring. Both showed relatively low levels of aggression and conduct disorder. In cases in which there were many parental problems in the adoptive family, however, the adopted children of birth parents with psychiatric disorders showed high levels of aggression and conduct disorder, whereas those with normal birth parents did not. In other words, the genetic liability for conduct problems became manifest only in adoptive families that had multiple psychiatric, marital, or legal problems themselves. A very similar finding has been obtained in a large adoption study of schizophrenia: the genetic liability for schizophrenia, as judged by the presence of schizophrenia in a birth parent, is manifest only in families with multiple problems (Tienari et al., 1985; Tienari et al., 1987; Tienari et al., 1994).

Gene-environment interaction is an important concept for understanding how genetic and environmental factors intertwine to influence psychological development, and so we will return to it later when we discuss our ideas about a "relationship code." But gene-environment interaction is also an important concept for the statistical analyses undertaken in this study. The most straightforward analytic tools estimating genetic and environmental influences on development assume that these two domains of influence are additive; that is, that genetic influences and environmental influences add up to account for individual differences. An interaction is "nonadditive" in the sense that it reflects not a simple sum of two distinct influences, but a more complex mix. We have elected not to include an estimate of this interaction in our study for two closely related reasons.

First, in nonclinical samples in which there is not a heavy concentration of children at risk for serious psychopathology, gene-environment interactions have been hard to detect (DeFries et al., 1994; Plomin, DeFries, and Fulker, 1988), and when they are found they are often small in magnitude, as in the study of separated monozygotic twins described above (Bergeman et al., 1988). We may, however, question the trustworthiness of these conclusions about rarity of gene-environment interactions by arguing that adoptive families must all be high functioning since they were screened by agencies for their suitability for adopting. If this is true, then the absence of substantial gene-environment interactions in nonclinical samples is an artifact attributable to this very restricted range of adoptive family environments. We cannot discover the magnitude of gene-environment interaction, according to this argument, unless the full range of environmental strengths and weaknesses is encompassed in the sample of adoptive families.

Although this is certainly an important argument, the data we have available suggest that, despite screening by adoption agencies, all adoptive families are not notably healthy or high functioning. For example, as mentioned in Chapter 3, one adoption study carefully compared personality features and reported family environment of birth parents rearing their own children with those of adoptive parents and found little difference (Rhea and Corley, 1994). Moreover, if the personality measures are compared with those from national samples, the adoptive parents show no substantial differences from these norms. Of special importance is the range between the highest- and lowest-functioning adoptive parents; it is the restriction in range or variance that would reduce the chances of finding gene-environment interactions. Findings on a range of scores, expressed as a variance, are quite comparable for adoptive families, birth families rearing their own children, and national norms. These findings are buttressed by more comprehensive clinical studies of adoptive families that show a broad range of marital and legal difficulties in adoptive families, as well as numerous instances of serious psychopathology in adoptive parents. For example, one recent study found that less than one-third of adoptive families were free of serious marital problems, legal problems engendered by parental misbehavior, or major psychiatric disorders in the parents (Cadoret et al., 1995).

A second reason for not including gene-environment interaction in our analysis is that most of the positive evidence we have for this effect comes from adoption and twin studies in which individuals are separated into those with high genetic risk and those with very low risk. For example, the most compelling data come from studies in which some of the birth parents were selected for having schizophrenia (Tienari et al., 1985; Tienari et al., 1987; Tienari et al., 1994), severe antisocial behavior, substance abuse (Cadoret and Cain, 1981; Cadoret et al., 1983; Cadoret et al., 1990; Cadoret et al., 1995), or major depression (Kendler et al., 1995). In our study there is no reason to believe that adolescents will fall into just these two extreme categories of genetic risk. In fact, they will be distributed all along the continua of genetic risks, as might any sample drawn from a large population.

Nonetheless, there are a few exceptions to the pattern of findings described above. We have already mentioned the interesting interaction between genetic factors and marital status on drinking behavior (Heath et al., 1989). There is also an intriguing single report on the apparent inter-action between genetic influence and a large-scale social change on educational opportunity (Heath et al., 1985). According to the data from that study, genetic influences on how far a child progresses in school were apparent in a very large Norwegian sample only after major societal changes had occurred; these changes opened up schooling according to individual abilities rather than social rank. Although marital status and large-scale societal change are not directly relevant to our study, these findings require us to be vigilant about the potential role of gene-environment interactions. Even though our sample is comparably large, it is too small to detect all but the most dramatic gene-environment interactions. Moreover, we cannot anticipate how their inclusion in our analyses—for those instances where they were very large—would affect our current estimates of heritability or environmentality. If substantial gene-environment interaction influences the psychological development of the adolescents we examined, then our estimates of heritability or environmentality or both might be influenced.

In sum, with some very important exceptions, genetics researchers have not been much more successful than psychosocial researchers in finding strong and replicable interactions between characteristics of an individual and environmental influences on the course of psychological development. One important reason may be that individuals are not sorted at random into all possible environments. The phenomenon of person-environment correlation is probably quite pervasive from both a psychosocial and a genetic perspective. For example, severely antisocial children are not equally distributed in rich and poor neighborhoods, nor in well or poorly functioning families. Children who have a personal risk for severe antisocial behavior grow up in environments that may add to that risk; in short, they suffer from a person-environment correlation.

Selective placement by adoption agencies presents special challenges for researchers conducting adoption studies. In placing children with adoptive parents who are like their birth parents, the adoption service in effect correlates the infant's genotype with its newly acquired environment, creating an unwanted form of person-environment correlation. Adoption research designs that can circumvent this difficulty are uniquely sensitive to person-environment interactions in general, and to gene-environment interactions in particular. The absence of selective placement does allow children to be, in effect, randomly assorted into a wide range of adoptive environments without reference to their genetic propensities. Evidence indicates that for most developing children, there is a high correlation between the genetic and the environmental influence on their development, and thus we return repeatedly to these correlations.

Adoption studies in which selective placement has been avoided may be very unusual. Nonetheless, they tell us something very important about the malleability of genetic influences. In particular, they make a powerful case for how effectively variations in the social environment can completely, or nearly so, eliminate the influences on genetic factors even when those influences, in certain environments, are very strong. It is surprising that not a single geneticist, or psychosocial researcher, for that matter, has ever attempted to induce positive variations in family environments to see if such changes might offset serious genetic risk for psychopathology in the offspring. For example, a good deal of evidence points to the importance of effective family functioning, particularly marital functioning, in suppressing adverse genetic influences on the offspring (Cadoret, Winokur, Langbehn, Troughton, et al., 1996; Cadoret, Yates, Troughton, Woodworth, and Stewart, 1996; Cadoret et al., 1995; Heath et al., 1989; Tienari et al., 1985; Tienari et al., 1987; Tienari et al., 1994).

Studies Using Longitudinal Data

The area of longitudinal studies that uses genetically informed designs is not precisely analogous to longitudinal studies in the purely psychosocial realm. Recall that in Chapter 2 we reviewed briefly two approaches to longitudinal designs. The first was temporal sequencing studies, in which the investigator seeks to establish order in developmental sequences. For example, does poor parental discipline precede or follow the development of antisocial behavior in the child? The second approach, disentangling group influences from assortative affiliation, is relatively unique to the study of voluntary associations, particularly peer groups.

Researchers using longitudinal data in genetically informed studies have the same overall aims as those using such data in psychosocial studies. As with psychosocial studies, the first order of business it to under-

stand the emerging characteristics of behavior across development. In genetic and psychosocial studies of antisocial behavior, for example, we
must pay attention to the fact that early in development antisocial behavior will show itself in family, peer, and school settings, but later in development will manifest itself not only in these areas but also in serious
substance abuse and distinctly criminal behavior. Thus, in tracking the
influence on changes in behavior across time, genetic studies must be as
sensitive as psychosocial studies to changes in the form of the behavior.

Once this common concern is addressed, however, the order of questions asked is different for genetic studies. As in cross-sectional genetic
studies, the first step in longitudinal genetic studies is to estimate the
influence of anonymous genetic, shared environmental, and nonshared
environmental factors on both change and stability in behavior across
time. There are many ways to define the terms "change" and "stability."
The most frequent focus, and the one we use in this book, is on change
and stability of an individual's rank order position versus that of others in
a sample. For example, we know that *in general* boys tend to show more
antisocial behavior in adolescence than in pre-adolescence. Thus, if we
measured the mean or average levels of misbehavior in a group of eight-
year-olds and then measured them again when these children were thirteen, the average level would almost certainly increase. This is definitely a
change, but one that reflects influences acting on many if not most boys
as they move into adolescence. Many psychosocial studies, and all behavioral genetics studies, are focused on factors that distinguish one person
from another. Studies in this genre are interested in factors that account
for why one boy might actually show a decline in antisocial behavior on
entry into adolescence whereas another might show a distinctively sharp
increase. The first boy will almost certainly move down in rank order: in
early adolescence he might have been in the middle of the pack, but by
adolescence he may have a very low rank on the scale of misbehavior. The
second boy might also start out in the middle of the pack but emerge in
adolescence with one of the highest ranks.

In many instances some individuals will show dramatic change in rank
across time and others will show little. In other instances most individuals
will show some change in rank. In both these cases we can say that there is
some stability and some change. We can ask, then, how significantly

anonymous genetic and environmental factors contribute to both stability and change. A more detailed explanation of how analyses of this kind are conducted is presented in Chapter 8. Here we give just a brief explanation of genetic influence on stability and change.

In studying how genetic influences might operate to influence stability, we could ask how genetic influences account for the fact that children who score high on intelligence tests in early adolescence, in comparison with other children their age, also score high later in adolescence. The most parsimonious explanation is that genetic factors influence the rank order of intelligence at both stages of development, and that the same genetic factors operate at both times. Alternatively, genetic factors might influence stability by having an impact only early in development, but the *residue* of this impact might remain relatively unchanged across time, even though genetic influences cease to operate later in development. For example, genetic influences may affect certain areas of brain functioning early in development. This brain functioning may not alter much over time, even though genetic influences no longer operate. The residue might also be environmental. Genetic factors might influence intelligence early in development. For example, an intellectually gifted child may develop both high self-confidence on intellectual tasks and a social reputation as being a bright child. These factors—self-confidence and a reputation as a bright child—may operate across time to maintain the child's high rank even though the original genetic influences have faded. Indeed, several studies have shown that genetic influence on the stability of cognitive achievement is important in young children (Plomin et al., 1993) as well as in individuals in their sixties (Plomin, Pedersen, Lichtenstein, and McClearn, 1994). The mechanisms underlying this remarkable stability have not yet been clarified.

Genetic factors can also account for change across time, for example, change in cognitive ability in young children (Cardon et al., 1992; Plomin et al., 1993). Genetic factors can account for change in many other domains as well, such as in antisocial behavior from adolescence to adulthood (Lyons et al., 1995) and in the expression of positive moods from fourteen to twenty months of age (Plomin et al., 1993). The simplest explanation for this influence is that genes can "switch on" and "switch off" at the molecular level. The influence of genetic factors on change in emotional expression might come about, for example, if certain genes that influence emotional expression in fourteen-month-old toddlers switch off in some toddlers by twenty months but not in others. Indeed, modern molecular biology is detailing a broad range of mechanisms that control the molecular expression of genes. These mechanisms can be induced by environmental factors such as the use of psychoactive drugs (see Hyman and Nestler, 1993, for a review) and environmental stress (Glaser et al., 1990; Platt et al., 1995).

In some instances, however, genetic factors may account for change in behavior in ways that are better understood at the social rather than the molecular level. For example, later in development, powerful environmental factors that "swamp out" the effect of genetic influences may arise for some individuals but not for others. As will be shown in Chapter 10, this may be the case for depression in adolescence. The data from our study suggest that relationships with fathers become very salient for older adolescents and may obscure the ongoing influences of some genetic factors on depression in adolescence. But we will not be able to elucidate the relative contribution of molecular and environmental mechanisms to changes in genetic influence across time until research yields a clearer picture of the role of particular genes in the ontogeny of complex behaviors.

Studies Using Experiments

Thanks to Mendel's experiment with smooth and wrinkled peas, experimental approaches in the field of genetics, even those relating to relatively simple behaviors, are now widespread. Mendel's experiments with vegetables have been supplemented with a variety of breeding experiments in animals that have led to a better understanding not only of genetic influences on behaviors such as fear and aggression, but also of some of the environmental circumstances that may moderate these genetic risks. More recently, advances in molecular biology have provided many new avenues for using the rigors of true experiments to explore the connections between genetic mechanisms and behavior. For example, experiments using rats show that psychoactive drugs may work to the extent that they can alter gene expression of proteins that regulate neuronal function. Even more dramatic are gene knockout techniques whereby a gene is made dysfunctional and the offspring are then studied. These offspring are literally "missing" a particular gene. Techniques of this sort have already been used to study the molecular basis of memory (Grant et al., 1992).

As intriguing as these techniques are becoming, they are not, as yet, clearly relevant to the development of complex behaviors and adjustment patterns in children and adults. More relevant to the concerns of this book are a series of studies attempting to illuminate the mechanisms by which genetic factors influence alcoholism. These studies have used a very restricted form of experiment, usually testing research subjects when they are fully sober and then, under conditions of careful supervision, after they have been given a standard amount of alcohol. A basic though simple developmental theory has been tested in these studies: the effects of genetic factors on alcoholism are mediated by their effects on sensitivity to alcohol ingestion. People are more inclined to become heavy drinkers if they show little sensitivity to alcohol; that is, if they feel less intoxicated than others after a standard dose of alcohol, or if they show less tendency to sway back and forth after a standardized dose. Although there are clear genetic influences on these hypothesized mediators, the data are less clear on the precise effect of genes on the ontogeny of alcoholism (Heath and Martin, 1991; Heath and Martin, 1992).

The logic of these alcohol studies bears important similarities to the logic of discovering gene-environment interactions as described earlier in this chapter. The straightforward hypothesis being tested in these studies is that there are genetic factors that operate only after the ingestion of a substantial amount of alcohol. Thus, the impact of these genetic factors on body sway, for example, would not be manifest or expressed in the sober state. The individual must be exposed to a particular "environment" in order for these genetic factors to be manifest: the "environment" is the consumption of alcohol and its subsequent influences.

The alcohol studies are an example of an experimentally induced "environment" that encourages the expression of an otherwise latent difficulty. There is a substantial amount of evidence, however, that some environments may suppress the expression of a latent problem. These data provide clues to an experiment of an entirely different sort: one that studies changes in the environment that may enhance development. Indeed, the research on gene-environment interaction provides an impressive body of clues for true preventive intervention experiments (see Chapter 2 for an outline of the logic of these interventions to test psychosocial theories of psychological development). For example, a preventive intervention that improved marital function, improved family communication, and reduced the impact of parental psychopathology on children might reduce or eliminate the heritability of some serious psychiatric disorders and thereby provide a rigorous test for the role of these factors in moderating genetic influence (Cadoret, Winokur, Langbehn, Troughton, et al., 1996; Cadoret, Yates, Troughton, Woodworth, and Stewart, et al., 1996; Cadoret et al., 1995; Heath et al., 1989; Tienari et al., 1985; Tienari et al., 1987; Tienari et al., 1994).

It is theoretically possible to use a twin study in pursuit of this strategy. We could, for example, seek to enhance the family environment of twins in an "experimental" group while leaving the environment unaltered in a "control group." We could provide assistance with any marital difficulties, reduce the impact of parental psychopathology on children, or enhance family communication. The heritability of the adverse condition should be reduced or eliminated in the experimental group if the theory about gene-environment interaction in the ontogeny of schizophrenia, substance abuse, or antisocial personality is correct. But an enormous sample of twins would be necessary to demonstrate an effect of this kind, and large sample sizes are just too costly in carefully planned and executed preventive interventions. Thus the preventive intervention would have to be carried out with adopted children and their adoptive parents, where a smaller sample can still yield clear results. The adoptive family environment could be enhanced in the experimental or treated group. The intervention would be targeted on an aspect of family environment that has been shown to have important gene-environment interactions. Not only should the offspring in the treated group of adoptive families show reduced frequency and severity of psychological problems, but the heritability of these problems should be reduced or eliminated by the intervention. We would infer that the latter had been accomplished if we found that there was little or no correlation between birth-parent psychopathology and psychological problems in the adopted child in the treated but not in the untreated group.

It is fair to say that psychosocial researchers are far ahead of genetics researchers in recognizing the power of the experimental design to test carefully crafted theories of development. This is because they have developed meticulous hypotheses about mechanisms of development that emphasize influential factors that are malleable. For nearly two decades genetic researchers have had, in widely known studies of gene-environment interaction, clear evidence that genetic influences may be malleable by particular variations in environmental circumstances. Except for rather restricted attempts in the field of alcoholism, we are unaware of any efforts to capitalize on these important leads. Addressing this surprising inaction is a particularly intriguing example of the value that could come from a genuine integration of genetic and psychosocial perspectives on development. It also constitutes an excellent transition to the next phase of our book, in which we make a major effort to propel this integrative process.

Chapter 5

STUDYING ADOLESCENT SIBLINGS

AND THEIR FAMILIES

This chapter begins our scientific account by describing the strategy and methods of our study. We designed the study with the aim of understanding the nonshared environment of sibling pairs. Our most fundamental concerns were how differently siblings are treated in their families and whether these differences are associated with their psychological adjustment. To answer these questions we selected a large sample of families with two sibs close in age and then carefully measured relationship patterns within these families.

From the genetic perspective, the most important question was whether these differential experiences of siblings in the same family were attributable to genetic differences between them. If the answer is "yes," then genetic mechanisms might be implicated in any associations found between differential family experiences and adjustment of the adolescents. To address this issue we selected a sample consisting of adolescent siblings who were identical twins, fraternal twins, full sibs in nondivorced families, and full, half-, and genetically unrelated sibs in step-families.

Adding step-families to our design enhances both its power and its generalizability. But it also raises questions that reflect both psychosocial and genetic issues. For example, blended sibs almost certainly reside in a common household for less time than half-sibs. Would this difference then confound our efforts to use step-families in a design that simultaneously examines the role of family processes and genetic factors in development? We examine this issue in detail in this chapter.

The concept of the nonshared environment is perhaps the best example of an important topic for the dialogue between geneticists and psychosocial researchers. As noted, genetics research has reported striking findings over the last two decades. For example, research has revealed a good deal of dissimilarity between monozygotic twins on most measures of psychological adjustment. In addition, it has shown that siblings who share no genes, such as siblings who have been adopted, are not much more similar than randomly paired children. Taken together, these data suggest that environmental factors in general are very important for development, but that what matter most are those environmental influences that lead to sibling differences in the same family. These findings, and their implications, remain a challenge to almost every prevailing view still held by psychosocial researchers about the impact of the environment on psychological development.

Findings of this kind were first explored by the behavioral geneticist John Loehlin (Loehlin and Nichols, 1976). Loehlin studied 850 twin pairs of high school seniors and assessed them on a broad range of abilities, personality, and interests. He noted that correlations between identical twins were no greater than .50. Although he was not the first to report these findings, he was the first to give strong emphasis to their implications for development. He correctly concluded that only environmental influences unique for each twin could account for these striking differences. He emphasized the marked contrast between the implications of these results and the fact that parents reported treating these identical twins similarly. He wondered just what kinds of unique environments these might be.

Ten years after Loehlin's path-breaking observations, Robert Plomin published a review article on the nonshared environment that sparked lively discussion (Plomin and Daniels, 1987). Plomin noted that many studies confirmed the importance for development of differences in experiences, not only between identical twins, but between all siblings. He further noted that the genetic evidence was compelling but that psychosocial researchers had few data on the differential experience of siblings. The study we report in this book was designed as a response to that review article and the genetic evidence it summarized.

At the time we designed our study, almost thirteen years ago, an adequate response to the behavioral genetic data would have been a thoughtful psychosocial study on the nonshared environment. Genetic data did not distinguish whether the nonshared environment relevant for psychological development was drawn from just the psychological and social environment or might include aspects of the physical environment, such as physical illness, as well. We decided to focus on the nonshared social environment and leave aspects of the physical environment, for example, the nonshared bacteriological environment, to others. Moreover, we felt there was a compelling case for focusing on the nonshared *family* environment.

Even if we had elected to design just another psychosocial study, Plomin's review, and the genetic data it summarized, would have spurred us to revise the traditional approach to studying the family environment. We had determined from the outset that our study should focus on the unique or differential experiences of siblings within families. This important objective, of course, cannot be accomplished without including at least two siblings from each family in a research design. Readers who are not familiar with the field of family research will find it difficult to believe that family researchers rarely include siblings in their research designs. Disappointingly, Plomin's review didn't initiate much of a response from such researchers. Thus we still don't know much about even the simplest aspects of the nonshared family environment—or of any other environment, for that matter. The number of psychosocial investigators who have explored this issue could fit in the hand of a very small child (see, for example, Daniels, 1986; Daniels and Plomin, 1985; Daniels et al., 1985; Deal et al., 1994; Dunn and McGuire, 1993; Dunn et al., 1990; McHale et al., 1990; Tejerina-Allen et al., 1994).

At the time we designed our study, very little was known about whether parents treat children differently in the same family, and if they do, whether those differences affect children's development. Had we opted to use traditional psychosocial designs, paying special attention to sibling differences, we could have made provisional estimates of these differences in parenting and their association with development. We would have added only one new feature in response to the genetic challenge: equal attention to at least two siblings in every family we studied. But to conduct a limited study of that kind would have been—in scientific terms—a tragic missed opportunity. Worse yet, it would have been seriously misleading.

Fortunately, we recognized from the start that inferences drawn from such a study would be limited and potentially deceptive. For example, differential parental treatment of siblings might play only a minor role in individual differences in development. Instead of influencing development, differential parenting might be a response to differences in the children. Ordinary siblings share only 50 percent of their segregating genes, on average, meaning that a full 50 percent are *not* shared. Thus differential treatment of siblings by their parents might simply reflect heritable differences between siblings in temperament and in the way they relate with others. If we did not use a genetically informative design, we might have seriously overestimated the independent, nongenetic role of differential parenting. To ensure that such a design was indeed employed, we created the first major study of psychological development to consider equally both the social environment and genetic influences.

Our first decision, to combine psychosocial and genetic approaches to development, spurred several other major decisions. The second was to select the most appropriate genetically informative design from those available. As reviewed in Chapter 3, there are two classes of genetically informative designs: parent-child designs and sibling designs. The former, usually conducted as adoption studies, provide only an indirect assessment of the effects of the nonshared environment on development. Thus we chose a sibling design as the best method available. Although the traditional choice would have been a sibling design that employed twins, we felt there were advantages to broadening the design to include nontwin siblings, particularly full sibs in nondivorced families and in step-families, along with half-siblings and unrelated siblings in step-families. We opted for this broadened design because it enabled us to examine possible genetic and environmental influences on development in three very different types of family systems—twin, nondivorced, and step—and thus might enhance the generality of both environmental and genetic findings.

As noted, however, adding step-families also poses a significant challenge. Although comparing three different types of families may make our findings more applicable to a full range of families, it also raises the possibility that there are differences among the groups of siblings that could make inferences difficult. The most conspicuous difference is that step-families are, by definition, families of remarriage. In some step-families the marriage has been relatively recent. By definition, full sibs in nondivorced families are living in family environments shaped by a marriage of long-standing. Moreover, most twins live in nondivorced households. It would have been extremely difficult to locate twins within stepfamilies, and so we could not avoid this lack of comparability of marital duration across the groups of siblings we studied. We sought as much comparability as possible using recent research that clarifies how stepfamilies readjust following the marriage of the step-parents. It is now clear that the process of readjustment may take at least three years, and perhaps many more, and have implications for both the parents and the children (Hetherington and Clingempeel, 1992). In order to achieve as much comparability across the sibling groups as possible, we required all parents in our study to have been married at least five years.

This requirement, however, introduced a new problem. Remarriages, we knew, are more likely to break up than first marriages. Thus, on the one hand, we had to match the groups of families—as much as possible—for marital duration. On the other hand, because remarried families break up more readily, remarried families with stable marriages are extremely hard to find. The addition of step-families required us to screen hundreds of thousands of households in order to fulfill our sampling requirements.

A third decision arose from our focus on individual differences in development. Specifically, we wanted to explore factors that were associated with individual differences in adjustment at a particular point in time and also those factors that account for both change and stability of adjustment across time. At a minimum, this objective required us to observe the siblings and their families at two different points in time.

A fourth decision concerned the phase of development to be studied. Adolescence seemed ideal. There is substantial change both in adolescent behavior across time and in the social influences that are associated with these changes, making it easier to study factors that are responsible for both change and stability. It is also becoming increasingly clear that measures of adjustment and psychopathology in adolescence predict corresponding areas of adjustment and psychopathology in adult life (see, for example, Kandel and Davies, 1986; Kandel et al., 1986).

Ideally, the adolescent siblings would have been approximately the same age across all the families we studied. Thus we might have studied a sample of early adolescents, say, ages twelve to thirteen, and noted both changes and stability in their development to mid-adolescence, at which point all subjects would then be about fifteen to sixteen years old. Unfortunately, we could not avoid a trade-off. Step-families with both stable marriages and half- or unrelated sibs are very rare; if we required the eligible sibs to be in the same phase of adolescent development, say, early adolescence, at the time of our first observation, the study would have been impossible. Instead of screening 675,000 households, we would have had to screen more than three million. Thus, of necessity, the siblings we studied ranged from early to late adolescents. At the time of our first observations the age range was ten to eighteen, although we kept the age differences between siblings within families to a maximum of four years. By the time of our second observation period, three years later, many families no longer qualified because the older children had grown up and left home. The subsample used for longitudinal analyses across time therefore consists mostly of children who were of roughly the same ages.

The fifth design decision involved gender. It would have been possible to include families in which one of the adolescent siblings was female and the other male. In these families we almost certainly would have seen substantial differences in parental treatment that reflected gender differences between the siblings. But these differences would be difficult to interpret. We would have wondered whether they arose from the biological differences between boys and girls or from social processes shaped by our culture's conception of male and female. The former are, of course, genetic in a crude sense: girls have two X chromosomes and boys have only one, whereas boys are unique in having a Y chromosome. Although researchers have studied individuals with extra or deleted X or Y chromosomes, there are no straightforward behavioral genetic designs that permit us to distinguish whether social or genetic factors account for differential treatment of boys and girls within the same family or for differences in their psychological development. Moreover, the relatively crude distinction between one and two X chromosomes (or the presence of Y in boys and its absence in girls) is not the kind of genetic difference we are interested in studying.

For these reasons we decided not to place a high priority on examining the effects of gender, but instead focused on pervasive genetic and nonshared influences that are likely to operate in the same way for both boys and girls. If we recruited for our sample an equal number of all-female and all-male sibling pairs, we could make some very imprecise comparisons between genders. For example, are some nonshared experiences associated with developmental differences in girls and not in boys (or vice versa)? Are some genetic factors influential in boys but not in girls (or vice versa)? Comparisons of this kind, in effect, cut our sample in half and severely curtail our power to detect meaningful environmental or genetic associations with development. Expanding the size of our sample to give us power to detect moderate, or even substantial, gender differences was out of the question. The sample we did recruit—720 families—was the largest ever employed for detailed studies of the family environment. For all these reasons, we knew in advance that we could say little about gender-specific effects.

Sixth, in order to address fully the questions that genetic research raises about the nonshared environment, we sought to measure the nonshared social environment as carefully as possible. Although we focused on the family as a major source of environmental influence on adolescent development, we included peers, teachers, and other important components of the social environment as well, though our measurement of these aspects of the social world of adolescents is not as thorough. Data on these other relationships are suggestive, but for simplicity and focus we restrict attention in this book to measures of the family. To explore thoroughly the nonshared family environment, we adopted two strategies. On the one hand, we measured all relationship subsystems in the family: the marital, father-child, mother-child, and sibling. On the other hand, we used a broad range of measures as well as many sources for our measurements. Included among our sources were father's reports, mother's reports, child's reports, and objectively coded videotaped records of interaction within each subsystem made in the family's home.

Our seventh decision concerned assessing adjustment. Since adolescence is a time when psychological development progresses at a rapid rate in many areas, we chose to measure a broad range of indicators of adolescent maladjustment and adjustment. Among the most frequent manifestations of maladjustment are antisocial behavior, depressive symptoms, or both, and so these domains were obvious ones to measure. But young people also develop a broad range of skills and competencies during adolescence. With this in mind, we included in our measures of adjustment cognitive skills and involvement in school; successful involvement with peer groups and other social activities; increased initiative in household responsibilities, outside activities, and leisure activities; awareness and respect for the rights and perspectives of others; and general levels of self-perceived competence.

Design and Sample

Our sample consisted of 720 families, 708 of which were used for genetic analyses. We included both families that had experienced no divorce since the birth of the oldest child in the study, and step-families in which the parents had been married at least five years. The former consisted of three different groups of families: those with MZ twins, those with DZ twins, and those with ordinary siblings. Of the 287 families in this group, 91 percent of mothers and 88 percent of fathers had no history of any previous divorce. These parents were evenly distributed across all three sibling groups.

The zygosity of twins was determined by ratings made by the research interviewers and by both parents using a standardized questionnaire about physical similarity (Nichols and Bilbro, 1966). When checked against blood tests using single gene markers, this questionnaire provides an accurate classification in more than 90 percent of cases. Six percent, or twelve twin pairs of our total sample, could not be classified, a rate of difficulty comparable to that of other samples of twins. The families of these twins were dropped from all genetic analyses.

The step-families we studied fell into three groups. First were step-families with full siblings in which the mother had brought both children to the new marriage from a previous marriage. Second were step-families in which the mother had brought a child from a first marriage and had had a second child with her current husband. The third group consisted of step-families with genetically unrelated siblings. In this group the father had brought to the family a child from a previous marriage, as had the mother. The number of families in each of the six groups is shown in Table 5.1, along with the amount of genetic overlap between the siblings. The table emphasizes that the full range of genetic relatedness is encompassed by these six groups, adding power to our genetic analyses.

Table 5.1 Characteristics of the sample at time 1 and time 2

Characteristics	Genetic relatedness (%)	Time 1	Time 2 (3 years later)
		Sample size	
MZ twins—nondivorced	100	93	63
DZ twins—nondivorced	50	99	75
Full sibs—nondivorced	50	95	58
Full sibs—step	50	182	95
Half-sibs—step	25	109	60
Unrelated sibs—step	0	130	44
Twin/sibling pairs total		*708*	*395*
		Adolescent characteristics[a]	
Age range (years)		10 to 18	13 to 21
Mean age for child 1 (years)		13.5 ± 2.0	16.2 ± 2.1
Mean age for child 2 (years)		12.1 ± 1.3	14.7 ± 1.9
Mean age difference (years)		1.61 ± 1.29	1.47 ± 1.34
Female sibling pairs (%)		48.4	49.4
		Parental characteristics[b]	
Existence of step-families			
Mean length of time (years)		8.5 ± 3	11.6 ± 3.5
Range (years)		5–18	8–21
Mother's age (years)		38.1 ± 5.2	40.5 ± 4.8
Father's age (years)		41.0 ± 6.5	43.7 ± 6.1
Mother's education (years)		13.8 ± 2.3	13.9 ± 2.4
Father's education (years)		13.9 ± 2.7	14.0 ± 2.6
Family income in thousands (median)		25 to 35	25 to 35

a. Sibling pairs were required to be the same sex and to reside in the home at least half the time.

b. Both parents were required to reside together in the home.

Source: Neiderhiser et al., 1999. Copyright © 1999 by the American Psychological Association. Adapted with permission.

Among the 708 families we studied, 238, or 33 percent, had only the two children; 141, or 20 percent, had an additional child older than those we tested; 272, or 38 percent, had one child younger than those we tested; and 69, or 10 percent, had at least one younger and one older child. As mentioned, the siblings were required to be of the same gender, no more than four years apart in age, and no younger than ten or older than eighteen. In the case of dual custody in step-families, the adolescents selected for study had to have resided in the household at least half of each week in order for the family to be included.

Because of its special characteristics and its size, as well as the rarity of some of its family types, this sample constitutes one of the most precious ever assembled in the history of behavioral science research. For example, according to 1980 census data—the most recent available when we began recruiting our sample—only .04 percent of all households met our criteria for step-families with unrelated siblings. Thus to find these rare families, we screened more than 675,000 households using two techniques. First, we randomly called households throughout the country. This procedure allowed us to recruit all the nondivorced families with ordinary siblings and some of the families with twins. Second, we used market panels to identify the remaining twin families and all step-families. Because they were so rare, we approached families with half- and unrelated siblings first. Of these, 44 percent and 48 percent, respectively, completed assessment. These rates of recruitment and data collection compare favorably with other family study designs that require families composed of three or four members to be assessed as a unit and that use community or national registers to identify them. For example, a large community study in California recruited 30.5 percent of eligible families (Fisher et al., 1992), and a national sample of families with children recruited 41 percent (Olson et al., 1989). Once we recruited these two rarest groups, we directed the recruitment of other groups in order to match age and age-spacing of children in the blended and half-sib families. Given that we did not approach all eligible families in these other groups, a true recruitment rate cannot be computed.

Three years after assessing all 720 families, we sought to retest all the families in which both siblings still resided in the house at least half of the time. Of the 434 families who qualified, 395, or 91 percent, completed data collection at time 2. For the great majority of families not eligible at time 2, one or both siblings had moved out of the house. Demographic characteristics of our sample at time 1 and time 2 are reported in Table 5.1. At time 1 there were only a few differences among the six groups. As expected, the marital duration in families of half-sibs is greater than that in the other two step-family groups, and, somewhat surprisingly, their education and total family income are somewhat lower. Finally, step-families show somewhat lower levels of social class than nonsteps, consistent with data showing an inverse relationship between social class and divorce. The sample is mostly white and middle-class: 94 percent of the men and 93 percent of the women are white. Some of the racial and economic skew of our sample is attributable to our requirement for long marriages; whites typically have lower divorce rates and higher remarriage rates than blacks. Despite this class and racial skew, the children in our sample scored about the same as age-matched comparison samples on measures of psychopathology (Glyshaw et al., 1989; Kovacs, 1985; Peterson and Zill, 1986; Wierzbicki, 1987).

As Table 5.1 shows, this is a well-educated, mostly white sample. Although there is no *a priori* reason to believe that these results would not generalize to other populations, we cannot assume that they would.

Methods of Measurement

We selected three areas relevant to adolescent development for careful measurement: adolescent psychopathology, adolescent competence, and measurements of all subsystems in the family. We also assessed other aspects of the environment, including the adolescents' relationships with peers, teachers, and best friends. In general, we don't focus on those measures in this book, though they are occasionally used for comparisons.

Psychopathology

Table 5.2 summarizes the measures we used to assess psychopathology. Some general features of our measurement strategy are displayed. First, we employed established and frequently used methods wherever possible. Second, since each parent had to assess psychopathology in two children (the adolescent siblings), we tried to cut down on total testing time on this topic; hence, we used the Zill-abbreviated form of the widely used but much more lengthy Child Behavior Checklist. Third, we sought to use measures that extend back in time to varying degrees from one week to three months in their inquiries about symptoms. Fourth, we used many sources for our data about children: their own reports, the reports

Table 5.2 Summary of measures of adolescent psychopathology

Construct	Measures	Source	Examples	Construct reliability (Chronbach's alpha)
Antisocial behavior	Behavior Problems Index (Zill, 1985)	M,F,C	Got into trouble at school, was a bully or was mean to others *during the past 3 months*	
	Behavior Events Inventory[a] (Hetherington, 1992)	M,F,C	Stole, lied, or cheated, skipped school *during the past week*	.85
	Global Coding Scales (Hetherington, 1992) —Antisocial behavior	O	Observed disruptive, rude, coercive, aggressive behavior (interrater coding reliability: ICC = .86)	
Depressive symptoms	Behavior Problems Index	M,F,C	Sudden changes in mood, felt sad or depressed *during the past 3 months*	
	Behavior Events Inventory[a]	M,F,C	Felt withdrawn, depressed, or lonely *during the past week*	.85
	Child Depression Inventory	M,F,C	Depressed affect, poor appetite, or poor sleep *during the past two weeks*	
	Global Coding Scales —Depressed mood	O	Depressed mood (interrater coding reliability: ICC = .70)	

a. Deleted from time 2 assessments.

Note: M = mother; F = father; C = child; O = observer; ICC = intraclass correlation coefficient.

of each parent, and observer ratings using video recordings of family discussions. We also obtained teachers' ratings of the subjects' psychopathology and competence, but because these reports were not available for some of the adolescent siblings, teacher data are not used in this book. Fifth, we sought to measure stable rather than transient psychopathology. Thus, for the Behavior Problem Index and Child Depression Inventory we obtained measures on two separate occasions, usually separated by a week, and then averaged the two. The correlations between measurement on the first and on the second week were invariably very high.

The measures of depression and antisocial behavior were aggregated for each observer to yield, for example, a depression score as reported by mother or an antisocial score as reported by father. Standard or z scores were computed for each measure for all 1,440 adolescents in the study. Time 1 raw scores were used for computing z scores for time 2 so that meaningful change could be observed across the two time periods. The z scores were added within each domain. To test the reliability of these composite scores, we computed Cronbach's coefficient alpha for the aggregated score for depression and antisocial behavior. The average reliability, cumulating data across all raters and across both adolescents, is shown in Table 5.2. The separate mother, father, child, and observer ratings of psychopathology and of competence are used for the phenotypic analyses in Chapter 7. Otherwise, we used only a single measure each for depression and for antisocial behavior: a composite of all measures across all raters. Also, to reduce the testing burden and expense of time 2 data collection, we eliminated the Behavior Events Inventory and the second assessment of antisocial behavior and depression a week after the first assessment.

Competence

We employed a number of measures to assess the adolescents' cognitive and social competence, self-esteem, and psychological autonomy. These are shown in Table 5.3. Here again we used well-established measures and derived data from mother, father, and child reports.

Because the conceptual domains and the measures are not as securely tied to one another as they are for psychopathology, we subjected all the

Table 5.3 Summary of measures of adolescent competence

Construct	Measures	Source	Examples	Construct reliability (Chronbach's alpha)
Cognitive agency	Child Behavior Checklist (Achenbach, 1983, no. 653) Cognitive competence subscale	M,F,C	School grades in several areas	.85
	Harter Perceived Competence Scale (Harter, 1982, no. 1131) Cognitive competence subscale	C	Degree of positive self-regard in cognitive areas	
	Child Competence Inventory (Hetherington, 1992, no. 126) Cognitive agency subscale	M,F	Parental rating of adolescent's industriousness and orientation toward schoolwork	
Sociability	Child Behavior Checklist Sociability subscale	M,F,C	Activities in social organizations, number of child's friends, and how well child gets along with others	.76
	Autonomous Functioning Checklist (Sigafoos, 1988, no. 313) Sociability subscale	M,F,C	Extent of positive peer activity, involvement in organized prosocial activity, quality of peer network	
Autonomy	Autonomous Functioning Checklist Maintenance, self-care, and recreation subscales	M,F,C	Engagement in independent and self-reliant activity: initiates activities and transactions with community services; takes care of self and belongings; independently pursues leisure and work activities	.80

Construct	Measures	Source	Examples	Construct reliability (Chronbach's alpha)
Social responsibility	Child Competence Inventory (Hetherington, 1992, no. 126) Social responsibility subscale	M,F	Adherence to adult norms	
	Behavior Events Inventory[a] (Hetherington, 1992, no. 126) Prosocial behavior subscale	M,F	Helping and sharing behavior	.73
	Behavior Events Inventory[a] (Hetherington, 1992, no. 126) Prosocial subscale	M,F	Helping and sharing behavior	
	California Psychological Inventory[a] (Gough, 1966, no. 114) Socialization and responsibility subscales	C	Social maturity, adherence to social norms, internalization of moral principles	
Global self-worth	Harter Perceived Competence Scale Global self-worth	C	General tendency to view self positively	

a. Deleted from time 2 assessments.

Note: M = mother; F = father; C = child.

measures of competence to factor analysis and rotation of the factors. This enabled us to group measures together based on the relatively high correlation among measures within a group and the relatively low correlation of measures across groups. The grouping of measures obtained through the factor analytic procedure is shown in Table 5.3. The reliability or internal consistencies of measures in each of these domains are also shown.

Family Subsystems

In recent years researchers have made considerable advances in thinking about the household-based family as a single unit. Clinical and research observations have led to typologies of family units based on the overall quality of relationships in the family such as levels of conflict and cohesion or the kinds of values, perceptions, and world views that organize family life (see, for example, Moos and Moos, 1981; Reiss and Klein, 1987). By definition, however, these attributes of the family are part of the shared environment for siblings. The same is true for the quality of the parents' marriage. Considerable progress has also been made in assessing marital quality, stability, and the mechanisms by which couples achieve satisfying marriages (see, for example, Booth et al., 1983; Gottman, 1979; Gottman, 1994; Spanier, 1976). In general, the quality and stability of the parental marriage are aspects of the environment that a child shares with other siblings. Since our main objective from the start was to measure the nonshared environment, these attributes of the family were tangential.

To be sure, siblings almost certainly have different views about their family and their parents' marriage. In this sense, the *perceived* family environment does constitute a unique social world for each child. Despite the importance of these perceptions, however, we did not want to ground our measurement strategy in their pursuit. If we were to find that differences in perceived family environment were indeed related to adolescent adjustment, we would be faced with the larger question of where those differences originate: do they come from the child or do they result from the family process itself?

To circumvent this question, we chose a measurement strategy that had two components. First, we focused on family subsystems, particularly parent-child and sibling subsystems. For example, we measured each child's relationship with his or her mother. This specific relationship might be quite different from one sibling to the next: one child might be favored and another scapegoated. A second component of our strategy was an effort to take several perspectives into account in evaluating the quality of a particular subsystem: the parent's, the target adolescent's, the sibling's, and that of a trained observer. In this way we sought to estimate essential features of family relationships that might be recognized by any observer. The adolescent's own perceptions can then be regarded as his or her subjective construction of the relationship subsystem and a potential *mediator* of the influence of that subsystem on development. Although a child's perceptions may modify the generally observed nature of a subsystem and vice versa, it is reasonable to assign some causal priority to the essential nature of the subsystem itself. Thus in this book we focus on composite scores of the relationship subsystem. Elsewhere we have paid close attention to the child's perception of relationships (Neiderhiser and Pike, 1995; Plomin, Reiss, Hetherington, and Howe, 1994).

Following this strategy, we used many questionnaire and coding measures to study the quality of the mother-child, father-child, and sibling subsystems in each of the families in our sample. We also include in this book, at time 1, one marital measure: the degree to which the couple reported disagreeing or fighting about a particular child. We included this aspect of marriage because it could vary from one child to another. In addition to this simple measure of the marital relationship, we included very detailed assessments of the parents' marriage, including many questionnaires and scales for coding videotaped marital interaction at both time 1 and time 2. One of our initial hypotheses was that severe marital conflict might result in parents' treating their children unequally. Our findings suggest, however, that this is not the case (Henderson et al., 1996). Thus these additional measures of marriage will not be considered in this book.

Videotaped records of family interactions were obtained during a home visit made to each of the families. Early in the home session each member of each dyad in the study filled out questionnaires inquiring about an area of difficulty, disagreement, or conflict with other family members. For example, mother and her older child each filled out a questionnaire asking them to select areas that were a source of significant arguments between them. Toward the mid-point of the session a member of our staff reviewed these questionnaires to locate topics of disagreement or conflict that were jointly rated by members of each salient dyad. For example, father and the younger adolescent sibling might each report, after responding independently, that they frequently argued over the adolescent's cleanup of his room. A "hot topic" such as this could be readily identified for each dyad in the family. Later in the session the dyad was presented with the hot topic and asked to discuss it and attempt to reach a resolution within ten minutes. The staff member then left the room for ten minutes after turning on a video recorder. In many cases the dyad seemed to have forgotten that they had provided the information about the hot topic, but they invariably acknowledged its salience for them. The videotaped interaction was typically lively and emotional.

Videotapes were rated by a large team of highly trained coders under the supervision of Mavis Hetherington at the University of Virginia. They used an adaptation of the Global Coding System developed by Hetherington to be equally applicable to nondivorced and step-families (Hetherington and Clingempeel, 1992). Using a very detailed manual, coders viewed the entire ten minutes of interaction and rated the segment on several five-point scales. Care was taken to assign separate coders to each of the two target siblings in the family. For example, one coder rated mother's interactions with the older adolescent and another rated her interactions with the younger adolescent. Coders were blind to one another's coding judgments and to the sibling status of the family: twin, full sib, half-sib, and so on. During the extensive training period for the coders, we used a criterion coder system for measuring how fastidiously the coder followed the coding manual. The criterion coder was the research team member who adapted the manual to the requirements of this study and was the most familiar with its requirements. All coders were trained to achieve 60 percent perfect agreement with the criterion coder. After this criterion was achieved at the conclusion of training, about 48 percent of each coder's work was double-checked at random. In the course of the actual coding we used a "round robin method" for assessing reliability whereby each coder's work was compared with that of every other coder.

When a coder fell below 60 percent agreement with either the criterion coder or the other coders, the work of that unreliable coder was redone and the coder retrained. If coders could not achieve and maintain high levels of agreement with the criterion coder or with other coders, they were dropped from the coding team.

The measures of the family relationship subsystems are summarized in Table 5.4. For questionnaire measures we provide the internal reliabilities across both adolescents (labeled "child") and mother and father (labeled "parents"). For scales used to code videotapes we provide the intraclass correlation coefficient (ICC), which is a measure of exact agreement between the coder being assessed and the criterion coder. The questionnaires and codes were assigned *a priori* to the domains of subsystem interaction as shown in Table 5.4. For example, we hypothesized that coercion as noted by an observer would measure the same underlying relationship quality of parent-child conflict/negativity as would punitiveness as reported on a questionnaire by mother, father, and child. We confirmed these groupings of measurements by confirmatory factor analysis (for details see Reiss et al., 1995).

Data-Collection Procedures

Because of the extreme rarity of half-sib step-families who met all criteria for inclusion in the study, and the even greater rarity of step-families with genetically unrelated sibs, we had to scour the entire country for our sample. In fact, all fifty states are represented in the sample, with the exception of Hawaii, Alaska, and South Dakota. For time 1, we employed thirty-seven interview teams under the general supervision of a survey research unit, NORC, at the University of Chicago. Interviewers resided all over the country. They came to Chicago for intensive training prior to beginning their work, which was closely supervised through checks of the quality of the questionnaire responses and videotapes. Interviewers were assigned families that lived in their region. In sparsely populated areas, however, interview teams had to travel great distances to reach the families. At time 1, two interviewers were assigned to each family in which one of the children was younger than fifteen. At time 2 we assigned only

Table 5.4 Summary of measures of family subsystems

Construct	Measures	Source	Examples	Scale or coding reliability
Parent-child conflict and negativity	Parent-child disagreement (Hetherington and Clingempeel, 1992)	M,F,C	Frequency of disagreement on specific issues (e.g., telephone use)	child = .80 parent = .84
	Punitiveness (Hetherington and Clingempeel, 1992)	M,F,C	Frequency of yelling and punishing	child = .90 parent = .90
	Yielding to coercion (Hetherington and Clingempeel, 1992)	M,F,C	Surrendering to noncompliant and disruptive behavior	child = .59 parent = .68
	Open conflict (Hetherington and Clingempeel, 1992)	M,F,C	Frequency of severe conflicts and fights	child = .71 parent = .74
	Verbal aggression (Straus, 1979)	M,F,C	Swearing, threatening, stomping out of the room	child = .74 parent = .74
	Global Coding Scale (Hetherington and Clingempeel, 1992)			
	—Anger and rejection	O	Verbal and nonverbal criticism and denigration	ICC = .81
	—Coercion	O	Manipulative control including whining and threats	ICC = .78
	—Transactional conflict	O	Reciprocated and escalated argument and conflict	ICC = .79
Parent-child warmth and support	Closeness and rapport (Hetherington and Clingempeel, 1992)	M,F,C	Pleasure in closeness and involvement	child = .91 parent = .88
	Expression of affection (Patterson, 1982)	M,F,C	Overt signs of affection and pleasure in joining parent-child activities	child = .77 parent = .81

Construct	Measures	Source	Examples	Scale or coding reliability
	Global Coding Scale (Hetherington and Clingempel, 1992)			
	—Warmth and support	O	Positive interpersonal feelings and affection	ICC = .79
	—Assertiveness	O	Positive expression of personal needs, self-confidence, and patience	ICC = .75
	—Communication	O	Ability to listen, explain, and elicit other's view	ICC = .62
	—Involvement	O	Feeling comfortable with one's own initiative and encouraging other's participation	ICC = .68
Parent-child monitoring and control	Monitoring/Control Scale (Hetherington and Clingempel, 1992)			
	—Knowledge	M,F,C	Shows evidence of having information about child's activities	child = .89 parent = .91
	—Attempted control	M,F,C	Unsuccessful efforts to control child's activities	child = .93 parent = .91
	—Actual control	M,F,C	Successful efforts to control child's activities	child = .95 parent = .94
	Global Coding Scale (Hetherington and Clingempel, 1992)			
	—Child monitoring	O	Evidence that parent knows about child's activities	ICC = .64
	—Attempted influence	O	Parental attempts at control	ICC = .69
	—Parental authority	O	Successful influences on child's behavior and opinions	ICC = .64

Table 5.4 *(continued)*

Construct	Measures	Source	Examples	Scale or coding reliability
Sibling conflict and negativity	Sibling Inventory of Behavior (Schaefer and Edgerton, 1981)			
	—Rivalry	M,F,C	Sibling is competitive with adolescent	child = .82 parent = .84
	—Aggressiveness	M,F,C	Sibling is angry with adolescent	child = .80 parent = .84
	—Avoidance	M,F,C	Sibling avoids adolescent	child = .78 parent = .85
	Conflict Tactics Scale (Straus, 1979)			
	—Symbolic aggression	C	Verbal threats and swearing by sibling	child = .79
	—Violence	C	Physical violence including hitting by sibling	child = .88
	Sibling disagreements	C	Sibling disagrees with adolescent; circumstances include taking things without permission, messing up adolescent's room, etc.	child = .72
	Relationship Quality Survey (Furman and Buhrmester, 1985, 1992; Hetherington and Clingempeel, 1992)	C	Sibling displays negative behavior such as nagging	child = .84
	Global Coding Scale (Hetherington and Clingempeel, 1992)			
	—Anger and rejection	O		ICC = .65, .73[a]
	—Coercion	O		ICC = .62, .71[a]
	—Transactional conflict	O		ICC = .66[b]

Construct	Measures	Source	Examples	Scale or coding reliability
Sibling warmth and support	Sibling Inventory of Behavior (Schaefer and Edgerton, 1981)			
	—Companionship	M,F,C	Sibling treats adolescent as a good friend	child = .86 parent = .90
	—Empathy	M,F,C	Sibling shows sympathy when times are hard	child = .86 parent = .92
	—Teaching	M,F,C	Sibling teaches new skills	child = .68 parent = .72
	Global Coding Scale (Hetherington and Clingempeel, 1992)			
	—Warmth	O		ICC = .51, .59[a]
	—Assertiveness	O		ICC = .60, .62[a]
	—Communication	O		ICC = .40, .60[a]
	—Involvement	O		ICC = .55, .59[a]
Parental conflict about adolescent		M,F		.88 (M) .87 (F)

a. Reliability coefficients were calculated for each sibling.
b. This code was given to the sibling pair, not to each sibling.
Note: M = mother; F = father; C = child; O = observer; ICC = intraclass correlation coefficient.

one interviewer per family to obtain questionnaire responses and to record videotaped dyadic interactions.

Analysis of Data

The Main Questions

We used three fundamental strategies to analyze the data we collected. These strategies are the foundation for each of the major sections of Part 2.

THESIS   Our thesis is that nonshared aspects of the family are closely linked to levels of adjustment in the adolescents we studied. In this section we must address three questions. First, how different is the family for adolescent siblings? For example, do mothers and fathers treat their children differently? Are sibling relationships perfectly reciprocal so that they constitute the same relationship for each sibling, or are they asymmetrical and hence functionally different for each sibling? For example, if one sibling is aggressive, is the other likely to be passive? In this example the sibling relationship for one would be exercising aggression and for the other, submitting to it—a form of nonshared sibling environment.

Second, does the unique family experience of the adolescent influence adjustment? In other words, if the family is different in important ways for each of the siblings, does this difference matter? To answer this question we use analytic approaches that delineate these unique effects and then associate them with all seven domains of adolescent adjustment.

Finally, are adolescents influenced by the unique family experience of their siblings? As we explain in the next chapter, there is substantial evidence to assume that children, from an early age, are keen observers of how their parents treat their siblings. What is unknown—particularly for adolescents—is how these unique experiences influence them. For example, suppose a father is particularly cruel to an adolescent's sibling but spares the adolescent himself. Does witnessing cruelty to another sibling have the same implications for the adolescent as being the victim of that cruelty himself? Does the adolescent say to himself, in effect, "this cruelty could have been meted out to me" and thus suffer when his own adjustment falters? Or does the adolescent feel uniquely privileged and protected, resulting in improved adjustment?

Our analytic preference is to develop approaches that answer several related questions simultaneously. In effect, we do not use simple statistical tests. Rather, we construct a mathematical representation of our entire thesis—differential family treatment, unique effects on adolescents, vicarious effects of sibling-unique experiences—and compare this model with our actual findings. Chapter 7 provides more detail on this model and presents the major findings relevant to our thesis.

ANTITHESIS   Our antithesis turns the spotlight away from the effects of the nonshared environment to the effects of genetics. The antithesis is that genetic factors, not the nonshared environment, have the strongest influence on adolescent adjustment. The antithesis centers on two main questions. First, what is the influence of genetic factors on adolescent adjustment? Genetic influence on antisocial behavior in adolescence has been demonstrated in other studies, and so it would not be surprising if we confirmed those findings here. There are many questions arising from our main query on genetic influence, however, for which there are no data. For example, though there are some data suggesting genetic influence on social responsibility and sociability (see, for example, Loehlin and Nichols, 1976), there are no previous data on autonomy and self-esteem. In addition, we have more searching questions that go beyond the magnitude of genetic influence on measures of adolescent adjustment. For example, what is the balance among genetic, shared, and nonshared influences on all seven domains of adolescent adjustment? Are there notable changes in this balance between early and mid-adolescence? And, perhaps most important for our interest in development, what is the balance among genetic, shared, and nonshared environments in accounting for both stability and change between early and mid-adolescence?

Second, our antithesis asks, what is the influence of genetic factors on adolescents' relationships in their families? This question poses a stiff challenge to the thesis. Our antithesis states that the associations we observe between nonshared environment (and shared environment, if any) and adolescent adjustment are illusions. This is because genetic factors influence not only adjustment but also the adolescent's behavior in the family and, as a consequence, the development of relationships with both parents and siblings. Thus, since genetic factors could account for both relationships and adjustment, they may account for the association between the two. In testing the validity of our antithesis, however, we stop short of pursuing this last question and examine systematically only the balance of genetic and environmental influences on parent-child and sibling relationships. We also examine these influences on parental conflict concerning the adolescent. Our inquiries about genetic influences on family relationships parallel our questions on adjustment. We want to ascertain the balance of genetic and environmental influences on family relationships earlier in adolescence and later in adolescence, as well as on stability and change in relationships across this developmental period.

As with our thesis, we test our antithesis with a model embracing all the component questions: genetic influences on early and mid-adolescence and genetic influences on change and stability. We present these models in Chapters 8 and 9.

SYNTHESIS   A synthesis, of course, depends on the apparent validity of the thesis and the antithesis. On the basis of currently available data, we can state both the thesis and the antithesis with clarity. If the evidence favors either the thesis or the antithesis—but not both—there isn't much need for a synthesis. Suppose it favors both? Our analytic models then must reconcile the two. We must wait, however, to see how our dramatically contending ideas fare when they are faced with actual data.

In order to assure ourselves that the thesis and antithesis are challenged to the limits currently available, we must examine two issues. First, how do we use the exceptional data we have collected? We need justifiable ground rules for ensuring that, once collected, the data are maximally exploited. The voluminous data from our study are of little use unless we can analyze and report our findings in a way that is useful for refashioning ideas about development. Second, how useful is the step-family design, anyway? Are the assumptions behind its use different from those for the twin design? Are these assumptions warranted? Does the step-family design yield results that are comparable to the twin design? In sum, does the step-family design enhance the power of our genetically informed design to detect both genetic and environmental influences on adolescent development and family relationships, or does it just muddy the waters?

Using Data Effectively

In writing this book, we adopted one fundamental ground rule concerning our data: present everything. The findings from our study, and their implications, should offer the clearest grounding for the next generation of developmental research, characterized, we hope, by geneticists' and psychosocial researchers' focusing together on emerging questions. There is little use, then, in presenting only a portion of the data: we could easily fool ourselves into thinking we had seen patterns that might be limited or contradicted by the data we did not examine. Thus, the reader is invited to examine the entire corpus of our findings with us. All the major conclusions of our study are offered here for the first time in response to the major patterns of findings that emerged in the process of writing this book.

Despite our good intentions, however, the decision to present all the data will be taxing to our readers, as it has been to us. For us, it has meant arranging the data so that we can see the entire terrain and not get lost in small crevices. The decision also posed a very heavy burden on explanation. Even if we could see clearly the main contours of the terrain, we had to be able to explain those contours. As an even stiffer challenge to ourselves, we held fast to a severe scientific aesthetic: our explanations should be as parsimonious as possible. We attempted to devise a few simply stated principles of development that could account for all our findings because we believed that these principles were what people would remember about our work; we also thought they would be useful to future researchers studying genetic and social factors in adolescents and families. We wanted, at all costs, to avoid two mournful and typical consequences of large, complicated studies. First, data are often used and presented in ways that obscure rather than reveal the large-scale patterns of findings. Second, explanations of data are complex, conditional, intellectually provincial, and mutilated by hesitations and appeals for the next study or the next wave of data to clarify matters. Either of these circumstances, we reasoned, was enough to sink a study.

To avoid drowning our study, we restrict our report here, as mentioned, to data about the family environment. Occasionally, we use data reflecting other aspects of the adolescents' social world; for example, peer relationships. As even a quick glance at Tables 5.2 through 5.4 makes clear, however, we use many different measures to assess numerous aspects of the family and the adolescent. Thus limiting our focus to the family seems like a modest restriction at best. Had we not devised these ground rules for the use of the data, we would still be facing a torrent of analyses and results too numerous to interpret meaningfully.

DATA REDUCTION   Even a sturdy thesis, antithesis, and synthesis could easily become buried under an avalanche of data. Thus we have devised several other ground rules to help us come to terms with these findings and help the reader follow our reasoning. First, we combine as many measures as possible. If we have reason to believe that many measures assess the same underlying attribute, we aggregate them. It may seem logical in such a situation to use the single best measure and drop the rest, but we believe that each measure we use has its own idiosyncrasies. For example, we ask each adolescent to respond to three different measures of depressive symptoms. These are distinguished from one another by the different periods of time they cover—three months to two weeks to one week—and by the specific questions asked about depression. By combining these measures we can identify adolescents who were depressed at the time of assessment. Suppose, however, that the adolescent had experienced a brief lightening of mood at the time of assessment but had suffered sustained symptoms for most of the previous three months. Because of the increased time coverage of one of the instruments, this young person, too, would be identified as depressed. Further, by using three different but overlapping lists of symptoms to inquire about depressive symptoms, we increase the likelihood that we will identify a depressed adolescent who may show signs of depression that are emphasized, let us say, on two of the instruments but not on the third. The use of the third instrument alone would have missed the depression. Indeed, with this strategy, the adolescent receiving high depression scores would have a sustained depression across three months and would have experienced a broad range of symptoms.

Second, wherever possible we combine ratings and data from all available observers: mothers, fathers, adolescents, and observers. There are pros and cons to this strategy. On the one hand, it vastly reduces the number of analyses we need to do and the number of findings we must synthesize. On the other hand, it obscures the important and particular nuances each observer brings in his or her perceptions of family relationships and adolescent adjustment. For the most part, however, we regard each observer as an equally valid source of data about underlying "true scores," attributes that can readily be observed from different vantage points.

VISUAL PRESENTATIONS We believe graphs are more effective than tables. Without exception, every important datum in this book is reported as either a bar graph or a "balloon diagram." This artifice helped us recognize patterns and integrate findings, and we hope it helps the reader as well. Nonetheless, this strategy also represents a compromise. First, the bar graphs reflect, for the most part, a statistical abstraction of the data. In effect, they summarize many correlations among the measures. We have already referred briefly to these complex models.

Second, in order to present our findings visually, we cannot report in the graphs themselves, or in the text, standard statistical estimates of how well our abstraction actually represents the myriads of simpler correlations. For the statistically minded reader, we present the correlations on which these models are based and indexes of the fit between the models and the correlations in Appendix B. Finally, in contrast to some researchers in this field, we use the same models, across different analyses, as often as we can. To be perfectly clear on this point: we construct analytic models using the same assumptions about the data across all our analyses. We have already stated one assumption of this kind: that there is no appreciable gene x environment interaction in any of the domains we analyze. The rationale for this assumption was provided in the last chapter. Another assumption is that the level of shared environment influencing the adjustment measures is the same for all six of the sibling groups. These assumptions may be accurate for many of our analyses but hold less well for others. The tactic of using the same analytic models across all analyses makes for easy visual scans of an array of data, although it means some models in an array don't fit the data as well as others. Again, the statistically minded reader can evaluate our strategy here by carefully examining the graphical presentations themselves as well as the data provided in Appendix B. Appendix B also shows that although some of our models fit the data better than others, almost all have a fit well above thresholds that are widely accepted in the field of developmental studies.

We anticipate that members of our own team who continue to work with these data, as well as other researchers who examine the data or comparable findings from future studies, may explore these simplifying assumptions in greater depth. This would be especially appropriate for a more intense analysis of particular domains of family process or adolescent adjustment. We hope a critical re-examination of these assumptions may yield a more refined conception of psychological development and thus bolster the dialogue this book is intended to stimulate.

The Step-Family Design

In Chapter 3 we discussed the advantages and limitations of the twin method. By using ordinary siblings in nondivorced families and, more particularly, by using step-families, we have an opportunity to test the generality of the twin findings. The step-families are particularly interesting because they provide siblings at three levels of genetic relatedness: full (50 percent); half (25 percent); and blended (0 percent). Our study represents the first time that step-families have been used in this way in behavioral genetics research. In order to determine whether they provide a useful addition to the twin method, we must ask two questions. First, what assumptions must be made before we can use step-families for estimating genetic and environmental effects, and are these assumptions valid? Second, do step-families provide estimates of genetic and environmental effects that are, generally speaking, comparable to those provided by the twin subsample?

Assumptions

Step-families have three distinct features that differentiate them from nondivorced twin families. These differences may influence estimates of genetic and environmental effects on adolescent development. First, step-families are by definition reassembled groups; the duration of the current marriage is variable, whereas it is uniformly of long duration in the other groups. Second, in the reassembled families, some parents are raising children who are not their "own" in a biological sense. Third, the set of factors influencing marital choice is broader in the step-families than in the nondivorced families. For example, when couples bring children with them into a relationship, it is reasonable to suppose that the characteristics of those children will influence both the likelihood of marriage and the duration of that marriage should it occur. These three distinctive characteristics of step-families would influence estimates of genetic and environmental influences if they influenced the correlations between siblings of our measures of the family subsystems, of adolescent adjustment, or of both.

Duration of Current Marriage and Co-Residence of Siblings

In our study design, the current marriage between parent and step-parent had to be shorter than the age of both children in the full- and blended-sib step-families, and shorter than the age of the oldest sib in the half-sib families. In contrast, in our twin sample, the parents were invariably married before the twins were born and before the ordinary siblings in nondivorced families were born. Thus children in the last three groups have resided together in families headed by the same marital partners for their entire lives. This co-residence in the same family system is one aspect of the equal-environment assumption that lies at the core of the twin method.

The situation is different in step-families. With the exception of half-siblings, siblings in step-families have not lived under the aegis of the same parental marriage for the duration of their lives together. Even more important, there are almost certainly differences among the three groups of step-families. In full-sib step-families the siblings—for the most part—

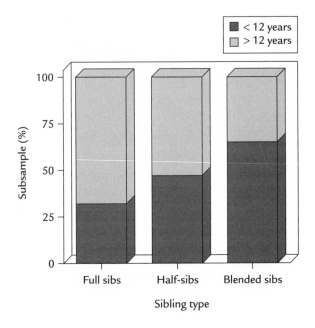

Figure 5.1 Comparison of co-residence among full, half-, and blended sibs

resided together in a household headed by mother and biological father. Then, after varying lengths of time and complex custody arrangements, they again resided together in a new household headed by their mother and step-father. If we add together the period of the first marriage and that of the second marriage, this sum will equal in many cases the time of co-residence under the aegis of the same parents of the three nondivorced groups. The big difference is that, in the course of time, the step-father replaces the father in this arrangement. For the group of half-sibs, the siblings have co-resided in a household under the aegis of the same marital couple for the duration of their sibship. Blended sibs began life in different households. The father's child lived with a different mother before the end of the father's prior marriage. The mother's child, of course, lived with a father who was not shared with the blended sib. Thus, the duration of co-residence in a common household will invariably be shortest for these groups.

This inequality across groups could pose a problem for our step-family model. Let us suppose that the duration of co-residence in a household headed by the same marital couple determines the magnitude of the correlation between sibs on measures of adjustment or on measures of their relationships with their sibs or parents. It could be that sibs who have a lifelong co-residence in such circumstances are much more highly correlated with each other, on any measure of interest, than are sibs who have co-resided in the same household for only five years. This would mean that, on the basis of co-residence alone, blended sibs' correlations would be lower than those of other groups. In effect the observed correlations between blended sibs would be lower than those of full sibs or half-sibs in step-families because of a difference in co-residence and not, as it would seem, because they differ in genetic relatedness. This spurious effect of co-residence would, in turn, lead to underestimates of the shared environment and perhaps overestimates of both the genetic and the nonshared environments.

Figure 5.1 shows that, indeed, blended sibs have co-resided in the same household and under the same marital aegis as each other much less than the other two groups. Full sibs show a little more co-residence than the others, but recall that this duration is the sum of co-residence under the aegis of at least two different marriages. Figures 5.2 and 5.3 provide reassuring data on all these step-siblings. They show that there is little difference in correlation between step-siblings co-residing together less than twelve years, the sample median, and greater than twelve years. There is only one significant difference among the seventeen comparisons shown in Figures 5.2 and 5.3; this is likely to be due to chance alone. Indeed, of the seventeen comparisons, seven show those co-residing less than twelve years with *higher* correlations than those co-residing for more than twelve years. Thus we may conclude that differences among step-family groups in duration of co-residence under a common marital aegis do not influence our estimates of genetic and environmental influences. This finding is quite consistent with other data suggesting that no aspect of the shared environment plays a major role in influencing most measures of adjustment.

Biological "Owness"

Another distinctive characteristic of step-families is that parents are rearing children who are not biologically related to them. For example, in our full-sib step-families neither child is the father's progeny, whereas both

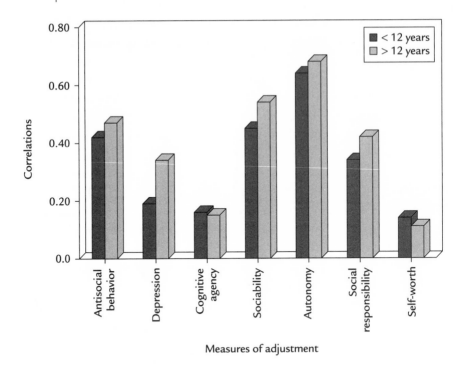

Figure 5.2 Co-residence and correlations among measures of adolescent adjustment

children are the mother's biological offspring. In contrast, in blended-sib families one child is the father's offspring whereas the other is not; the same is true for mothers. It is conceivable, as we suggested in Chapter 3, that parents may feel more warmly toward children biologically related to them but also feel a greater responsibility to set limits on their unruly behavior. This might mean, in blended steps, say, that mothers would be both warmer and firmer to their "own" children than to their husbands' children. This discrepancy might reduce the correlation between the mother's behavior toward her "own" child and her behavior toward her step-child. The same may be true for fathers. Once again we are in a situation where the blended-sib families may show lower correlations than the full-sib families because of contrasts in biological relatedness. And, once again, this would lead to underestimates of the shared environment and overestimates of the genetic and nonshared environment.

Fortunately, our design includes 104 half-sib families, the largest sample of families of this kind ever studied with careful measurements of family process. In this group both children are the mother's "own," whereas only one child is the father's "own." Let us assume that similarity between children in "owness" is an important influence on correlations of parenting directed toward the sibs. We should regularly find correlations to be higher for mothers than for fathers. We computed correlations for each parent, in half-sib families, between the parenting of one child compared with the parenting of the other child. We then compared mothers and fathers on twenty separate measures of parenting (see Table 5.4 for a list of these measures): three parental self-reports of conflict/negativity, warmth and support, and monitoring and control; three child reports on these same variables; and fourteen scales for coding the videotaped parental behavior toward the child. Of these twenty, eleven showed fathers with

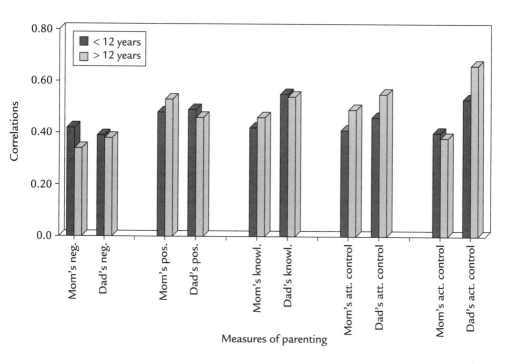

Figure 5.3 Co-residence and correlations among measures of parenting

higher correlations. Thus, variations in owness are unlikely to play a role in comparisons across the three groups of step-families.

Our findings here do not imply that parents make no distinctions between their "own" children and their step-children. In fact, we have reported elsewhere (Henderson et al., 1996) that both mothers and fathers are warmer toward their "own" children than toward their step-children. Mothers also show more conflict and negativity toward their "own" children and fathers show more monitoring and control. What our findings here do imply is that a parent's behavior toward his or her step-child is just as *predictable* from his or her behavior toward his "own" child as it would be if the first child were also an "own" child or the second child a step-child. For example, a father who treats his step-child wretchedly is likely to treat his own child a bit better but not nearly so well as a father who treats his step-child decently. The latter father will treat his "own" child even better. This constancy of parental predictability across sibship types is probably governed to some extent by the parents' personality and temperament, which shaped consistent responses to children despite their intrinsic differences. It may also be governed by the parents' high valuation of consistent treatment of their children.

Marital Choice

It is possible that children play a substantial role in influencing their mothers or fathers during courtships that lead to remarriage. We know that once the remarriage occurs, step-children have a substantial effect on parental behavior, whereas the reverse does not seem to occur (Hetherington and Clingempeel, 1992). This possible effect is most salient in blended step-families. It is conceivable that a father's child will raise strenuous objections to the remarriage if that child objects to either the prospective new mother *or* the prospective new step-sibling. This same process is likely to go on in the initial phases of the remarriage and even contribute to shortening it. Since there are a number of similarities between biological parents and their offspring, children's effect on remarriage is likely to be manifest in a greater degree of similarity between the parents themselves. In effect, children may participate in enhancing assortative mating in parents—particularly in blended step-families.

Differential assortative mating across the three step-family groups would have an effect opposite from the potential artifacts described above: it would *inflate* the correlation between measures of blended siblings' adjustment and hence lead to an underestimate of genetic and nonshared influences and an overestimate of shared environmental influences. Our design does not allow a direct test of whether there was "assortative mating" among sibs. That is, parents tended to marry and stay married if their children were similar, and didn't marry in the first place or stay married if their children were different. It would seem unlikely, however, that parents would give priority to child similarity over their similarity with each other. That is, if there is no evidence of differentially high assortative mating in the parents in this group, it would also seem improbable that there would be "assortative mating" of children.

We were able to analyze for differential assortative mating of parents across the three step-family groups using two pertinent measures: verbal IQ (corrected for parental age) and depression as measured by the Center for Epidemiological Studies of Depression Scale (Radloff, 1977). For IQ, the blended sibs ranked fourth from the highest and the step-families showed somewhat lower scores than nondivorced groups. Among steps, the half-sibs were highest and the full sibs the lowest. For depression, there was little difference across the six groups; again the blended sibs ranked fourth from the highest. This time the full-step sibs were the highest and the half-sibs the lowest. Parents in the three step-sib groups had, again, slightly lower correlations than parents in nondivorced families. We conclude that the presence of step-children has little or no effect on assortative mating of parents and that, within the step-families, blended sibs have no more effect on assortative mating of parents than do other types of sibs.

Comparing Step-Families and Twins

As we have seen, the step-family design conforms well to our requirements. Three possible sources of invalidity of this design—unequal marital environments, differential prediction of parents' behavior toward their "own" children from their parenting of step-children, and differential assortative mating—are unlikely to influence our findings in this study.

A more positive evaluative test of the step-family design is to determine how effectively this design replicates the results of the twin design. We have chosen data on height, weight, and vocabulary. These are particularly useful measures because they have been studied frequently, allowing us to compare the data from our study with those from many other studies. Figure 5.4 shows the genetic cascade for these three measures. In general the cascades are steeper for height and weight, implying higher heritabilities than for vocabulary. For height and vocabulary the slope for twin families is roughly comparable to that of the steps, except for the higher-than-expected correlations of the blended sibs. For weight, the non-divorced slope is somewhat greater for twins than for steps. Rather than relying exclusively on visual inspection here, however, we used standard model testing, which not only provided heritability and environmentality estimates, but also systematically compared the step-family results with the twin results. This model testing will be described in more detail in

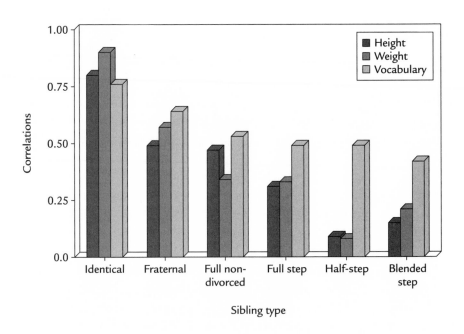

Figure 5.4 Genetic cascade for height, weight, and vocabulary

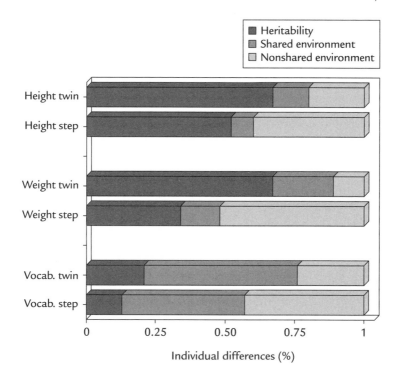

Figure 5.5 Univariate models comparing twins and steps on measures of height, weight, and vocabulary

subsequent chapters. Figure 5.5 shows the results of applying these models. As predicted from inspecting the correlations, the heritabilities for twins and steps are roughly the same for height and vocabulary but not for weight. Formal statistical tests of these comparisons confirm the visual impressions (Neiderhiser, O'Connor, Chipuer, Reiss, Hetherington, and Plomin, submitted). Across the three measures, heritability estimates for step-families are a bit lower than those for twins, and nonshared-environment estimates are a bit higher. Additional model testing, directed at comparing results between twins and steps, confirms the equalities and inequalities that are evident in the graphs.

As Figure 5.4 shows, the correlations from our study compare quite well with those from other studies of youngsters of the same age. The twin correlations for height, weight, and vocabulary are similar to those reported by other researchers (Bouchard and McGue, 1981; Fischbein and Nordquist, 1978; Grilo and Pogue-Geile, 1991), as are full-sibling correlations for height and weight (Martin et al., 1973; Mueller, 1977; Mueller and Malina, 1980). Further, for vocabulary ability the correlations for the blended sibs we studied are similar to those reported for other genetically unrelated sibs in adoptive families (Bouchard and McGue, 1981; McGue, Bouchard, Iacono, and Lykken, 1993). Our rather high sibling correlations for height and weight in the blended-sibling group exceeded those reported elsewhere (Biron et al., 1977).

The statistical tests we used to compare genetic and environmental estimates of height, weight, and vocabulary can be refined and extended. We can use them for other measures in our study that have not been examined nearly as extensively as height, weight, or vocabulary. The refinement is to ask a narrower question: Do shared or nonshared factors differ across the six groups of families in our sample? Perhaps children growing up in nondivorced families will experience more shared environment than those in step-families owing to the continuous, nondisrupted nature of their parents' marriage. As a consequence, either nonshared or genetic influences may be less apparent in nondivorced families than in step-families.

In the conduct of our study we have repeatedly examined not only whether nondivorced families differ from step-families, but whether twins differ in shared or nonshared environment from nontwins. We have made these comparisons for a very broad range of measures: adolescents' self-worth as rated by themselves (McGuire et al., 1994); teacher, parent, and child reports of social and cognitive competence (Neiderhiser, McGuire, Plomin, Hetherington, and Reiss, unpublished); parent and child reports of parent-child and sibling relationships (Plomin, Reiss, Hetherington, and Howe, 1994); observer reports of parent-child interaction (O'Connor et al., 1995); and child and parent reports of child relationships with teacher, best friend, and peers (Manke et al., 1995). In the majority of these comparisons there was no statistically different pattern of findings between twin and nontwin or nondivorced and step-families. What findings there were proved to be either small or inconsistent across comparisons. This exceptional range of analyses suggests that there are not important differences in shared or nonshared environment across the six groups. Nonetheless, we have continued to examine this issue for many of the analyses reported for the first time in this book. They confirm findings we have already published: differences between subgroups in shared or nonshared environments of the families in our sample are small and not consistent in direction across analyses.

Conclusion

This brief analysis of comparability to other samples suggests that in many cases the step-family model yields estimates of heritability and environmentality that are comparable to those provided by the twin model. Where steps differ they may provide a useful dampening or corrective on twin-model estimates of both heritability and shared environment. This role for the step-family model is strengthened by evidence supporting the basic assumptions underlying the inclusion of step-families in our sample: equal parenting and marital environments, equal assortative mating, and comparable shared and nonshared environment across the three groups. In general, our estimates of both heritability and environmentality should be reasonably accurate and more generalizable to a broader range of families than are estimates derived from the more highly specialized samples of traditional twin and traditional adoption studies. Nonetheless, we are introducing the use of step-family estimates of genetic and environmental analyses for the first time. The assumptions underlying their inclusion should, and almost certainly will, be examined carefully. Indeed, as Chapters 2, 3, and 4 have clarified, a thoughtful examination of the assumptions underlying both psychosocial and genetic research is a major opportunity for collaboration across the behavioral sciences.

Part 2

GENES AND RELATIONSHIPS:

THESIS, ANTITHESIS, SYNTHESIS

THESIS I: A THEORY OF ADOLESCENTS' SHARED AND NONSHARED FAMILY RELATIONSHIPS

We have laid out some logical principles of both psychosocial and genetic studies in an attempt to intensify the dialogue between these two perspectives. Genetic research has suggested that the influences unique to each sibling in the family are the most powerful environmental determinants of psychological development. This same line of research has also shown that genetic influences can be very strong on measures of the unique or nonshared environment. Unfortunately, neither genetic nor psychosocial research provides much guidance as to what these nonshared factors might be. Thus, we decided to design a study that not only focused on the differential experience of siblings but also assessed the potential role of genetic factors in influencing the nonshared environment.

We described the logic and methodology of our study in the last chapter. To measure aspects of the nonshared environment we recruited families with at least two same-sexed adolescent siblings close in age and designed a comprehensive set of assessments for them. In order to evaluate the role of genetic factors in the nonshared environment we recruited siblings who differed in genetic relatedness—from genetically identical twins to sibs who share none of their segregating genes.

We have structured our analyses of data, and the organization of this book, around three steps. The first is a thesis about the nonshared environment. We explore evidence from previous research that supports the importance of this phenomenon and its relevance to our own work. Although the intellectual stimulus for the concepts of the nonshared environment derives from genetics, we present our thesis as a psychosocial concept. The plausibility of this idea comes from a mixture of behavioral genetics and psychosocial research, which we summarize in this chapter. In Chapters 8–10 we examine the antithesis of this premise: that genetic factors not only play a major role in psychological development but also may explain the association of nonshared factors with measures of development. The inferences that can be drawn from this section challenge those that can be drawn from the exploration of our thesis. As a consequence, these chapters reflect if not recreate a fundamental tension in the field of developmental studies. We attempt a resolution of this tension in Chapters 11–13: the synthesis.

This chapter lays out our thesis, stating the strongest case yet for the importance of the nonshared environment in adolescent development. It also reviews data that support several hypotheses of how nonshared factors might influence development. We present four important concepts about the nonshared environment in this review. First, measurements of the nonshared environment need to assess many domains, both within the family and without. Second, direct environmental influences on the adolescent must be measured along with indirect effects that estimate how the unique environment of the sib influences the development of the adolescent. Third, adolescents develop their own interpretations of people and of relationships in their social environment. Fourth, adolescents play a major role in shaping the social environments unique to them. Thus, any associations between nonshared environment and adjustment may reflect the impact of the latter on the former.

Sibling relationships are a pervasive feature of family life, and are among our most enduring ties. Only recently, however, have they been the object of serious study among psychosocial researchers, who are now turning their attention to these relationships as a key to understanding psychological development. Much of this recent work has been stimulated by findings from behavioral genetics. As Chapters 3 and 4 made clear, genetics research has helped focus psychosocial family research on siblings. Indeed, sibling research has been a central tool for drawing inferences about the weight of genetic and environmental influences on psychological development. Further, it has, with some exceptions, consistently drawn the same conclusion: siblings in the same family are different from one another on a broad range of measures. The similarities they do show are more likely the result of shared genes than of shared environmental experiences. Sibling differences may also be due to genetic differences (among all sibs but identical twins, of course), but genetics research tells us that their differences are almost certainly due to differences in their experience of the environment as well. Such findings have focused researchers' attention on a central question: Which aspects of the environment make siblings so different?

A search for these factors motivated our team for the last decade. On the basis of a substantial set of findings from genetic studies and a few preliminary psychosocial studies, we expected to find differences among the adolescents we studied in their experiences with their parents, with each other, and with their peers; such differences, we surmised, would account for differences in their development. But in our effort to identify important nonshared environmental factors, heretofore known primarily by their anonymous effects in genetic analyses, we had to make sure that our findings did not reflect genetic influences masquerading as environmental effects. For example, parents might treat adolescent children differently because the children have different heritable traits that elicit or shape their parents' behavior toward them. One child might be irritable and easily upset; another might be good-humored and affectionate. These same heritable differences in the children may also underlie differences in their levels of adjustment. The first child might have a strong tendency to break rules and get into fights, whereas the second child might develop warm relationships with peers and adults.

In this chapter and the next we cover four major topics. First, we review briefly the evidence supporting the importance of the nonshared environment. Second, we review the social and psychological mechanisms by which nonshared factors may work. Third, drawing on these ideas, we formulate two straightforward strategies for analyzing the data from our study. Fourth, we present the full range of our findings on the association of nonshared factors with adjustment in the adolescents we observed. These data are presented as they would have appeared in an ordinary psychosocial study, one that was not designed to be genetically informative. That is, we pool the twins, the siblings from nondivorced families, and the step-families and report data from all of them as an aggregate.

Evidence for the Nonshared Environment

Evidence for the importance of the nonshared environment has been re-viewed extensively (Plomin, Chipuer, and Neiderhiser, 1994; Plomin and Daniels, 1987), and so we review only the main lines of evidence here.

Informative Designs

Data supporting the importance of environments unique to each sibling come primarily from two sources. The first source consists of genetically informed designs, particularly those using twins and siblings. These stud-ies are designed to explore the heritability of a broad range of skills, personality features, and psychological problems. They provide good esti-mates of genetic influences but also yield implications about the environ-ment, as we shall see. Adoption studies can also provide indirect estimates of the importance of the nonshared environment.

Data obtained from monozygotic twins provide a good direct estimate of nonshared-environmental influences. Indeed, subtracting from 1 the correlation between monozygotic twins on any trait is a very direct esti-mate. Given that MZ twins are genetically identical, any factors that make them different must be their unique environmental experiences. Thus, if we find that MZ twins are correlated .45 on any trait, we can es-timate that (1 − .45), or 55 percent, of the variation in that trait is ac-counted for by anonymous nonshared factors. Always included in this estimate are unwanted errors of measurement. To the extent that our measures do not reliably measure what we intend them to, they will un-derestimate any correlation. This unwanted error cannot readily be ac-counted for statistically, although we present some statistical approaches in the next chapter. The best way to keep this unwanted factor to a mini-mum is to use very reliable measures.

Genetically unrelated step-siblings, such as those used in our study, provide a direct estimate of the shared environment. Any factors that ac-count for the similarity of these genetically distinctive siblings must be environmental experiences that they share. If we know the heritability of a trait, then we can add to it the shared environment; the difference be-tween the total of these two and 1 is an indirect estimate of the nonshared environment. For example, let us say we find that genetically unrelated sibs have a correlation of .07 for a particular trait, and that we know the heritability is 42 percent. The nonshared environment is then estimated as follows: $(1 - .42 - .07) = 51\%$. This simple subtraction procedure assumes there is little or no interaction between genetic and environmental factors, a topic covered in detail in Chapter 4.

Twins raised apart share no rearing environment. Thus, it is interesting to compare correlations on measures of adjustment in this group with a group reared together; such a comparison is a direct estimate of the shared environment. For example, if twins reared apart are correlated .40 on a particular trait and those reared together correlate .51, we can say that the shared environment accounts for 11 percent of the variation in that trait. The correlation of .40 is also a direct estimate of heritability.

In adoption designs the nonshared environment cannot be assessed directly. But the shared environment can be directly estimated by the correlation between adoptive parents and their children. It is important to note, however, that the adoption design is the least direct of the genetically informed designs for estimating nonshared environment. Parent-child correlations are almost always reduced by noncomparability of measures used to assess particular traits in adults and children. For example, we are still not certain whether depressed mood in a young child is the same as clinical depression in a parent, and, more to the point, whether our assessment measures of child mood and adult depression are measuring comparable phenomena. Thus, adoption designs may underestimate the importance of the shared environment. For the same reason, they can also seriously underestimate genetic influences, particularly in young children. That is, the correlation between a trait of the birth parents and a trait of the adopted child—the key to estimating genetic influences—may be reduced by noncomparability of measures of parents and children.

These genetically informed approaches can estimate the cumulated total of nonshared influences, but that total cannot tell us which particular environmental factors are unique to siblings and are salient for their psychological development. Indeed, results of genetic analyses such as these are entirely consistent with nonshared factors' being of one sort in one family and quite different in another.

Thus a second source, studies measuring the environment directly, is essential. Such studies typically examine family dynamics, focusing on differential experiences of siblings. They directly assess these experiences but cannot determine the extent to which they may be attributable to genetic factors. Recall that full siblings share approximately 50 percent of their segregating genes, meaning that they *don't* share 50 percent of these genetic factors. Thus, differences in genetic influences between them can account for observed differences in their adjustment; these genetic differences can also account for differences in how they are treated.

Before we began our study, only one previous study had combined direct assessments of the environment with a genetically informed design that could rule out the role of genetic factors in the association between environmental differences and psychological adjustment (Baker and Daniels, 1990). Recently, a second study of this kind has been published (Vernon et al., 1997).

Genetic Studies

Because findings from genetic studies have been reviewed so extensively elsewhere (Plomin, Chipuer, and Neiderhiser, 1994; Plomin and Daniels, 1987; Plomin et al., 1977), we confine our review to a brief summary of the main conclusions from many different studies.

First, in the area of personality traits—such as extroversion, impulsivity, and shyness—results are surprisingly consistent across studies of twins reared together, two major studies of twins reared apart, and adoption studies. With very few exceptions, researchers have found substantial effects of both heritability and the nonshared environment. In the vast majority of cases, data provide little evidence for the importance of the shared environment in influencing personality traits.

Second, in the area of psychopathology, results are almost as consistent. Virtually all studies on depression and schizophrenia point to the importance of genetic factors and of the nonshared environment. A similar pattern has been found for most disorders that typically begin in early childhood, such as autism, hyperactivity, reading disabilities, and speech disruptions. There are two notable exceptions, however. Delinquency in adolescents typically shows the importance of shared environment. The same is true in twin studies of young alcoholics. It is possible to understand both of these findings as evidence for the influence of twins on each other. Twins can be partners in crime and can share the bottle. But common rearing environments—environments that undermine self-control in children—might also be particularly important for these disorders (Patterson, 1982; Patterson, 1989; Patterson, Reid, and Dishion, 1992).

Third, studies of IQ are particularly revealing. Such studies show that there is clear evidence to support the importance of the shared environment on development, particularly in childhood. As children become adolescents, however, evidence for the shared environment begins to diminish, so that by the end of adolescence heritability and nonshared environment are the major influences.

In sum, for almost all dimensions of behavior, at least by adolescence, there is evidence that nonshared environmental influences are by far the most important. On the basis of this literature, we expected to see this same pattern of findings in our own work.

Studies That Directly Measure the Environment

Only a few studies have directly measured nonshared influences on adolescent development. One study examined personality differences in adolescents (Daniels, 1986). In that study data revealed an association between personality differences such as levels of sociability and differential experiences with siblings and peers. For example, siblings who scored high on measures of sociability also reported being involved with popular peers and experienced more closeness from their siblings than those who scored low (Daniels, 1986). Similar results were found for adult twins (Baker and Daniels, 1990).

In the area of psychopathology, no studies have addressed diagnosed syndromes in relationship to differential experiences of the environment. A small sample study did examine the association of differences in mother's parenting, as reported by mother, and psychological symptoms such as depressive feelings or impulsive behavior, as reported by mother and teacher (McGuire and Dunn, 1994). For example, the study found that siblings who received more warmth, acceptance, and attention from their mothers showed less misbehavior and impulsiveness—symptoms referred to as "externalizing"—than siblings who received less warmth, acceptance, and attention. Similar findings have been reported for children's self-esteem (Dunn and McGuire, 1993). In one study of adolescents, children's experience of affectionate bonding with their mothers was associated with differences in oppositional behavior on the part of the children; the greater the bond, the less the problematic behavior. Interestingly, adolescents who received more punishment than their siblings were more likely to have suicidal thoughts (Tejerina-Allen et al., 1994). There is no direct evidence linking nonshared environments to cognitive abilities in children.

Studies That Combine Genetic Analyses and Direct Measures of the Nonshared Environment

Aside from publications derived from our study, we are aware of only two studies that have attempted to control or partial out genetic differences between siblings as an explanation for the relationship between non-shared measures and measures of adjustment. The first of these studies focused on a subsample of identical twins (Baker and Daniels, 1990). Any association between differences in the nonshared environment and differences in adjustment of identical twins can only be attributed to psychosocial mechanisms, not genetic ones. The correlation between nonshared experiences and adjustment in MZ twins directly measured the associations between nonshared influences on the measure of the environment and on the measure of adjustment. In the Baker and Daniels study several correlations were found between nonshared factors and personality. More recently, a larger-scale study found substantial associations between nonshared experiences of the family, the school, and the physical environment, on the one hand, and measures of personality, on the other (Vernon et al., 1997).

It should be noted that in both these studies the research subjects were the sole source of information about their environmental experience and their adjustment. Thus it is possible that these findings are the result of simple rater bias. For example, an aggressive subject is likely to feel that his siblings and parents treat him more aggressively than they treat others (Dodge and Tomlin, 1987). Indeed, in our own previously published analyses of the correlation of specific measures of the nonshared experience with differences in adjustment in identical twins, we could only obtain meaningful associations within reporter (Pike, Reiss, Hetherington, and Plomin, 1996). For example, the adolescent's report of differential conflict and negativity from father correlated with the adolescent's own report of differences with the sib in antisocial behavior and depression. Similarly, the father's report of his own differential conflict and negativity with his adolescent children correlated with his report of the adolescent's antisocial behavior. We did not obtain any significant correlation across reporters, however. For example, there was no association of differential negativity as reported by the father with differences in antisocial behavior as reported by the adolescents. In order to determine whether there is an association between specifically measured aspects of the nonshared environment and adjustment, we must pool information about the environment and adjustment from several different sources. This will avoid results that could be attributable to rater bias. Pooling results across raters is a major analytic tack in our study.

In sum, genetic data on the influence of the nonshared environment are very strong. The few studies that have directly measured nonshared experiences provide promising leads, suggesting the importance of experiences that may arise from differences in parental treatment of siblings, from differences in how siblings treat each other, and from differences in the nature of siblings' peer groups. It is not clear from most of these studies, however, whether these associations involve genetic influences. So far, the associations are independent of genetic influences only when the measures come from the same reporter.

Nonshared Environments and Psychological Development

The evidence we have reviewed thus far suggests that environments unique to each sibling in the family are powerful influences on psychological development. These data, however, do not tell us how such experiences exert their influence. In designing our own study, we relied on prior research to construct reasonable hypotheses to explain how differential experience has its impact. Indeed, on the basis of previous research, we can formulate two hypotheses that are not mutually exclusive.

The first postulate we call the "single-system hypothesis." According to this theory, there are substantial and verifiable differences in environmental circumstances for each adolescent. Differential parenting as well as differential sibling and peer relationships are included in this set, but other differential aspects of the social environment may also be important. Moreover, it is the variation in the magnitude or quality of these unique environments that directly impacts the growing child. This impact does not depend on the adolescents' comparing themselves with their sibs. Indeed, what happens to the adolescent's sib is not salient for the adolescent.

The second postulate we call the "multiple-system hypothesis." According to this theory, adolescents always consider their siblings in responding to questions about their environment; for example, "I am being treated worse than my sibling."

Such comparisons may extend beyond the family as well. For example, an adolescent may have a keen sense that her father's relationship to her is abusive only if she knows that other fathers do not treat their daughters in this fashion or that her father does not treat her sister in the same manner.

Single-System Nonshared Environments

Some characteristics of social relationships may directly activate or influence adolescent functioning without the need for comparisons. The best example may be adolescents' perceptions of facial movements or tones of voice that convey others' emotions (Dunn and Brown, 1994; Wiggers and Vanlishout, 1985). Since these facial and voice cues to emotions are understood in the same ways by people in many different cultures, it is likely that an adolescent or an adult does not need to make elaborate inferences, including social comparisons with siblings and peers, in order to detect emotions in another. Of central importance here are the emotions that refer directly to the perceiving adolescent. For example, most individuals can recognize anger, disgust, joy, and other emotions in themselves and use the same cues to identify these feelings in others (Ekman, 1994; Ekman and Friesen, 1971; Ekman and Heider, 1988). These cues are contained mainly in facial expressions, but can also be detected in tone of voice and posture. Given that the capacity to perceive these basic emotions is so widespread in literate and preliterate societies, it is reasonable to assume that both the expression and the perception of such emotions are built-in mechanisms.

Despite these universals, however, children and adolescents vary in their ability to respond to emotional cues in others. Several studies are documenting characteristics of children that may interfere with their capacity to identify emotions accurately. These studies suggest that states of high and persistent emotionality in children themselves (Walker and Leister, 1994) or persistent impulsive behavior (McCown et al., 1985) may interfere with their accurate reading of emotional states in others. Thus, an adolescent's perception of affect in a parent or a sibling may be inaccurate and idiosyncratic.

Within this domain, nonshared environmental effects may be due either to differences between siblings in the emotional signals that are directed at them, or to differences between them in their ability to recognize these cues accurately and respond to them.

Adolescents may require some information about the context or setting of a relationship in order to interpret accurately the meaning of another's words, gestures, and facial expressions. This context or setting has been called a "frame" (Goffman, 1974). Goffman distinguishes between primary frames and their transformations. Primary frames are implicit cultural meanings ascribed to human action that help all members of a culture judge whether actions are excessively intimate, disrespectful, deceitful, or psychotic. These frames can be transformed by a process Goffman calls "keying." For example, many behaviors that would be threatening in a primary frame can be transformed if all participants in the social transaction understand that they are involved in a "play" or a "game." Behaviors in this transformed context are interpreted very differently from those in a primary frame. Adults and children can play, for example, at being aggressive without eliciting hostile responses from others in the game.

Following Goffman (Goffman, 1974), we assume that most adolescent members of a culture or a peer group can utilize these frames and their transformations to form judgments about the affect and intent of the important agents in their social community. Indeed, research on much younger children suggests that they become quite adept at using their knowledge of other people's expectations as part of their responses to a parent (Grusec and Goodnow, 1994). Because these frames are pervasive and implicit, they can guide adolescents in forming perceptions of the quality of their primary relationships. They provide such guidance without specific reference to other relationships and so are placed under the rubric of "single system."

Research is beginning to delineate differences among adolescents in their ability to "read" these common social frames, particularly those that clarify general or broad expectations for their desirable or approved behavior. For example, some adolescents are typically so preoccupied with expressing or enacting their own feelings that they are inattentive to either primary frames or their quick shifts and transformations. Often such children are shunned by others and develop severe and abiding loneliness (Carr and Schellenbach, 1993; Graziano et al., 1987). A particularly productive line of research has focused on aggressive children. These children typically misread the intent of other children and see them as hostile. Further, they are much more likely than other children to retaliate against this misperceived aggression. Many of these children, particularly those who use aggression to achieve personal gain rather than to respond to frustration, overvalue the success that their aggression is likely to achieve with others (Dodge and Crick, 1990).

Within this domain, nonshared environmental experiences can be accounted for in several ways. When siblings' responses to their social world are shaped by the same frame, verifiable differences in the conditions of the social world may be responsible for differences between them. For example, one sibling may have a warm, supportive, and skillful teacher, and the other a belittling and authoritarian one. Both siblings, however, use the same primary frame for judging teacher behavior in the classroom, and hence the actual differences between the teachers, as filtered by a primary frame common to both children and equally well read by both, may have a substantial impact on the children's scholastic achievements and related areas of development. Differences between siblings may also arise if they differ in their ability or motivation to "read" primary frameworks or their transformations and thus reach very different conclusions about environmental circumstances that are relatively similar for both of them.

Multiple-System Nonshared Environments

According to the multiple-system hypothesis, the influence of an adolescent's primary social relationship on his or her adjustment is moderated by the qualities of other primary or secondary relationships. For example, an adolescent may regard her mother as abusive to her in part because she observes that her other siblings are not treated as harshly. This kind of social comparison may take one of three forms.

At the simplest level, adolescents make some distinction between a social agent's behavior toward them and that aimed at a sibling. For example, adolescents may know when warm, supportive, and affectionate behavior from mother is directed at them, when affection is directed at both them and their siblings, and when behavior is intended primarily for a sibling. The adolescents attend mostly to the behavior directed at them and are relatively unaffected by that directed toward their siblings.

It is also possible, as mentioned, that an adolescent may be affected by how a sibling is treated. There are two possibilities here. First, adolescents may respond to how siblings are treated in the same way that they respond to how they are treated. For example, an adolescent may develop impulsive or depressive behavior if she is treated harshly or seductively by her father, and she may be further affected, in a similar way, by the same sort of paternal behavior directed at her sibling. A second possibility is that the adolescent may respond in the opposite way. If she notes that another sibling is being treated harshly but that, for the moment, she is not, she may respond positively to that paternal treatment of her sibling even if, under comparable circumstances, she responds negatively to the same paternal treatment of her.

Finally, these social comparisons may, like single-system processes, be influenced by factors in the social context. Evidence suggests that families develop their own frames for interpreting social cues and qualities of relationships both inside and outside the family (Fitzpatrick and Ritchie, 1994; Reiss, 1981; Reiss and Klein, 1987; Reiss et al., 1983). For example, in efforts to classify families, researchers consistently note distinctions between families who stress achievement and independence and those who either are organized by values of "just getting by" or are dominated by conflict or disorganization (Billings and Moos, 1982; Marjoribanks,

1988; Mink and Nihira, 1986). In effect, each family functions as a miniculture with its own frames and transformations.

From these investigations of differences among families, it is reasonable to argue that the qualities of parent-child relationships may be associated with different outcomes in different types of families. For example, in families geared toward achievement, both warmth and control from the parent will be perceived as encouraging independence and achievement. Thus, if one sibling receives more warmth and control than the other, that sibling is likely to show more evidence of scholastic and related achievements. The reverse may be true, however, in families that are fearful or resigned to living in a world that they feel they cannot master. Warmth and control may be understood as encouraging the child to keep ambitions low and remain in the family. In such cases, the child who receives less warmth and control is more likely to venture out of the family and may show greater scholastic and related achievements than the child who receives more warmth and control.

To summarize, the sources of multiple-system nonshared experience are shaped by two intersecting circumstances: actual differential treatment of the siblings by parents and others in the family, as well as a frame that is characteristic of each family. This frame, if fully grasped by each child, shapes the meaning the child attributes to this differential treatment.

Adolescents' own attributes might contribute to differences in their apprehension of the social world. It is worth emphasizing once again that different experiences of the environment by siblings in the same family are unlikely to arise only from differences in the objective or verifiable aspects of their social world. One particularly important source of these differences is the siblings' unique social histories. Each sibling in a family accumulates a series of salient experiences with family members and with others outside the family. In many instances these experiences continue to resonate and shape the adolescents' current perceptions, their tendencies to react in certain ways, and their beliefs about the impact of their actions on those closest to them.

One intriguing example comes from studies done a number of years ago on schizophrenic patients and their nonschizophrenic siblings. Researchers noted that, in contrast to their siblings, schizophrenic patients tended to feel extraordinarily attached to their mothers. The attachment seemed strong enough to merit the term "symbiosis." In a separate study, the same research group discovered that a large proportion of schizophrenic patients in one sample had experienced the death of a grandparent within two years of their birth, in contrast with other psychiatric patients and with individuals without psychiatric difficulties. It is possible that the bereaved parent included the infant in the mourning process and, in part, saw the child as some sort of replacement or restitution for the loss of the parent. This would be in contrast to the sibling who was born at a time more distant from the grandparent's death (Summers and Walsh, 1980; Walsh, 1978). If one sibling is born during a time of extreme parental bereavement but another is not, it is likely that this experience of a grieving and needy parent will influence the first child's experience of that parent and perhaps of other important social agents for many years to come.

Distinctive and even traumatic past experience may profoundly shape the way children and adolescents view relationships, not only with their parents, but with many individuals in their lives. For example, physical or sexual abuse by a caretaker can have a lasting impact on a child's ability to detect social cues accurately or to understand the intent of others. Research has shown that four domains of children's experience are frequently associated with past abuse. Abused children tend to have difficulty discriminating their own needs from those of others, see their relationships with others as painful and malevolent, have difficulty investing themselves in relationships, and cannot accurately define the motives of others (Freedenfeld et al., 1995; Ordnuff et al., 1994; Westen et al., 1991).

How Single- and Multiple-System Environments Affect Psychological Development

The study of the nonshared environment is in its earliest stages, and so previous research provided little guidance for our own study. There are ample data, however, to suggest that multiple systems play a central role in the social world of the developing child from infancy onward. The importance that past research has attributed to maternal or paternal influences from an early age needs no additional documentation here, although conclusions from such research may need revision in the light of recent genetic data. More pertinent to our concerns is a research literature on siblings that may be less well known. This research suggests that even very young children are responsive, not just to their parents, but to their siblings as well. Research data suggest that even infants are responsive to their older siblings.

Infants are affected by their siblings in three ways. First, an older sibling can influence the experience of a younger one indirectly. When an older sibling joins an interaction between a mother and an infant, the mother-infant interaction shows important changes. When the older sibling approaches a mother who is interacting with an infant, the mother typically directs more talk to the older sibling than to the younger child and becomes less responsive to the younger one (Woollett, 1986). As evidence that this change is registered by younger children, the younger children show a corresponding reduction in their talk as well as in their responsiveness to their mothers. Second, infants are capable of imitating and learning from older siblings (Wishart, 1986). Third, older siblings provide important cognitive (Teti et al., 1986) and care-taking (Stewart, 1983) environments for their infant siblings.

During childhood it becomes clear that children not only are responsive to several subsystems within the family, but also enlarge their social network and begin to integrate their experiences of all its subsystems: with siblings, each parent, peers, and other family members. For example, children as young as three can recognize and use the quality of their play with peers to buffer the effects that the birth of a new sibling has on them: positive quality of play with peers is associated with greater acceptance by the young child of his or her new sibling (Kramer and Gottman, 1992). As children develop, their social world widens and there are progressive changes in the people to whom they turn for support (Furman and Buhrmester, 1985; Furman and Buhrmester, 1992). For example, whereas parents are important early in development, peers are important later on. In addition, young children intervene—effectively and ineffectively—in family fights that do not, initially, involve them (Vuchinich et al., 1988). Further, they develop an understanding of how other family members can displace anger that is intended for one member on to another (Covell and Abramovitch, 1988). By the time they reach adolescence, they understand how families can create scapegoats (Arnold, 1985).

Thus, as they enter adolescence, children have a good deal of experience mastering and integrating many different relationships and are developing their own understanding of how these relationships are interconnected. At this stage they are likely to be highly sensitive to multisystem sources of the nonshared environment—particularly to comparisons between their own status in the family and that of their siblings.

By contrast, it is less likely that single-system nonshared environments, unmodified in the minds of adolescents by their registration and representation of other relationships, play an enduring role in development. It seems unlikely that adolescents would reimpose blinders on themselves and hence be insensitive to the many social systems to which they were so attuned during infancy and childhood.

But this may not be the case in early phases of development or in some limited circumstances later on. For example, when they find themselves in situations that they consider potentially dangerous, such as standing at the edge of an artificially construed cliff or facing strange toys, infants and young children will often glance quickly at a parent (Hirshberg, 1990; Sorce et al., 1985) or a familiar adult (Klinnert et al., 1986). They are looking for and receive quick emotional information about what to do; almost all this information is contained in emotional signals from the adults' facial expressions. This is termed "social referencing" and, much later in development, may have its analogues in parent-adolescent relationships. For example, the emotional tone of a parent's discussion of sex with an adolescent is associated with actual sexual behavior and contraceptive use by that child (Mueller and Powers, 1990). The prospect of sexual behavior in early adolescence may be analogous to the cliff or threatening toy in infancy. Parental tone and affect may continue to provide the adolescent with important guidance in a tense and ambiguous situation just as facial expressions do for infants. It seems likely, however, that social referencing in adolescents is influenced substantially by their registration and representation of other relationships in the family, particularly how they perceive their parents' emotional signals to other siblings.

Similarly, by adolescence a variety of primary frames and their transformations may play some role in orienting young adults to expectations and meaning in peer groups (Stein and Reichert, 1990) and schools (Marsh, 1990). A more complete knowledge of both these frames and the experiences they organize may provide a rich clue to sources of nonshared experience outside the family, though for the purposes of this study we restrict our focus to the family. Research suggests that each family serves as its own miniculture and helps adolescent siblings interpret the meaning of parents' and siblings' behavior toward them.

As our focus in this study is on nonshared environments in the family, we are concerned less with single-systems sources of the nonshared environment and more with multiple systems, particularly social comparisons with siblings. The multisystem nonshared environment that consists of the influence of past relationships on current ones requires specialized study and techniques. Because it may influence some of the findings from our study, we return to it later.

Adolescents' Contributions to Their Unique Experiences

From the earliest phases of development, children are proactive. For example, children who are treated harshly or aggressively by parents or siblings often start fights with them and seem particularly insensitive to limit-setting efforts in their families (Patterson, 1982; Patterson, 1989; Patterson et al., 1992b). This is conspicuously the case for friendship and peer relationships, as reviewed in Chapter 2. Adolescents pick peers who are very much like themselves and hence create distinctive social worlds for themselves that may then exert continuing influence on their behavior. Thus the nonshared experience is not something that "happens" to a child, but rather something that the child plays a major role in eliciting. Indeed, at any point in time, measures of the nonshared environment reflect—in all likelihood—a bi-directional effect: the influences of parents, siblings, peers, and others on the developing child and the continuing impact on that child of family and friends.

Guidelines for Assessing the Nonshared Worlds of Adolescents

Hypotheses about the nonshared environment can provide researchers with guidelines for developing measures and analytic strategies, as well as for interpreting data. They can also clarify limits on the inferences that can be drawn from research findings. We focus on four guidelines that can be derived from the available evidence on the nonshared environment.

First, there is reasonable evidence that important components of the nonshared environment reside in siblings' differential experience of the *social* as opposed to the *physical* environment. Preliminary research suggests, however, that this social environment needs to be broadly sampled. Although parents are, of course, important in the lives of developing children, siblings themselves are also important. They provide, in all likelihood, an important reference point for adolescents, helping them to gauge and interpret the meaning of parenting directed toward them. In addition, differential aspects of the sibling relationship itself might be a powerful influence on psychological development. For example, if one sibling is warm and supportive to the other, but the latter is hostile in return, both experience a different sibling environment. Whereas hostile behavior may contribute to maladjustment in the sibling who experiences it, the experience of warmth by the second sibling may contribute to healthy adjustment.

Friendships and peer groups are another potential source for the nonshared environment, as is the classroom for school-age children. Obviously, the broader the range of measures used in any study, the more shallow and insubstantial each measurement will be, since research subjects cannot be asked to endure hours and days of measurement procedures. As noted, we concentrate our attention on the family but have included measures related to extended family, peers, friends, and teachers.

Second, measures have to allow for direct as well as indirect effects of specific environments on the adolescent. Direct effects are environmental circumstances unique to or directed specifically at the target adolescent. For our purposes, the adolescent's environment can be divided into three parts: that experienced only by the sib, that which is common to both adolescent and sib, and that experienced only by the adolescent after the first two have been partialed out of any analysis. Our review of single-system mechanisms strongly supports measuring the last component, although multiple-system mechanisms may influence how an adolescent responds to this component. Our review of multiple-system mechanisms provides equally strong support for measuring the first component. That is, we want to understand how treatment of the sibling affects the adolescent's adjustment. To our knowledge, an analysis of this kind has never been undertaken before now. To anticipate the analytic approach we take in the next chapter, associations between measures of the environment unique to adolescents, on the one hand, and their adjustment, on the other, are called *specific correlations.* Associations between environments unique to the sib, on the one hand, and the adolescent's adjustment, on the other, are called *cross-correlations.*

Third, the adolescents' own views of their social worlds are likely to play a large role in linking the environment with adolescent adjustment. It is entirely conceivable that much or most of the so-called nonshared environment actually resides in these differential interpretive processes. It is not unreasonable to suppose that genetic factors may influence how adolescents interpret their environment, given that such factors are known to affect a broad range of perceptual, attentional, cognitive, and personality functions.

Unfortunately, here is where breadth and depth collide. It is within the scope of a major study such as this one to ask meticulously how the adolescent perceives relationships with each parent, with the sibling, and with peers and teachers. But it is entirely uncertain if the adolescents' interpretations are distorted by traumatic experiences, shame, jealousy, and rage; nor is it clear how well they pick up on the cues provided by family and cultural frames. In our study we have had to emphasize the assessment of the objective or verifiable unique environments of adolescents and give much less attention to the subtlety of their subjective interpretations.

Fourth, this review reminds us of the active role adolescents play in shaping their own unique environments, both inside the family and outside it. The adolescents' own, active role is another route for genetic influences to play a major role in nonshared environments that may appear, in typical psychosocial studies, to exert their influence purely by psychosocial means. But even in the absence of genetic influences on these associations, we must be very cautious in interpreting observed relationships between the nonshared environment and adjustment: causality can flow in either direction between these two. Longitudinal and even experimental data are essential for establishing the causal role of any particular nonshared influence.

In sum, despite our strenuous efforts to design a comprehensive assessment strategy that has both range and depth, we will miss many of the crucial nuances of the nonshared environment, nuances in the environment itself and in the way adolescents interpret it. Thus, any failure to find substantial and systematic nonshared influences on development may be partially attributable to our inability to capture these nuances. We return to this theme in the final two chapters.

Chapter 7

THESIS II: MAJOR FINDINGS
ON ADOLESCENTS' FAMILY RELATIONSHIPS

We began by summarizing the logical principles that underlie many psychosocial studies of adolescent development. We outlined a comparable summary for genetic studies. We argued for the importance of dialogue between the two perspectives and suggested a study design that would allow us to present the psychosocial perspective, the genetic perspective, and a synthesis of the two. The psychosocial perspective focuses on the importance of social environments unique to siblings and constitutes the thesis of this book. We present the genetic data as an antithesis, and the resolution between the two perspectives as a synthesis. In Chapter 6 we began to articulate the thesis: we presented evidence for the importance of social environments specific to each sibling in the family, as well as evidence suggesting how these might have an impact on adolescent development.

In this chapter we continue to support our thesis. In particular, we present data that show very strong associations between the nonshared environment and the seven domains of adolescent adjustment that we measured: antisocial behavior, depressive symptoms, cognitive agency, sociability, autonomy, social responsibility, and self-worth. Many of the associations are very strong and were predictable from previous psychosocial research. For example, the positive associations of parental conflict and negativity with psychopathology confirm many previous findings. The inverse relationship of such conflict and negativity with positive areas of adjustment was predictable as well. We also find substantial associations between the quality of sibling relationships and most areas of adjustment.

Among the most intriguing findings discussed in this chapter is that cross-correlations reflect very substantial associations between the adolescent's adjustment and the sibling's family experience. We found, for example, that ineffective parenting directed at the sibling, such as failed attempts at control, is associated with very positive outcomes in the adolescent, defined as less antisocial behavior, fewer depressive symptoms, and more cognitive agency and autonomy. Similarly, we found that when the adolescent is the victim of aggression in the family, he or she also shows evidence of poor adjustment. In contrast, the adolescent seems to benefit if the sibling is the victim and the adolescent is the aggressor. Whether these findings can be attributed to intriguing family dynamics or to genetics remains to be seen.

In this chapter we turn to the most important questions raised by behavioral genetics research: How different are the family environments of adolescent siblings, and are these differences associated with levels of adjustment in adolescents? We give special attention to contrasts among types of individuals who provide data to us: parents, children, and trained observers. We ask whether parents see the same differences in their families as do observers or adolescents, or whether there are important distinctions among these three sources of information about family relationships. If there are distinctions, we ask how they might affect adolescent development.

Answers to these questions provide a bedrock for a second set of questions: Are these differences in the social environment associated with adolescent adjustment? If they are, what are the psychosocial mechanisms that might account for these associations?

Comparing Siblings' Social Environments

Figure 7.1 shows the correlations between siblings in the same family on five different measures of parenting. As in previous graphs, the height of each bar reflects the magnitude of the correlation. For example, consider the very first bar on the left. This reaches the level of approximately .60, meaning that the correlation of mom's report of the conflict and negativity of her relationship with one child correlates .60 with her report on the same parenting toward her other child. These associations were evident

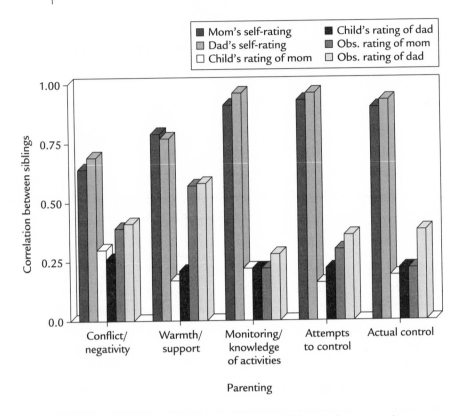

Figure 7.1   Sibling correlations of parent, child, and observer ratings

both at time 1, using data from the full sample of 720, and again at time 2, using data from the smaller sample of 376. As explained in Chapter 5, we sought to study at time 2, three years after time 1, only those families in which both adolescents were still living at home; 92 percent of the sample that was eligible at that time participated in the second data collection. The correlations are remarkably similar across this three-year time period. Moreover, a striking pattern is clear: parents report that they treat their children quite similarly in terms of knowledge of their activities, attempts to control their behavior, and successes in controlling them. Children, by contrast, report that their parents treat them differently. The contrast is notable for all five dimensions of parenting, but is particularly dramatic in the area of monitoring.

There is a simple explanation for this difference between parents and children that derives from the way we measure parenting. The correlations for mothers and fathers, on the one hand, reflect just one rater; for example, they reflect a comparison of how father reports he treats his older child with how he reports treating his younger one. The children's correlations, on the other hand, come from two raters. We asked the older child and the younger child how they were treated by their parents, and then we correlated these two separate reports. Perhaps the difference between parents and children is a relatively trivial reflection of this one-rater versus two-rater distinction. Figure 7.2 suggests that this factor is unlikely to account for most of the difference between parents' reports and children's reports. Figure 7.2 compares parents' ratings of their parenting, of the siblings' behavior toward each other, and of the quality of each child's peer group. As soon as the ratings focus outside the family, the correlation between parents' ratings of relationships with child 1 and with child 2 drops substantially. This is true for mothers and fathers and at time 1 and time 2. Thus, it seems likely that parents see their treatment of their children as consistent and their children's treatment of each other as reciprocal. But they see their children as quite different in their relationships outside the family. In other words, parents' consistency of rating depends on what they are rating, not just on the fact that they are a single rater.

A return to Figure 7.1 shows that for conflict/negativity and warmth/support, the correlations between observer ratings for child 1 and child 2 are midway between those of parents and those of children. It is important to recall that we always used separate observers to rate the relationships of the two siblings in order to avoid a single-rater problem. In the area of monitoring, the observers tend to agree more with the children and to code the parents as treating the two siblings differently. The observers are in the middle, between the generations, for conflict/negativity and warmth/support. They should not, however, be regarded as arbiters judging whether parents or children are closer to an objective truth. Observers view short segments of videotaped interaction, and they may be rating behaviors that are different from those that influence parent or child judgments.

One possible source of the lower correlations between children is age difference, an average of 2.2 years between siblings in the nontwin por-

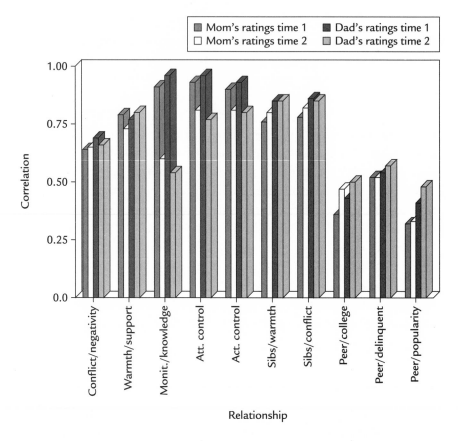

Figure 7.2 Parental ratings of family and peer relationships

tion of the sample. This simple difference might lead to children's different perspectives on their families and hence a much lower correlation between their ratings of family life. We can bring them, artificially, much closer in age by correlating the parenting directed toward the older sibling at time 1 with the parenting directed toward the younger sibling at time 2, three years later. This narrows the age gap between the two siblings from 2.2 years to about 8 months, but do we still get the same contrast between parents and children?

Figure 7.3 shows correlations that exclude twins and also those sibs who were the same calendar age, in years, at time 1; these data suggest that age differences between children are not a major factor in their low correlations and particularly not the chief factors in the striking contrast between parent and child reports.

As will be shown, the similarities and differences in siblings' experiences of their own relationship are an important part of our findings. Figure 7.4 examines these in greater detail. Parents, as was evident in Figure 7.2, tend to rate the sibling relationship as very reciprocal. The correlations of parental ratings of sibling conflict/negativity and sibling warmth/support average .81 for time 1 and .83 for time 2. This means, for example, that parents who consider one child to be very negative toward his sibling see that sibling as very negative in turn. The same is true for sibling positivity. As Figure 7.4 shows, however, the siblings see it somewhat differently. Their ratings of each other show a lower correlation than their parents' ratings for both conflict/negativity and warmth/support (positiv-

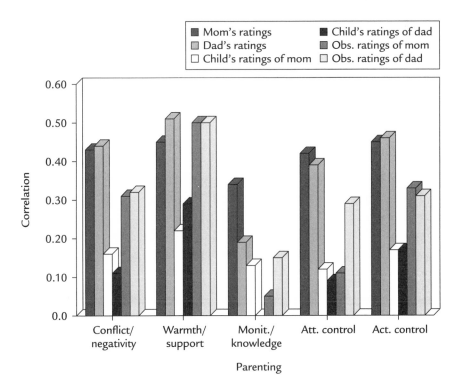

Figure 7.3 Correlations among ratings of sibs across time

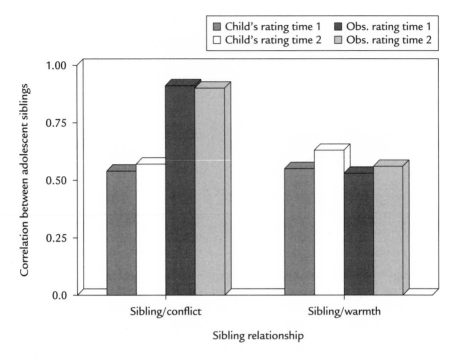

Figure 7.4   Self-report of sibling behavior versus observer report

ity). Observers show a similar pattern of findings for warmth but not for conflict. Here, one must note that the same observer makes both ratings using a videotape of siblings interacting with each other.

Figure 7.5 gives us a final look at child similarities and differences. Figures 7.1 through 7.4 tell us that there are moderate associations, across the whole sample, in the parent-child and sibling relationships. These could reflect modest differences across most or all of the families, or it could reflect a mix of both families who provide very discrepant environments for their adolescents and families in which these environments are rendered nearly equal.

To address this issue we used a *cluster analysis.* First, we combined all our measures to produce six scores for each family. The first two scores were simply the inverse of conflict; we call this inverse variable "conflict avoidance": the lower the score on the conflict/negativity scale, the higher the score (by definition and simple computation) on conflict avoidance. We computed these scores for parents (combining mother and father scores, since these were highly correlated) with the older child and parents with the younger child, and we computed analogous scores for parent-child warmth, sibling warmth, and marital satisfaction. The cluster analysis constructs an imaginary space defined by these six dimensions and then defines groups of families that are more like one another than they are like other families on these same six variables. Theoretically, families

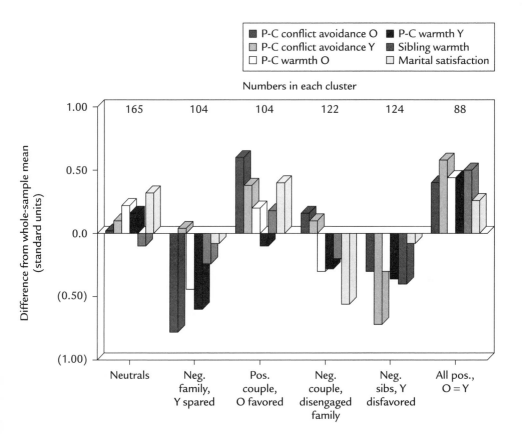

Figure 7.5   Patterns of parent-child and marital relationships (P = parent; C = child; O = older sibling; Y = younger sibling)

could be evenly distributed throughout this space. In this instance the cluster analysis would fail to delineate distinctive clumps of families. If they tended to clump or group together in the imaginary six-dimensional space, however, this would suggest that there may be distinct types of families, and that the attributes of each type are worth attending to.

Indeed, families did tend to clump together into more or less distinct clusters. Figure 7.5 shows that we identified six clusters and that the families were fairly evenly distributed across all six. All scores above the zero line, the mean for the whole sample of families, are regarded as positive, and those below it, negative. To make this graph easier to read, we have turned conflict/negativity on its head, as mentioned above: the more conflict/negativity, the lower the bar. Thus, the height of the bar measures "conflict avoidance." We have constructed a brief epigram to characterize each cluster by the profile of its scores. The first group is called the "neutrals" because the six scores are very close to zero. The second epigram, "negative families, younger (child) spared (by parents)," calls attention to the negative scores on all scores except for the slightly positive conflict-avoidance score for the younger sibling. The third group has the most satisfied marital unit but shows notable favoring of the older child on both conflict avoidance and warmth; hence the label "positive couple, older favored." The fourth group is the most dissatisfied in terms of marriage but, surprisingly, does not show problems in the other relationship subsystems. The sibling and parent-child relationships are noticeably cool, but there is also little conflict. Hence we have called this cluster "negative couple, disengaged family." The fifth group has the most distant siblings as well as the least conflict avoidance for the younger sib, and so it is labeled "negative sibs, younger disfavored." Finally, the last group, with the fewest families, is labeled "all positive."

Although there are many interesting aspects to these six clusters of families, for present purposes they illustrate one important finding: only three of the six clusters show any substantial difference in family relationships between older and younger sibs. In the second cluster, the younger child is favored and, in the third and fifth clusters, the older is favored. In the remaining clusters children are treated more or less equally. These data suggest that families do differ a good deal in the extent to which parents treat their children differently.

Shared and Nonshared Influences of Parenting

Analytic Principles

We now examine the association between these differences in family environment and the adjustment of the adolescents in seven domains: antisocial behavior, depressive symptoms, cognitive agency, sociability, autonomy, social responsibility, and global self-worth. We have developed a uniform framework for data analysis that examines closely both social and genetic influences on development. To make the data comprehensible, we sought to use or modify existing approaches to data analysis, although some innovation was required.

In order to evaluate fully the overall pattern of our findings, we must make sure that all our analyses fit together: the phenotypic and genetic analyses must be comparable, and the cross-sectional and longitudinal analyses must complement the phenotypic and genetic analyses in readily understandable ways. Moreover, we wanted our analytic procedures to remain as close to the raw data as possible, because we want readers to be able to estimate for themselves the results of our analytic models from the simplest correlations we present. Indeed, correlations are easy to compute using a desk calculator and can be estimated just from looking at an array of data.

This integrated framework for analysis starts with simple correlations. As a first step of almost every analysis, we ask whether changes in one variable are associated with changes in a second variable. Associations of this kind are often linear in the sense that the same degree of increase in one variable, say, parent-child conflict, is associated with a similar degree of increase in a second variable, say, adolescent antisocial behavior. For the sake of both simplicity and maintaining the integrity of our overall framework of analysis, we examine only associations of this kind. Many nonlinear relationships are also possible, however. For example, parent-child conflict may have a linear association with antisocial behavior in some kinds of families but not in others; this moderated or conditional association is known in statistics as an "interaction." We encountered this concept in Chapter 4 when we spoke of gene-environment interaction. We also discussed the topic in Chapter 6 when we talked about qualities of the family that might alter the relationship between nonshared factors in the environment and adolescent adjustment. Alternatively, parent-child conflict may have no association with adolescent antisocial behavior at lower or middle levels, but have a strong association with antisocial behavior when conflict is extreme. This "threshold effect" will not be detected very efficiently by the simple linear analytic procedures we employ here. Parent-child conflict might have a "curvilinear" relationship with antisocial behavior: at low levels it is associated with antisocial behavior, at mid-levels with high antisocial behavior, and at the highest levels with low levels of antisocial behavior. Thus, the data may be more complex, subtle, and interesting than our linear, correlational analyses will detect. Nonetheless, we stick with these because nothing else will serve our integrative aims as well.

The simplest use of a correlation would have been to estimate the association between an environmental variable and an adjustment variable. This could have been done once for the older siblings and then repeated for the younger siblings. Yet we had no compelling theoretical reason to track the differences between older and younger siblings. Keep in mind, for example, that the younger sibling in some of the families is a good deal older than the older sibling in some others. Moreover, the older sibling in our study is not, necessarily, the oldest child in the family. Thus, there are no straightforward theories about the effects of the age of the child or of the birth order that would apply to the heterogeneous age and birth-order mix in the sample. As a consequence, we combine the two siblings together with appropriate statistical adjustments to reflect that, unlike most studies, we are using two individuals from each family rather than only one (Reiss et al., 1995).

The most important initial result of our analyses is an estimate of the simple correlation between parenting toward the adolescent and that adolescent's adjustment. This can be thought of as an average of the association across the younger and older siblings in each family. This simple correlation is not easy to interpret by itself. We learn only whether there is notable association and whether it is positive (the higher the value of the environmental variable, the higher the value of the adjustment variable) or negative (the higher the value of the environmental variable, the lower the value of the adjustment variable). We have set a level of .20 as "notable." With our large sample this value is very significant in a statistical sense: the true association is very unlikely to be zero. Also, a correlation of .2 means that 4 percent or more of the differences in adjustment among the adolescents are accounted for, in some way, by the social environment. Although this is a small effect, it is not negligible. Further, correlations of .20 or greater can be explored reasonably by our genetic models.

The simple correlation, of course, tells us nothing about the role of genetic or environmental mechanisms that might explain the observed association between the measures of the social environment and the measures of adjustment. Indeed, we must compare these associations across the six subsamples to make inferences at the genetic level. The simple association also fails to make use of half the data we collected: the possible impact on each adolescent of the social behavior direct toward the sibling. This association is a clear reflection of the multiple-system nonshared environment we hypothesized in the last chapter. We examine associations of this type by use of a *covariance model.* The covariance model parses the two types of parental influence: the association between the parenting of the *adolescent* and the *adolescent's* adjustment, and the association between the parenting of the adolescent's *sibling* and the *adolescent's* adjustment. We refer to the first as the specific correlation and the second as the cross-correlation. (The term "cross-correlation" is used, following semantically confusing traditions in statistics, in three different senses in this book; we will make every effort to keep these three uses distinct.) Following from computational features of the covariance model (see Reiss et al., 1995, for details), the cross-correlation is highly specific for each child and constitutes a precise estimate of a component of the nonshared environment; that is, the computational method can estimate the unique effects of the parenting of the sibling on the adolescent's adjustment that are statistically independent of any other parenting effects on the adolescent or siblings.

The specific correlation, despite its name, is a good but imperfect estimate of the direct effects of parenting on the adolescent. It contains all the parenting influences that are statistically independent of parenting directed at the other sibling, so that, in this sense, it also estimates the nonshared environment. It also includes, however, estimates of the parenting that is correlated between the two siblings; that is, across all the families we studied, parenting that is directed at a specific adolescent can be predicted, to some extent, from parenting directed toward the sibling. It is debatable whether we should regard this component of parenting effects on adjustment as part of the shared or the nonshared environment. In other work, we have attempted to parse out this component of parenting, the component that is predictable between sibs, and estimate its unique association with adolescent adjustment (Anderson et al., 1994). For current purposes, however, the specific correlation can be regarded as a reasonable estimate of the second component of the nonshared environment. It reflects parenting that is directed or targeted at the adolescent, even though it is predictable in varying degrees from parenting directed toward the sibling.

It is possible that the specific correlations and cross-correlations can be opposite in sign; one may be positive and the other negative (see our explanation of negative and positive correlations earlier in this chapter). For example, aggressive parenting directed at a sibling may actually have positive associations with adolescent adjustment, even though aggressive parenting directed toward the adolescent may have negative effects. Delineation of these contrasting effects is a unique yield of including siblings in our design.

Figure 7.6 presents a schematic summary of the covariance model. It shows the distinction between specific and cross-correlations. It also illustrates a second important feature of the model: its treatment of biases that may be unique to each source of the data. For example, parents may have a tendency or bias to report their behavior toward their adolescents as less negative than it really is. Variation in this bias among parents is an important source of error or inaccuracy in measuring parenting; it is represented by the symbol "e" in Figure 7.6. Parents may also see their children as less troubled than they really are; variations in this bias contribute to the errors ("e") related to measures of adjustment. A significant distortion in our results might occur if these two errors are correlated: parents who substantially under-report their own negative behavior toward their children also under-report their children's psychological difficulties. This effect is known as "correlated error."

Research designs that use only one source of data can produce spurious findings because computed correlations can reflect only this correlated error, not a "true correlation" between two variables. When there are two or

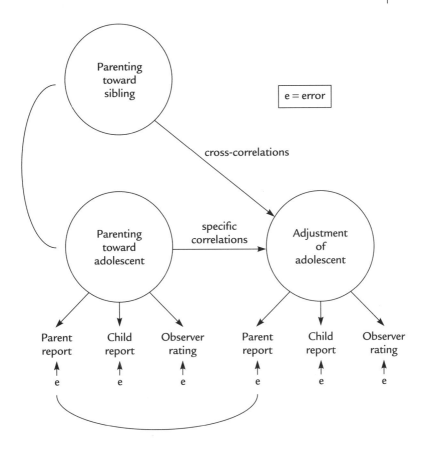

Figure 7.6 The covariance model

more separate sources of data (here we have used three: parent, child, and observer), the correlated error can be partialed out of the computed association between parenting and adjustment. This increases the precision of our estimate of both the cross-correlations and the specific correlations. In almost all our analyses the correlated error was high for each of the three sources of data, confirming the wisdom of partialing these out from our estimates of cross- and specific correlations. As noted in Chapter 6, this feature of our data analysis constitutes a considerable improvement over the few previous efforts to assess the relationship between the non-shared environment and psychological development (Baker and Daniels, 1990; Daniels, 1986; Vernon et al., 1997).

One more special feature of this model needs emphasis. The model treats each of two children in the family both as the "adolescent" and as the "sibling." There are two ways of thinking about how this is done, each roughly approximating the actual statistical procedures. First, one might think of the model as randomly assigning one adolescent to the "adolescent" role and the other to the "sibling" role. Second, a closer approximation is to think of the model as conducting two analyses for each family, one considering a particular child the "adolescent" and the other the "sibling," pooling the results, and then correcting for this "double analysis" so

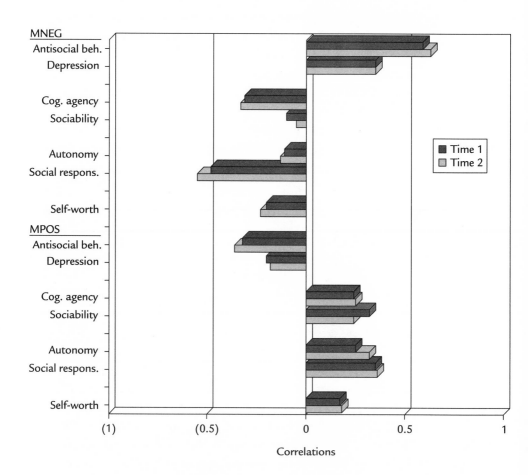

Figure 7.7 Simple correlations between mother's negativity and positivity and adolescent's adjustment

that the true sample size is 720 families and not the 1,440 "double" analyses. Details of this procedure have been reported elsewhere (see Reiss et al., 1995).

Simple Correlations between Parenting and Adolescent Adjustment

Figures 7.7 through 7.10 show the simple correlations between parenting and adolescent adjustment. These simple correlations cannot systematically partial out the effect of correlated error, as the covariance models can. Nonetheless, we do use scores that aggregate mother's, father's, observer's, and child's reports. Thus, mother's conflict/negativity ("MNEG" in Figure 7.7) reflects mother's and child's reports of her harsh discipline, fighting with the child, severe criticism, verbal abuse, and punitiveness. It also reflects the observer's coding of these features of mother's interaction with her child. Note also that in these analyses we have pooled data from the younger and older sibs in each family.

Figure 7.7 shows the associations between mother's negativity and positivity and the seven domains of adolescent adjustment. There are no surprises here. Mother's negativity has a strong positive relationship with antisocial behavior and depressive symptoms and a strong negative association with the five domains of positive adjustment, particularly social responsibility and cognitive agency. Conversely, mother's positivity shows modest negative relationships with antisocial behavior and depressive symptoms and modest positive relationships with the five positive domains of adjustment. The findings for time 2 are almost identical to those at time 1.

Figure 7.8 shows the same association for fathers, which are nearly identical to those for mothers.

Figure 7.9 shows associations between mother's knowledge of her adolescent's activities and the seven domains of adjustment. The findings look somewhat similar to those for mother's positivity. But mother's positivity and mother's knowledge are not measuring identical features of parenting. Their correlation with each other is .41, which is substantial but very far from 1.0. There is a distinction, however, between the pattern of findings for mother's knowledge and mother's positivity: the associations for knowledge go up in time 2, whereas for positivity there was

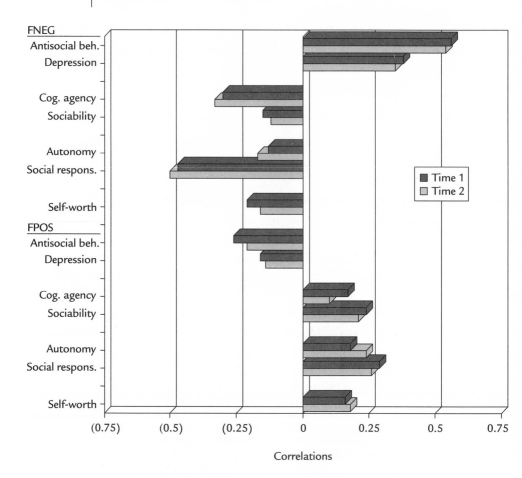

Figure 7.8   Simple correlations between father's negativity and positivity and adolescent's adjustment

little change across time. The increase is particularly notable for antisocial behavior and social responsibility. As will be shown, a very interesting story emerges concerning surveillance and limit-setting by parents of their adolescent children—a story that is unimaginable without integrating careful measurements of changes in family relationships over time with changes in genetic influence across time. The increase in association between mother's knowledge or surveillance of her children and their antisocial behavior (a negative association) and their social responsibility (a positive association) is just one clue to a fascinating interweaving of genetic and family processes in families of adolescents.

Figure 7.10 shows the association between father's knowledge about his adolescent's activities and adolescent adjustment. The pattern is similar to that shown for mothers. Fathers also show some increases in these correlations from time 1 to time 2. Interestingly, measures of parental attempted and actual control did not show notable correlations with measures of adjustment.

Covariance Model: Specific and Cross-Correlations

Figure 7.11 is the first in a series of bar graphs reporting the results of the covariance model. It reports the linear associations between mother's negativity and each of the seven domains of adolescent adjustment. For each domain we report first the specific correlations at times 1 and 2 and then the cross-correlations for times 1 and 2. The asterisks indicate coef-

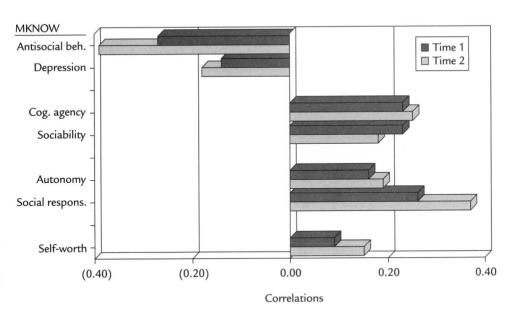

Figure 7.9 Simple correlations between mother's knowledge of adolescent's activities and adolescent's adjustment

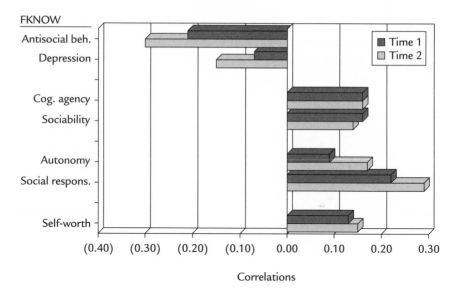

Figure 7.10   Simple correlations between father's knowledge of adolescent's activities and adolescent's adjustment

ficients that are statistically significant at or beyond the .05 level. For illustrative purposes, consider the first four coefficients. The specific correlations are quite high and the cross-correlations are negligible. Indeed, there are no cross-correlations that are consistently significant, in a statistical sense, across time 1 and time 2. For that reason Figure 7.11 looks quite similar to the top half of Figure 7.7 showing the simple associations. The correlations are somewhat larger in magnitude because we have partialed out the effects of correlated observer bias, which, in this case, has obscured or suppressed the magnitude of the direct associations. Some of the correlations shown here are very substantial, particularly for antisocial behavior, depression, and social responsibility. In the absence of genetic information, most researchers would be tempted to conclude that powerful environmental mechanisms must link mother's parenting and adolescent adjustment, with the former being the cause, the effect, or both.

Figure 7.12 shows comparable data for father's negativity, and these are quite similar to the findings for mother's. Figures 7.13 and 7.14 show the association of mother's and father's positivity with the seven domains of adjustment. These results are, to a large extent, the mirror image of the findings for negativity. The cross-correlations are neither large nor consistent across time 1 and time 2. Positivity, for both parents, is negatively or inversely associated with antisocial behavior and depressive symptoms and positively associated with the five domains of adaptive adjustment. Among the measures of adaptive adjustment, social responsibility shows the most substantial associations with both mother's and father's positivity. Indeed, these associations with positivity are very substantial for any behavioral science study.

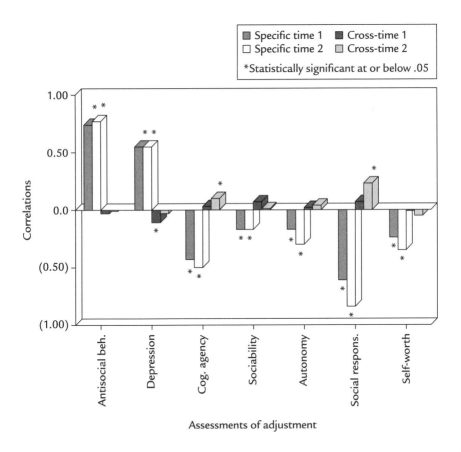

Figure 7.11 Associations between mother's negativity and areas of adolescent development using the covariance model

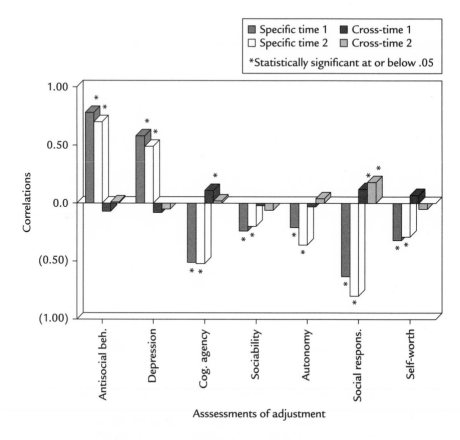

Figure 7.12 Associations between father's negativity and areas of adolescent development using the covariance model

Thus far, the covariance model has provided little information beyond that provided by the simple associations reported in Figures 7.7 through 7.10. This is because there is, apparently, very little association between positive or negative parenting directed at the sibling and the adjustment of the adolescent. In the area of control, however, the covariance analyses do provide fresh insight into family processes. As we review the results of these analyses, we can see that the cross-correlations become much more important. Turning to the associations with mother's knowledge in Figure 7.15, we note two patterns in the data.

First, the cross-correlations are in some cases quite large, particularly for antisocial behaviors, depressive symptoms, and social responsibility. Second, these large cross-correlations appear at time 2 only. Thus, the amount of knowledge the parent has about the *sibling* has adverse associations with the adjustment of the *adolescent:* it is associated with more antisocial behavior and depressive symptoms and less social responsibility. The contrast in direction of effect between the cross-correlation and the specific correlation—for antisocial behavior, depressive symptoms, and social responsibility—has an important statistical effect: partialing this effect out, using our covariance model, leads to dramatically greater specific

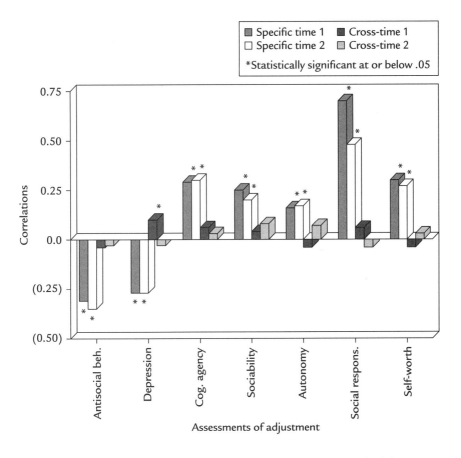

Figure 7.13 Associations between mother's positivity and areas of adolescent development using the covariance model

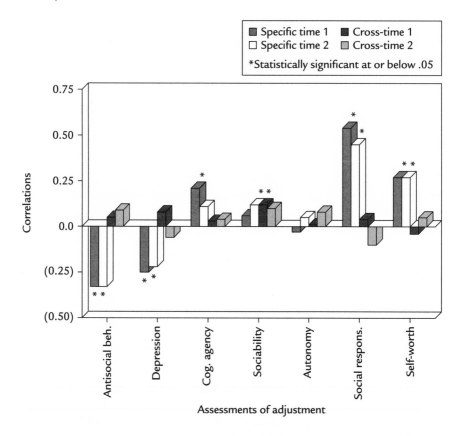

Figure 7.14 Associations between father's positivity and areas of adolescent development using the covariance model

correlations for mother's and father's knowledge (see Figure 7.16) with measures of these three domains of adjustment than is shown in the simple correlations in Figures 7.9 and 7.10.

When we examine attempted and actual control, using the covariance model in comparison with the simple correlation, we see an entirely new picture: the simple correlations failed to show *any* significant correlation between these variables and adjustment. By contrast, the cross-correlations are substantial for both attempted control and actual control. The results for attempted control are shown in Figure 7.17 for mothers and 7.18 for fathers. The more attempted control directed at the *sibling*, the better the adjustment of the *adolescent:* the adolescent shows less antisocial behavior, fewer depressive symptoms, and more cognitive agency, autonomy, and social responsibility. Again, the substantial cross-correlations clearly serve to suppress the simple correlations of mother's and father's attempted control adjustment. Without partialing out the effects of these cross-correlations, the simple correlations were so small we did not even report them in the series of graphs in Figures 7.7 to 7.10. Here, not only can we see a number of significant associations between attempted control and adjustment at times 1 and 2, but that attempted control is associ-

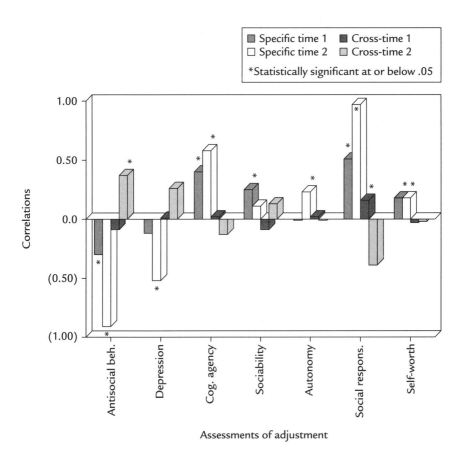

Figure 7.15 Associations between mother's knowledge of adolescent's activities and adolescent's adjustment using the covariance model

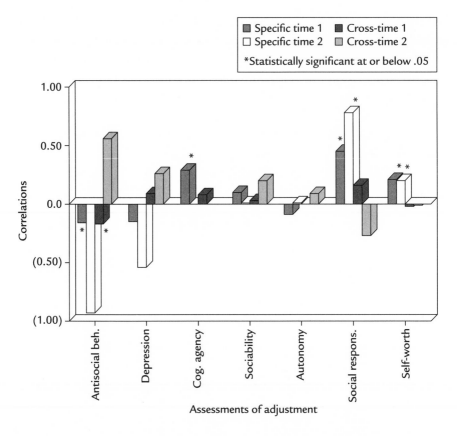

Figure 7.16 Associations between father's knowledge of adolescent's activities and adolescent's adjustment using the covariance model

ated with *mal*adjustment. It is positively associated with antisocial behavior and depressive symptoms and negatively associated with measures of adaptive adjustment, particularly cognitive agency and autonomy.

These correlations look very much like those for parental negativity. As will be shown, adding genetic analyses to our tools for understanding families provides a fascinating insight into the links between parental negativity and attempted control. The fact that the associations between parental negativity and adjustment, on the one hand, and parental attempted control and adjustment, on the other, look so similar is another surface clue to a story about parental control that will emerge throughout this book.

Cross-correlations play an important role for actual control as well (see Figures 7.19 and 7.20). Here they function much as they do for parental knowledge. First, they are substantial for sociability and autonomy for mother, and for antisocial behavior and autonomy for father. Second, they are almost invariably in the opposite direction from the specific correlations. Third, they are more prominent at time 2. When we partial out

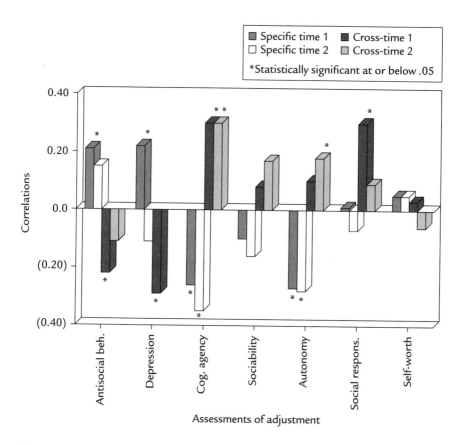

Figure 7.17 Associations between mother's attempted control of adolescent and adolescent's adjustment using the covariance model

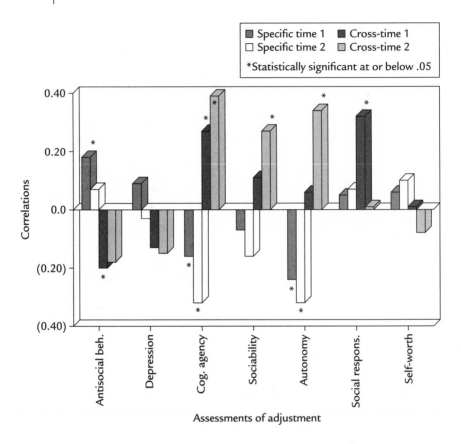

Figure 7.18 Associations between father's attempted control of adolescent and adolescent's adjustment using the covariance model

the effects of the cross-correlation in our covariance model, we see substantial specific correlations for actual control and adolescent adjustment. As with parents' knowledge about their adolescents' activities, these specific associations are negative for antisocial behavior and depressive symptoms and are positive for most of the five measures of adaptive adjustment. One difference stands out: the association between actual control and autonomy is negative for both mother's and father's actual control but positive for mother's knowledge of the adolescent's activities and insignificant for father's knowledge.

In sum, the simple associations show few surprises. Attempted and actual control have trivial associations with the seven domains of adjustment. Positivity and negativity have almost mirror-image effects. Negativity has a positive association with psychopathology and a negative association with the five areas of adaptive adjustment. Positivity shows the reverse pattern. As we shall see, genetic analysis helps to clarify that parental positivity and negativity are not just opposites but also quite distinct social processes. Moreover, they help us understand that mother's and father's positivity are also very distinct from each other: they have different

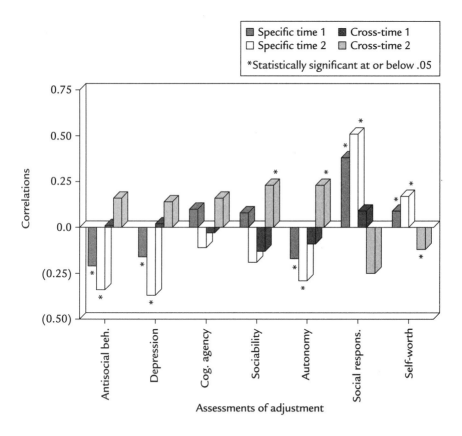

Figure 7.19 Associations between mother's actual control of adolescent and adolescent's adjustment using the covariance model

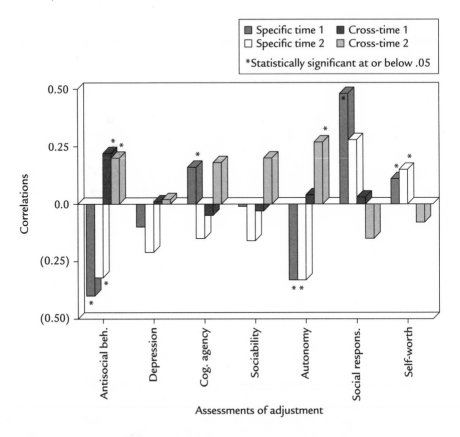

Figure 7.20 Associations between father's actual control of adolescent and adolescent's adjustment using the covariance model

origins and different functions in the family system. The covariance models, while revealing more substantial relationships between family systems and adolescent adjustment than the simple associations, do not add substantially to such associations. In the area of parental control, there are substantial cross-correlations for knowledge of child's activities, attempted control, and actual control. By isolating these, we also uncover substantial specific associations between parental control and adjustment. Attempted control seems inversely correlated with adaptive adjustment and positively correlated with maladjustment; the reverse is the case for knowledge and actual control.

Shared and Nonshared Influences of Sibling Relationships

Analytic Principles

As reported in Chapter 5, we pooled several measures to delineate two qualities of sibling relationships: conflict/negativity and warmth/support. The separate measures, and the way they were aggregated, make them analogous to measures of the same qualities in parent-child relationships. Recall that, in general, the correlations of siblings' relationships with each other were quite high—higher than they were for parent-child relationships. The average correlation between parent reports of sibling relationships was .83 at time 1 and at time 2. The average correlation between self-report of siblings was .55 at time 1 and .60 at time 2, and the between-sibling correlation of observer ratings averaged .72 and .73 at times 1 and 2, respectively. Thus, according to all three sources of information, siblings tended to act and feel similarly toward each other.

But they were not identical. In the area of conflict and negativity, some sibling behavior tended to be proactive: it was dished out without having been received. Other sibling behavior was passively receptive; some adolescents submitted to negative behavior from sibs that they did not reciprocate. Similarly, some siblings conveyed warmth and support that they did not receive, behavior that we classify as care-giving. We refer to the opposite of such behavior as care-receiving. The specific and cross-correlations in our covariance model can be used to index these different forms of sibling relationships. Recall that the specific correlations assess the behavior or relationship quality of which the adolescent is the target. Thus for conflict/negativity, the specific correlation assesses the extent to which the adolescent is a victim of negative behavior from his or her sib. For warmth/support, the specific correlation reflects the associations with warmth received. The cross-correlation reflects behavior that targets the sibling. In this case, however, the initiator is not a parent but the adolescent. Thus, the cross-correlation measures not just what the sib receives

Table 7.1 Summary of sibling relationships indexed by the specific and cross-correlations of the covariance model

	Type of correlation	
Quality of relationship	*Specific*	*Cross*
Conflict/negativity	Victimization	Aggression
Warmth/support	Care-receiving	Care-giving

but also what the adolescent actively gives: for conflict/negativity, the cross-correlation is the extent of aggression and for warmth/support, the extent of care-giving. The relationship between the covariance model and sibling behavior is summarized in Table 7.1.

The distinction between single-system and multiple-system mechanisms of the nonshared environment is somewhat different for siblings than it is for parent-child relationships. In a single-system mechanism, adolescents register or represent only how they are treated by their siblings. In the multiple-system analogue to parent-child relationships, adolescents also register or represent how they behave toward their siblings. If our results show significant specific and cross-correlations, they suggest that adolescent siblings register both the impact of their sibs on them and the impact of their behavior on their siblings. In the latter case, we cannot know from these analyses which matters most to the adolescents: their more distal observations of the impact of their aggression or care-giving on their sibs or the more proximal experience of venting their aggression or giving their support, or both. In any case, it does not seem accurate to regard specific and cross-correlations as evidence of a multisystem mechanism because giving and receiving here refer to one system, the sibling system. A more accurate distinction is between receiver-oriented and reciprocity-oriented sibling systems. If we found only simple associations among the sibling measures, all of which measure what adolescents *receive* from their sibs, we would be justified in concluding that these sibling systems are mainly "receiver-oriented." Wherever we find significant specific *and* cross-correlations in our covariance model, we can say that these systems are "reciprocity-oriented." It matters to adolescents both what they give and what they receive.

Simple Correlations between Sibling Relationships and Adolescent Adjustment

As with parenting, we begin our review of the data on sibling relationships by presenting the simple associations between the quality of sibling relationships and the seven major measures of adolescent adjustment. Here we have pooled all measures and all sources of information for assessing both sibling relationships and adjustment. Further, we have combined the data from both siblings in the family. We have shown all correlations that are above .10 for both time 1 and time 2. As Figure 7.21 shows, the pattern of results is clear and consistent across the two time periods. Sibling conflict/negativity is positively associated with measures of maladjustment, antisocial behavior, and depression, whereas sibling warmth/support is positively associated with all five measures of positive adjustment.

Covariance Model: Receiver- and Reciprocity-Oriented Sibling Systems

We turn next to the analysis of the same data using the covariance model. We begin with sibling conflict/negativity, shown in Figure 7.22. Note that, as occasionally happens with complex models, we could not successfully fit the covariance model to the data for social responsibility at time 2. In Figure 7.22 we see that a simple, receiver-oriented explanation for the sibling data is not tenable. There are striking cross-correlations for many of these associations. The most consistent of these are for antisocial behavior, autonomy, and social responsibility. In sum, being victimized by negative behavior from one's sibling is associated with increased antisocial behavior and depressive symptoms. Being the aggressor in these relationships, however, makes it *less* likely that the adolescent will describe him- or herself or be seen by others as antisocial and more likely that the adolescent will be characterized as socially responsible. Nonetheless, showing active aggression is not invariably associated with good outcomes; indeed, aggressors show less autonomy. Recall that our autonomy

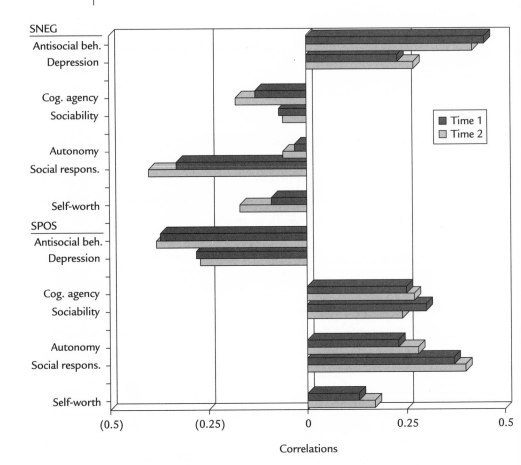

Figure 7.21 Simple correlations between sibling's negativity and positivity and adolescent's adjustment

instrument measures an adolescent's initiative in three areas: self-care in the home, self-maintenance outside of school and home, and recreation. Initiative here is not the same as "initiative" in aggression with one's sib; in fact, it appears to be just the opposite.

An earlier study of young boys explores the distinction between active aggression and victimization that may be relevant here. According to the study, children who are actively aggressive and children who are passively aggressive both have adjustment problems, but active aggressors also have some positive qualities as well. They are less likely than their passive counterparts to distort the intent of others, and their peers are more likely to rate them as leaders and as having a sense of humor (Dodge and Coie, 1987). A more complete interpretation of these findings must await re-examination using our genetically informative design (see below).

Figure 7.23 shows the associations between sibling warmth/support and major outcome domains. Again, substantial cross-correlations suggest that in the positive area adolescent siblings are reciprocity-oriented. The most consistent cross-correlations are for cognitive agency and autonomy. Overall, these data suggest favorable implications for being a care-receiver. There is a substantial inverse association between receiving care

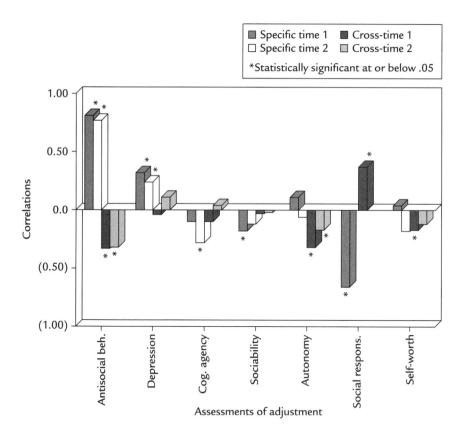

Figure 7.22 Associations between sibling's negativity and adolescent's adjustment using the covariance model

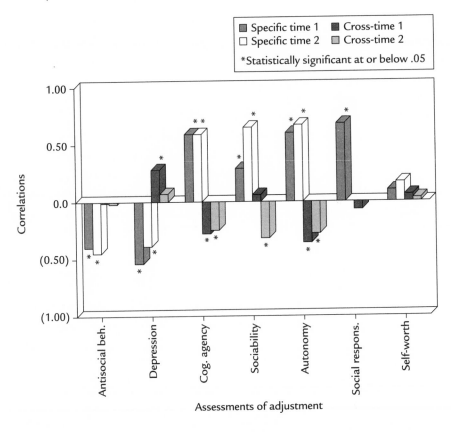

Figure 7.23 Associations between sibling's positivity and areas of adolescent development using the covariance model

from one's sibling and antisocial behavior and depressive symptoms and positive associations between care-receiving with four of the five positive adjustment values. Being a caregiver in a sibling relationship has a negative implication because it is marginally associated with having some depressive symptoms and inversely associated with cognitive agency and autonomy at both time 1 and time 2 and with sociability at time 2. Here again, initiative in the sibling relationship has negative implications.

How can we understand the inverse associations between care-giving and adjustment? Care-giving by one sibling to another may reflect withdrawal or emotional unavailability of one or both parents (Bryant, 1992).

Clinicians have coined the term "parentification" to describe a young child's taking over parental duties because of psychological or medical disabilities either in the parents or in a sibling. Clinical studies have suggested that this role thrusts children into responsibilities for which they are not ready and may impair the caregiver's psychological development (Seligman, 1988).

Psychosocial Theses and Genetic Information

Thus far we have presented data in a format that is familiar to psychosocial research. Using concepts summarized in Chapter 2, we have conducted two sets of correlational studies, one set at time 1 and another set at time 2. These have involved direct associations between assessments of family processes and adolescent adjustment (using specific correlations) and indirect associations concerning the influence of the sibling's relationships (using cross-correlations). We have given particular weight to findings that are substantial at both time 1 and time 2 because they confirm that the observed association is both important and relatively stable across three years. In conventional psychosocial studies the next logical step would be to ask about causal priority: do the adolescents' adjustment levels influence their own relationships and those of their siblings, or is it the other way around? Here longitudinal and intervention studies can be helpful.

But these follow-up studies, now quite commonplace in psychosocial research, might not clarify the causal mechanisms underlying these observed associations. Indeed, they might lead to erroneous conclusions because they might fail to delineate a crucial "third variable" (see Chapter 2), in this case, genetic influences. Genetic factors might explain some or most of our findings, but they cannot be uncovered by more conventional longitudinal or intervention studies. We need to draw on the genetic information in our design to evaluate more critically the findings we have reported so far. To appreciate the value of this information, it is helpful to anticipate what it might reveal.

Thus far we have seen what seem to be strong and direct "effects" of both parenting and sibling relationships on adolescents. Even more intriguing are the indirect associations between the relationships of one sibling and the adjustment of the adolescent, or between the parenting of a sibling and the adjustment of the adolescent. How might genetic mechanisms account for both the specific correlations and the cross-correlations we have noted in our report of covariance models?

For the specific associations, it is possible that the same genetic factors that influence measures of maladjustment or of positive adjustment may also influence the measures of parenting or sibling relationships. For example, consider the association between passive sibling negativity and antisocial behavior. On average, the siblings we studied share about 50 percent of their genes (from 0 percent in blended siblings to 100 percent in identical twins). Perhaps a set of genes influences the *sibling* to be aggressive and negative in his or her relationship with the adolescent; those same genes in the *adolescent,* at least 50 percent of them, anyway, influence antisocial behavior. Thus what appears to be an "effect" of sibling behavior on the adolescent is, in fact, an epiphenomenon. This genetic effect common to both siblings is a version of passive gene-environment correlation as explained in Chapter 4. Another explanation could be that the same genetic factor in the adolescents that influences their antisocial behavior might also make them irritable and aggressive with their siblings, which stimulates negative behavior toward the adolescents by their siblings.

Genetic mechanisms are easier to conceive in the sibling data than in the parent-child data. For example, we have reported that care-taking behavior is associated with reduced cognitive agency and reduced autonomy. These associations may be due to genetic correlation, that is, the same gene or genes that influence a sibling to adopt a care-taking role toward the adolescent may also diminish academic and extracurricular initiative. Geneticists refer to a correlation of different traits for genetic reasons as a "pleiotropism."

With respect to the parent-child data, we reported, for example, a substantial positive association between actual control directed at the *sibling* and level of antisocial behavior in the *adolescent.* Passive genetic mechanisms are harder to conceive of here: we would have to posit that some genes, shared by parent and adolescent, influence antisocial behavior in the adolescent and influence the parent to control the sibling. It seems very implausible that genetic factors influencing parenting would be sibling-specific. That is, genetic factors in the parents are more likely to influence parental style that would be more or less consistent across siblings than to influence differential parenting. A more likely explanation for the cross-correlations, from a genetic perspective, would rely on evocative, or active gene-environment, correlations. For example, genetic differences between siblings may be responsible for the adolescent's being more antisocial than the sibling. As a consequence of this genetically influenced difference between two sibs in the same family, the father—for example—gives up any attempt to discipline the antisocial child and turns his attention to the more tractable child in hopes of having some beneficial effect where positive outcomes still seem possible. We refer to this scenario as the "parental salvage effect." That is, the parent senses a looming psychological disaster in one child and moves to protect another child with a better chance at a good long-term outcome by using more effective parenting strategies. The parental salvage effect, if it exists, is a form of evocative gene-environment correlation, though in this case the parent is responding not to a heritable characteristic of a single child but rather to differences between two children caused by the genetic differences between them.

In sum, we have posited three closely related mechanisms that might explain the associations we have observed: gene-environment correlation, behavioral pleiotropism, and parental salvage. If any of these are operating, they would reframe our question of causal priority. We would want to know more than whether family processes influenced adolescent adjustment or the reverse. We would want to know specifically where genetic influence entered this sequence. Thus, it makes sense to examine the data for evidence of these and similar effects before proceeding to our longitudinal analyses.

Chapter 8

ANTITHESIS I: INFLUENCES ON STABILITY

AND CHANGE IN ADOLESCENT ADJUSTMENT

We have stated a strong thesis. Many previous research findings support the importance of environments unique to each sibling in the family. These nonshared environments, according to earlier studies, are the most significant environmental influences on psychological development. Evidence suggests that many of these nonshared influences may arise with the family, and so for our study we focused on two questions: How different is the family experience of two siblings? Are these differences associated with differences among adolescents in their psychological symptoms and in their adjustment? In the last chapter we presented very strong evidence supporting the thesis that parental behavior unique to each sib was very strongly associated with adolescent adjustment. In some cases, such as the relationships between parental negativity and antisocial behavior, these associations accounted for 50 percent of the variation among adolescents. We also found that parental treatment of the adolescent's sibling has a powerful influence on the adolescent, particularly in the areas of surveillance and parents' attempted control of behavior. We also found compelling evidence for the role of nonshared sibling experiences, with siblings who are victims of aggression or who become caregivers faring the worst.

The findings presented in the last chapter look like strong evidence for the argument that psychosocial mechanisms link the social environment to adolescent adjustment. But this may not be the case. Genetic factors could influence measures of adjustment as well as measures of the social environment. Moreover, the same genetic factors could affect both these sets of measures, hence accounting for the association. Such a circumstance would be a sharp challenge to the environmental thesis. In this chapter we take the first step toward exploring this antithesis, asking if genetic factors play a large role in differences among adolescents in the seven domains of adjustment we studied.

We find that genetic factors do have this importance. Moreover, they are the most significant influence on the stability of adjustment across time, with shared environment also playing an important role. For change across time, nonshared and genetic factors were both important.

This chapter begins to lay out what we term the "antithesis." Here for the first time we use genetic analyses to estimate the role of anonymous genetic, shared, and nonshared environmental factors on the seven domains of adjustment. By themselves, these data are neither supportive of nor antithetical to the findings discussed thus far. But they are a crucial first step in exploring whether the associations reported in the last chapter can be attributed to psychosocial or genetic mechanisms. The latter is possible only if genetic factors affect the measures of adjustment and the measures of the environment involved, and if the genetic factors affecting both are the same. Here we report the results of three separate analyses: an analysis of genetic and environmental influences on adjustment at time 1, the same analysis at time 2, and an analysis of the genetic and environmental contributions to both stability and change from time 1 to time 2.

Influences on Individual Differences among Adolescents at Time 1 and Time 2

Analytic Principles

The analytic tasks for estimating genetic and environmental contributions to individual differences at a single point in time are less imposing than those used for analyzing change across time. Indeed, in Chapter 4 we introduced the standard approaches behavioral genetics takes to estimating these influences. Recall that we use statistical models that compare correlations between siblings across the six groups of siblings. If genetic influences are important for any trait, we would expect correlations between siblings on that trait to be highest in MZ twins and lowest in genetically unrelated blended siblings. Indeed, the patterns of relationships among the intersibling correlations in the six groups should be as follows: MZ > DZ = full sibs (nondivorced families) = full sibs (step-families) > half-sibs (step-families) > blended sibs (step-families). We call a pattern of findings such as this one a "genetic cascade."

The balance between the shared and the nonshared environment is reflected in the overall level of the intersib correlations. Consistently high correlations, particularly in the blended sibs, suggest the importance of shared environments, and consistently low correlations (particularly in the MZ twins) suggest an important role for nonshared factors. The statistical procedures or models that we use serve to integrate two aspects of the data: (a) patterns, particularly the genetic cascade, of correlations across the six groups; and (b) absolute levels of correlations. These analyses permit us to estimate the proportion of differences among individuals (known as the "variance" of their scores on our tests) that are attributable to genetic factors as well as to the shared and nonshared environment. The use of these models rests on a number of different assumptions, including relatively small or significant gene-environment interactions and equal shared environments across the six groups. We have examined these and other assumptions in Chapters 3, 4, and 5. A general rule holds for all results in statistical analyses: the results are no more valid than the assumptions underlying the analyses that produced them.

In Chapter 3 we presented some simple examples of how comparisons of correlations across groups could be used to draw inferences about genetic and environmental influences. In Chapter 4 we presented several examples of correlations from the six groups, in the analysis of adolescents' self-perceived competence, alongside the results of our model-testing analyses. In this chapter we present only the results of the model-testing analyses. The reader should be able to imagine the pattern of correlations that gave rise to these results; if not, reference to Chapter 4 may be helpful. The actual correlations on which our models are based may be found in Appendix A.

Genetic and Environmental Influences at Early and Mid-Adolescence

Figure 8.1 presents our analyses of genetic and environmental influences on the seven major measures of adolescent adjustment for the full sample at time 1 and for the eligible and participating siblings at time 2. Recall that our sample consisted of 720 families at time 1; 12 of these families had twins of uncertain zygosity and so were excluded from all genetic analyses. The average age of the oldest child was then 13.5, and that of the youngest was 12.1. At time 2, 434 families were eligible for participation because the adolescent siblings were both still living at home. Ninety-one percent of these, or 395, participated. At that time the average age of the oldest child in the participating sample was 16.2 and the average age of the youngest child was 14.7. Several patterns in the data of Figure 8.1 are worth noting.

First, Figure 8.1 shows substantial genetic influences for three of the seven measures: antisocial behavior, cognitive agency, and social responsibility. Indeed, the heritability of each of these three dimensions of adjustment is in excess of 65 percent at both time 1 and time 2. In our study, cognitive agency is a broad-band measure of the adolescent's involvement in schoolwork as well as his or her scholastic achievement. It may be closely related to a range of mental abilities not measured in our study, including general intelligence. Its high heritability is consistent with previous studies of genetic influence on both mental abilities and scholastic achievement (Loehlin et al., 1994; Martin, 1975; Petrill and Thompson, 1993; Thompson et al., 1993). Genetic influence on antisocial behavior and related problems in adolescents has also been reported frequently (Graham and Stevenson, 1985; Rowe, 1983b; Silberg et al., 1994; Stevenson and Graham, 1988; Willerman et al., 1992). The magnitude of the heritability we report, however, is generally larger than previously reported, although a recent study also found very substantial heritability for impulsivity, conduct disorders, and oppositional behavior, all of which are closely related to antisocial behavior as we define it (Eaves et al., 1997).

Differences between our results and those of others may be due to differences in our sample or differences in how we assessed antisocial behavior. For example, no other genetically informed study has combined data from reports of both parents, from the child, and from direct observation. In effect, our study cumulates data on antisocial behavior in a broader range of settings than do previous studies: settings that can be observed by father, by mother, by observer, and by child. Thus, our aggregate score for antisocial behavior—which cumulates data from all these settings— may be more sensitive to differences among children and, consequently, more sensitive to genetic factors accounting for these differences. The nu-

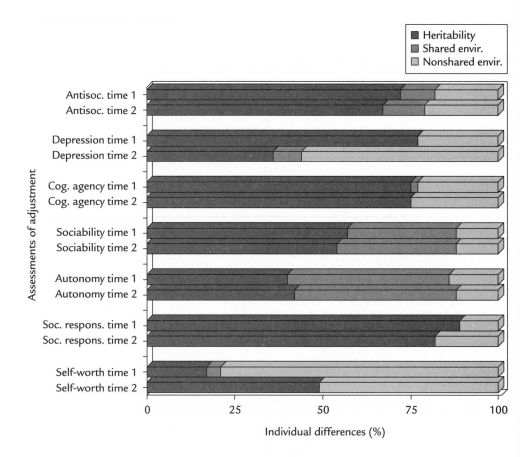

Figure 8.1 Heritability of adjustment

merous studies of social responsibility, and closely related aspects of adjustment in adolescents and adults, almost all use single measures or at most several measures from a single source (usually the adolescent or the adult). All showed much lower heritabilities than those reported here (Loehlin, 1992a).

Second, Figure 8.1 provides information on the magnitude and type of environmental effects. Note that for five measures nonshared environmental effects are the exclusive or preponderant environmental effects. The largest nonshared effects are for global self-worth, particularly at time 1. But the findings for global self-worth cannot be compared strictly with the other six measures of adjustment. Global self-worth is measured only by adolescent self-report and does not aggregate data from different measures or from different sources. Thus, there may be more error or uncertainty in this measure than in the others. As discussed, error or uncertainty in measurement influences the magnitude of nonshared effects and probably contributes to the magnitude of nonshared effects for global self-worth. Figure 8.1 shows that shared environmental factors are the preponderant environmental influences for both sociability and autonomy; they are also substantial, relative to nonshared effects, for antisocial behavior. Although the importance of the shared environment has been observed in prior studies of antisocial behavior (Plomin, Chipuer, and Neiderhiser, 1994), its substantial influence on sociability and autonomy in adolescence is a new finding.

A third finding concerns changes in the balance of influence among shared and nonshared genetic environments from time 1 to time 2. The most notable changes are for depression and global self-worth. For the former there is a sharp decline in the influence of genetic factors and a corresponding increase in the importance of nonshared factors. For the former the changes are just the reverse. Patterns of changes for cognitive agency are similar to those for depression, although they are much smaller. The remaining four measures show almost identical findings for time 1 and time 2. However, a great deal of change can be concealed behind this apparent stability across three years. The next section presents results of our efforts to examine underlying changes not apparent in these data.

Influences on Stability and Change in Adolescent Adjustment

Analytic Principles

Genetic and environmental factors can contribute not only to differences among individuals in adjustment at distinct points in time, but also to stability or change in any measure of adjustment across time. To study such effects we use a cross-correlation, illustrated in Figure 8.2. This figure shows contrived data for two time periods (comparable to the data in this book) and data for only three MZ and three DZ twins. The data presented could represent antisocial behavior or any other measure obtained at times 1 and 2. For purposes of simplification, we label the child with higher values for this variable at time 1 the "adolescent" and the one with lower values the "sibling."

Consider first the MZ twins and observe the rank order of the adolescent: adolescent A is highest, adolescent B is next, and adolescent C is the lowest. Note that all three adolescents change absolute values from time 1 to time 2. Adolescent A goes up, adolescent B stays about the same, and adolescent C goes down. The most important feature of this illustration is that the sibling changes in the same direction as the adolescent and by about the same amount. This means we can predict the rank of the "siblings" vis-à-vis each other at time 2 very precisely from the ranks of the adolescents at time 1. Regrettably, we use the term "cross-correlation" here, the same term used to express indirect effects of sibling relationships on adolescent adjustment in the last chapter. Even more unfortunately, we will encounter yet a third meaning of this term later in the book. Convention has, alas, fixed this term into scientific language with three distinct meanings.

Now let us compare the three MZ twins with the three DZ twins. The adolescents in all three DZ twin pairs also show a change from time 1 to time 2. The siblings' change, which is also considerable, does not, however, parallel the adolescents' change. Thus, we cannot predict very well the rank order of the sibling at time 2 from the rank order of the adolescent at time 1. What we have illustrated is that the cross-correlation is very high (perfect, or 1.0 in this illustration) for genetically identical siblings but much lower for siblings who share only 50 percent of their segregating genes. This is evidence that genetic factors play an important role in relationships between antisocial behavior measured at time 1 and at time 2. This relationship is known as stability, and we can say that Figure 8.2 provides strong evidence that genetic factors play a major role in the stability of the measured variable.

We can use cross-correlations to estimate the role of genetic and envionmental factors in this stability. To do so we extend the procedures we used for estimating heritabilities and environmentalities at a single point in time: we compare cross-correlations across groups of siblings who differ in genetic relatedness. Example data are shown in Figure 8.3. Here we use contrived data to illustrate an ideal genetic cascade. Since the statistics shown in Figure 8.2 are cross-correlations across time, they estimate not heritabilities but a relationship we term the "heritability of stability." At the same time they also assess the "environmentality of stability." The data in Figure 8.3 show that the difference between MZ, on the one hand, and DZ and other full sibs, on the other, is .20. The same difference holds between DZ twins and full sibs, on the one hand, and

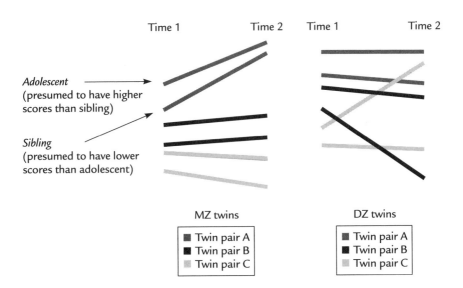

Figure 8.2 Data contrived to illustrate genetic influences on the stability of scores of antisocial behavior

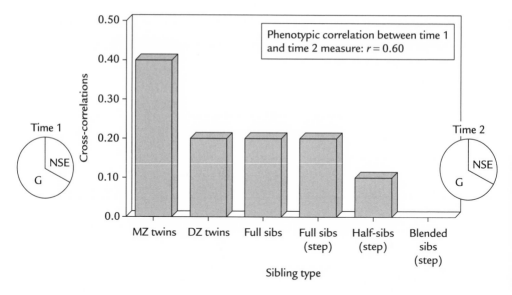

Figure 8.3 Data contrived to illustrate genetic and environmental effects on stability and change (G = genetic; NSE = nonshared environmental)

blended sibs, on the other. This is exactly what would be expected if the only genetic factors that were important in stability were strictly additive (the degree of genetic similarity of a pair of siblings is precisely related to their similarity on a measure of behavior).

If we double either of these differences we get .40. This is the extent of stability influenced by genetic factors. There are two ways of understanding this number. As in heritabilities computed at a single point in time, this number reflects a proportion of total differences among individuals in a sample. If this were a computation of heritability at a single point in time, we would say that 40 percent of differences among individuals are attributable to genetic factors. The .40 here means something different: 40 percent of the differences among individuals, on some measure of behavior at time 2, can be predicted from measures on those same individuals at time 1 for genetic reasons (or vice versa). Alternatively, since the overall phenotypic stability is .60 (see "*r*" in Figure 8.3), we can say that genetic factors account for two-thirds (.40 divided by .60) of the stability in these illustrative data.

Environmental factors must account for the remaining one-third of the stability. A close inspection of the data suggests that it must be nearly all nonshared factors that are correlated across time. The cross-correlations for blended sibs, a good measure of shared environmental contributions to stability, is zero and the cross-correlation for MZ twins is only .40. Indeed, the difference between the MZ cross-correlation (.40 in this example) and the overall stability (.60 in this example) gives a rough estimate of the contribution of nonshared factors to stability across time. The statistical model we use integrates data from all six groups simultaneously for a more precise estimate of these heritabilities and environmentalities of stability.

The pie diagrams at each end of Figure 8.3 provide a clue as to the factors responsible for change. Note that heritabilities at a single point in time, both at time 1 and at time 2, are quite high in this example (70 percent). The heritability of stability, however, is only 40 percent. Another way of thinking about this is to say that only 57 percent of the genetic influence (.40/.70) affecting behavior at time 1 continues to affect that behavior at time 2. Thus, 43 percent of all genetic factors influencing individual differences at time 1 operate uniquely at that time (that is, uniquely from time 2 but not necessarily uniquely from other time periods in development unmeasured in this example and in our study). Figure 8.3 also illustrates that there are comparable genetic effects at time 2 that do not contribute to stability and hence were not operating at time 1. We can regard these unique genetic effects at both time 1 and time 2 as those that are effective at one point in time but not at another and hence must be responsible for change in behavior across time. In this case, these genetic factors are the preponderant influence on change (in contrast to all environmental variables).

Note, too, that in Figure 8.3 nonshared environmental effects are also important at times 1 and 2. The nonshared influence on stability was 20 percent. We can estimate the unique nonshared influences at time 1 and time 2 just as we estimate the unique genetic effects: by subtraction. Thus the total influence of nonshared factors at time 1 was 30 percent. Of this, 20 percent is involved in stability. The remaining 10 percent is unique to time 1.

In sum, the stability of this variable is .60. Genetic factors account for

66 percent of this stability (.40/.60). Nonshared factors account for 33 percent of the stability (.20/.60). The coefficient of change is .40 (1 − .60). The unique heritabilities at each time point are 30 percent (.70 − .40). Thus we can say that .30/.40 of the coefficient of change, or 75 percent, is accounted for by genetic factors. Similarly, the unique nonshared environmental influences operating at time 1 and time 2 are 10 percent (.30 − .20). Thus we can say that .10/.40 of the coefficient of change, or 25 percent, is accounted for by change in nonshared factors. Genetic influence on change might be due to the switching on and off of particular genes in the course of development. Genes operating in early adolescence may switch off, to be replaced by newly activated genes. Similarly, certain nonshared factors salient in early adolescence, say, differential junior high school environments, may be replaced by new ones, for example, differential high school environments, in later adolescence. There are many other ways of understanding how genetic factors and nonshared factors might contribute to change, and we turn to these shortly.

In this and subsequent chapters of the book we represent the results of analyses like these by using a "balloon" diagram. Figure 8.4 illustrates how this diagram represents the simple correlational data presented in Figure 8.3. First, this diagram represents genetic, shared, and nonshared influences as *latent variables*. The influence of these latent variables on change and stability is not directly measured but inferred from the pattern of correlations at time 1 and at time 2 and the pattern of cross-correlations across times 1 and 2.

Second, factors that operate uniquely at time 1 or time 2 (and hence are responsible for change) are represented by smaller balloons that, defying their buoyancy, are suspended beneath the time 1 and time 2 measures. Their unique contribution to individual differences at either time point can be estimated by squaring the coefficients alongside the "strings" of the balloons; these are called *path coefficients*. In more technical parlance the balloon diagrams are referred to as "path diagrams." The unique heritabilities and environmentalities at each time point can be estimated by squaring these path coefficients. As we have seen by direct examination of the data in Figure 8.3, the unique genetic influences at each time point do indeed equal the square of the path coefficient $.55^2$ (30 percent), as do the unique nonshared influences, 33^2 (10 percent). Change reflects the

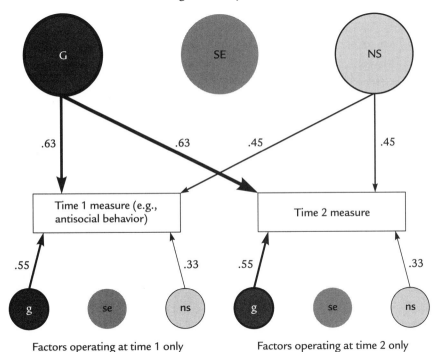

Figure 8.4   Balloon diagram showing results of model testing for factors influencing stability and change

disappearance of some influences across time and the addition of new ones. It is best expressed by averaging the squares of these unique influences. This is easy to do because, owing to the simplicity of the data, the unique paths are identical for times 1 and 2; usually they are not equal.

Third, factors that operate at time 1 and time 2, and are, therefore, responsible for stability, are reflected in balloons above and by the two path coefficients, reflecting their relative influence on time 1 and time 2 measures. The role of any latent variable in stability can be estimated by multiplying these two paths (in this example, .63 × .63 = .40 for genetic influence on stability, and .45 × .45 = .20 for nonshared environmental influence). The sums of these products should always equal the overall stability coefficient: .60 in this case. In actual practice, however, they may not add up, in part because, in an effort to simplify data presentation, we do not report paths that are not significantly different from zero. Also, the two paths from a single latent construct may not always be the same numerically. Our analytic model constrains these two paths to be equal, but occasionally there is a substantial rounding error.

Fourth, one more numerical feature of the model should be noted. The factors that account for change and those that account for stability should add up to 1 at time 1 and again at time 2. That is, differences among individuals at each single point in time consist of those factors that account for stability and those that account for change. Note that $.63^2 + .45^2 + .55^2 + .33^2 = 1$.

Finally, we simplify these balloon models to make them easier to read and to compare with one another. As noted, we eliminate from our diagrams those paths that are not significantly different from zero. Further, where paths are not statistically significant, we do not put any border around the corresponding latent variables (for example, "SE" among those factors accounting for stability in Figure 8.6). In addition, the thickness of borders around the circles as well as the arrows are proportional to the magnitude of the significant path coefficients.

Let us return to the specific "findings" in these illustrative, contrived data. First, how can we understand how changing nonshared environments lead to changes in the measure of adjustment illustrated here? Unique nonshared environmental effects in these analyses do not necessarily mean that siblings in all or most families experience the same or comparable shifts in their nonshared environment. Some, to continue our example, may change school districts, others might have experienced a major change in family economic conditions with differential effects on siblings in the same family, and still others may have experienced the death of a grandparent whose loss meant more to one child than to the other.

In the example we have been following, the preponderant influence on change is not environmental but genetic. There are three broad classes of psychobiological events that could account for change in genetic influence across time. The first group consists of intracellular mechanisms. We now know that a variety of intracellular events control gene expression, often by influencing the transcription of information in DNA amino acid sequences in the gene itself to RNA and subsequent protein synthesis. Recent research on animals and humans suggests that these intracellular gene-control processes can be influenced by a wide range of circumstances: the administering of psychiatric drugs, chronic and acute stress, conditioned fear, and sleep deprivation (Glaser et al., 1990; Hyman and Nestler, 1993; O'Hara et al., 1993; Platt et al., 1995). Thus far these findings are only heuristics for understanding the role of genetic factors in accounting for individual differences in behavioral change among adolescents across a three-year period. Nonetheless, these early findings do suggest that phase-specific environmental events may influence the expression of genetic differences in adolescents: as adolescents get older, some of the genes influencing behavior may be "switched off" by these phase-specific environmental circumstances and others may be "switched on." These phase-specific switches may affect some adolescents more than others, depending on the differences among them in their possession of certain genes that are responsive to these environmental triggers.

A second class of factors consists of major changes in environmental challenges. That is, genetic influences may be expressed in some environments but not in others. In this book we report only environmental changes that occur in the course of ordinary development. For example, as children move from more controlled school settings in the earlier grades to less structured settings in later grades, heritable differences among them in autonomy, self-control, and antisocial behavior may become more manifest. The genetic differences among these children were present all along, but they acquire greater significance, and are thus easier to detect, as the challenges of the environment shift.

A third class of events might account for differences among individuals in genetic change over time. These are genetic conditions that produce a slow accrual of changes not detected by ordinary measures until they are of major proportions. This is probably what happens, for example, in some cases of Alzheimer's disease—an example drawn from much later in development than the phase we study here. In some cases of Alzheimer's, genetic factors may exert a constant influence, leading to a very slow loss of brain cells and, consequently, a slow accrual of subtle loss of memory and other mental functions across many years. These genetic influences might not be apparent in ordinary clinical assessments of mental abilities until the disease has progressed significantly. Thus it might appear that, if measured at mid-life, mental abilities related to Alzheimer's disease are not influenced by genetics—even if the sample contains a large number of subclinical Alzheimer's patients whose subtle disorders are genetic in origin. Some years later this same sample might be measured again when the deficits attributable to genetically influenced Alzheimer's are more manifest; this sample might show large differences attributable to genetic influence. In analyses of this kind conducted on the adolescent data in our study, these late-appearing effects of constant genetic influence would appear as unique genetic influences, operating only at a second measurement occasion later in development.

Influences on Stability and Change in Adolescent Adjustment

ANTISOCIAL BEHAVIOR   Figure 8.5 presents an analysis of genetic and environmental influences on stability and change in antisocial behavior. It shows the cross-correlations (as illustrated in Figure 8.2) computed for each of six groups. For a full appreciation of this analysis it is helpful to turn to our analyses at two single points in time in Figure 8.1. Note that genetic factors accounted for more than 65 percent of individual differences at times 1 and 2, with the preponderant environmental influence being nonshared. Figures 8.5 and 8.6 unpack these findings.

First, note in Figure 8.5 that the stability of antisocial behavior across three years in adolescents is fairly high ($r = .61$). The pattern of relations

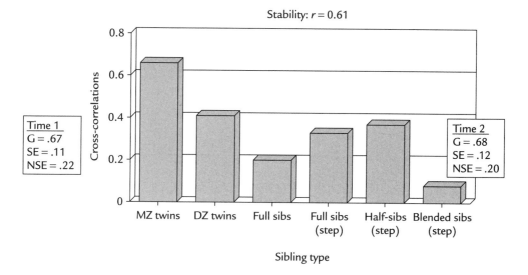

Figure 8.5 Cross-correlations for estimating the role of genetic and environmental factors in the change and stability of antisocial behavior (G = genetics; SE = shared environment; NSE = nonshared environment)

of the cross-correlations across the six groups does not fit neatly into the cascade expected for very high levels of additive genetic influence. Nonetheless, many of the two-way comparisons between the various sibling groups do meet genetic expectations. For example, MZs are greater than DZs, full sibs, full sibs–step, half-sibs, and blended sibs. Similarly, DZs are greater than blended sibs and half-sibs, and full sibs and half-sibs are greater than blended sibs. Thus, nine of the fifteen comparisons are consistent with an influence of genetic factors on stability, and some of the differences between cross-correlations in these comparisons are substantial. The most aberrant cross-correlation here is that for half-sibs, which is higher than we would expect if genetic factors were the only influence on the stability of antisocial behavior; similarly, the correlation for full sibs is lower than expected. The overall pattern, however, would lead us to expect genetic factors to play a substantial role in stability.

With respect to environmental influences on stability, it is important to note that the correlation for MZ twins exceeds the stability coefficient, and the correlation of blended sibs is substantially greater than zero. Thus, we would expect at least a modest role for the shared environment in stability as well. Given the very high heritabilities at time 1 and time 2 and the moderate role for the nonshared environment at these time periods, we might also expect a role for genes and nonshared environment in change in antisocial behavior. Nonshared factors are unlikely to account for much, if any, stability because of the very high MZ correlation (in comparison with the stability coefficient) and the notable correlation of blended sibs.

Figure 8.6, the result of multivariate model testing (see Appendix B) that takes all these comparisons into account simultaneously, confirms our expectations. Genetic factors account for 69 percent ($.66 \times .64/.60$) of the stability of antisocial behavior across three years of adolescent development. Shared environmental factors are the preponderant environmental influences on stability. Interestingly, genetic factors are also the predominant influence on change across time; changes in shared environmental factors play no role whatsoever.

Let us compare these results with those in Figure 8.1. The substantial genetic influence shown at time 1 and time 2 in Figure 8.1 really has two components; most of the genetic influences at these two time points come from genetic factors that are constant across the span of development from early to mid-adolescence, but another component of genetic influence changes across the two time periods. That is, even though the *magnitude* of genetic influence is the same at times 1 and 2, there are actually a substantial number of such influences that change over time. But this is not the case for shared environment. All of this influence can be designated as belonging to only one component: factors that account for stability. Thus, we can say with some confidence that changes in shared environment from early to mid-adolescence play little or no role in initiating changes in antisocial behavior during this period. The reverse is true for nonshared factors. At time 1, 70 percent ($.38^2/(.38^2 + .25^2)$) of nonshared influences are involved, as cause or as effect, in change in antisocial behavior, and a slightly greater proportion is involved in change at time 2. Another way of summarizing these analyses is to say that of all the environmental influences associated with antisocial behavior for nongenetic reasons, only about half are involved in developmental change.

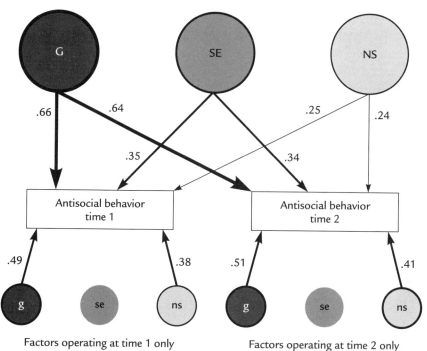

Factors accounting for stability from time 1 to time 2

Factors operating at time 1 only — Factors operating at time 2 only

Factors accounting for change from time 1 to time 2

Figure 8.6 Balloon diagram showing results of model testing for factors influencing stability and change in antisocial behavior

DEPRESSIVE SYMPTOMS Figure 8.7 presents cross-correlations for depression for time 1 and time 2. Note that the stability ($r = .60$) is nearly identical to that of antisocial behavior. The across-group comparison of cross-correlations is a bit different. To begin with, the correlations are, in general, lower—particularly for the MZ, full, and half-sibs in step-families. By contrast, these correlations conform a bit more closely to genetic expectations. The exceptions are the relatively high correlation for blended sibs and DZ twins. Thus, we would expect our model-testing analyses (Appendix B) to show substantial influence of genetic factors on stability in depression. Note, however, that there is a dramatic decline in heritability from time 1 to time 2 and a substantial increase in the influence of nonshared environment (we saw this also in Figure 8.1).

Findings from the model-testing analyses are shown in Figure 8.7. In terms of stability, we find the same balance as for antisocial behavior. Genetic factors account for the preponderant component (.67 × .66/.60 =

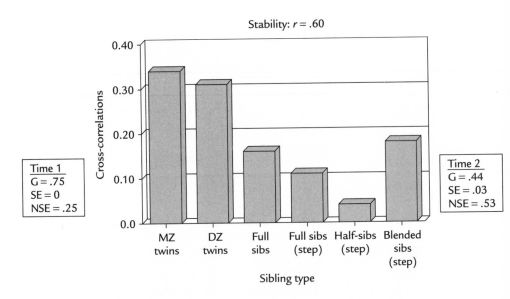

Figure 8.7 Cross-correlations for estimating the role of genetic and environmental factors in the change and stability of depressive symptoms (G = genetics; SE = shared environment; NSE = nonshared environment)

73 percent). Some of the factors accounting for change in depression, however, are quite different from those for antisocial behavior. As for antisocial behavior, genetic factors do account for change but in a special way: a substantial genetic influence on depressive symptoms notable in early adolescence does not influence depressive symptoms at mid-adolescence; indeed, there is no unique genetic influence at all at mid-adolescence. Yet new nonshared factors, which had no influence in earlier adolescence, now account for 38 percent ($.62^2$) of individual differences by later adolescence. The shared environment plays no role in stability or change in depression.

With reference to Figure 8.1 we can recognize, even from analyses at two single or cross-sectional points in time, that genetic influence declines substantially from earlier to later adolescence and that there is an increase in nonshared influences in later adolescence. Indeed, Figure 8.8 clarifies that the majority of genetic influences during this developmental period account for stability, but that most of the nonshared influences account for change.

COGNITIVE AGENCY   Figures 8.9 and 8.10 show results for cognitive agency with some similarities and important differences from antisocial behavior and depressive symptoms. Like the latter two, cognitive agency is very stable between earlier adolescence and later adolescence ($r = .71$). Further, as in the previous examples, analyses at a single point in time show substantial genetic influence, a modest nonshared effect, and no influence of shared environment whatsoever. The most striking difference, however, is that the cross-correlations conform almost precisely to genetic expectations.

As the pattern of correlations suggested, model-testing results show that genetic influences on stability are particularly strong, accounting for ($.81 \times .82/.71$) 94 percent of the stability. Changes in the nonshared environment from earlier to later adolescence account for 67 percent of the change (($(.35^2 + .52^2)/2/(1 - .71)$). Genetic factors account for 33 percent of the change (($(.44^2 + .02^2)/2/(1 - .71)$).

In Figure 8.10 we can draw roughly the same conclusions as we did for depression in Figure 8.1: most of the genetic effects shown as important at each time period are involved in maintaining stability in cognitive

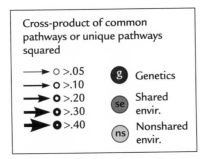

Factors accounting for stability from time 1 to time 2

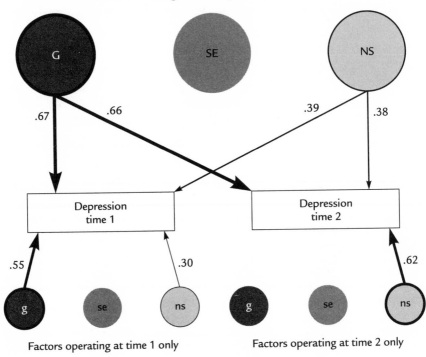

Factors accounting for change from time 1 to time 2

Figure 8.8 Balloon diagram showing results of model testing for factors influencing stability and change in depressive symptoms

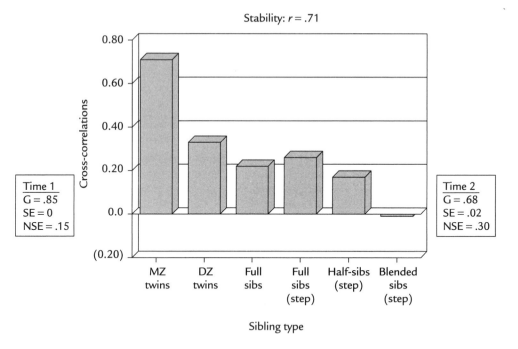

Figure 8.9 Cross-correlations for estimating genetic and environmental factors in the change and stability of cognitive agency (G = genetics; SE = shared environment; NSE = nonshared environment)

agency across time. Virtually all the nonshared effects change over time and are associated with change in cognitive agency from earlier to later adolescence.

SOCIABILITY For the remaining areas of adjustment we show only the balloon diagrams. Numerical values for the cross-correlations can be found in Appendix A. The importance of the shared environment in sociability was already suggested in our analyses of genetic and environmental influences at single points in time and shown in Figure 8.1. At time 1 and again at time 2, shared environment is the prevailing environmental influence on individual differences in sociability. These analyses do not, however, tell us whether shared environment is important for stability or change in sociability. Figure 8.11 provides a much clearer picture: all the

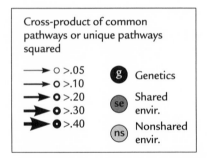

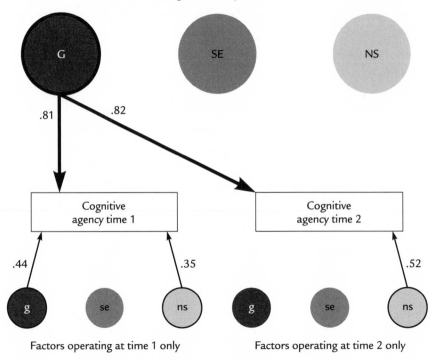

Figure 8.10 Balloon diagram showing results of model testing for factors influencing stability and change in cognitive agency

shared environmental influences on sociability are responsible for stability, not change. This means that the shared environmental influences operative at time 1 must be the same as those operative at time 2, or that the two sets of influences are highly correlated. Indeed, shared environment is slightly more important for the stability of sociability than are genetic factors, although the latter are the preponderant influence on change in sociability from earlier to later adolescence. The stability coefficient is .65; shared environmental influences account for 47 percent of this stability. Genetic characteristics account for about 41 percent. The remainder is accounted for by weak nonshared environmental effects. Genetic influences account for 77 percent of the change in sociability from time 1 to time 2.

With respect to Figure 8.1 these results, on the surface, could be read as the most disappointing for theories that posit strong environmental influences on sociability, independent of genetic mechanisms. Figure 8.1 raises the possibility that shared environmental factors, in particular, might be quite important to adolescent sociability. They account for about 30 percent of individual differences in sociability, as either cause or effect, at both time points. The analyses summarized in Figure 8.13 suggest, however, that none of these influences are associated with change in sociability during the early phases of adolescence. Only the smaller nonshared influences are implicated in environmental effects on change.

AUTONOMY   Figure 8.12 shows the results of model fitting for autonomy. The results are similar to those for sociability, which would be expected, since autonomy and sociability are moderately correlated ($r = .54$ averaged across both children in each family and across both time periods). The difference from sociability is that shared environment influences change as well as stability from earlier to later adolescence. The stability coefficient is .60 and shared environment accounts for 52 percent of it. Changes in shared environment of the adolescents we studied account for 32 percent of the changes in their autonomy between the earlier and later phases of their adolescence.

With reference to Figure 8.1, we can see that the small amount of nonshared influence apparent at time 1 and at time 2 is involved in change. However, most of the very substantial shared environmental asso-

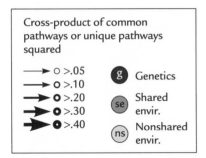

Cross-product of common
pathways or unique pathways
squared

⟶ ○ >.05
⟶ ● >.10
➤ ● >.20
➤ ● >.30
➤ ● >.40

g Genetics

se Shared
   envir.

ns Nonshared
   envir.

Factors accounting for stability from time 1 to time 2

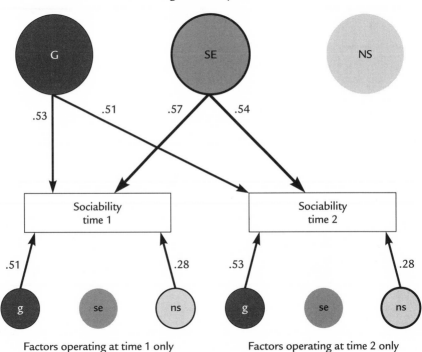

Factors operating at time 1 only   Factors operating at time 2 only

Factors accounting for change from time 1 to time 2

Figure 8.11 Balloon diagram showing results of model testing for factors
influencing stability and change in sociability

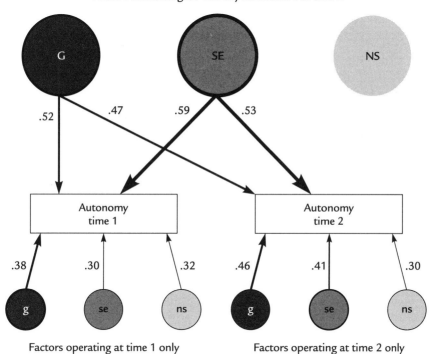

Figure 8.12 Balloon diagram showing results of model testing for factors influencing stability and change in autonomy

ciations with individual differences at time 1 and time 2 constitute stable influences on autonomy across time.

SOCIAL RESPONSIBILITY   Figure 8.13, depicting the results of model fitting, shows a pattern for social responsibility that is similar, in some respects, to that for antisocial behavior, with which it is highly correlated ($r = -.70$ averaged across both children and both time periods). The coefficient of stability for social responsibility from time 1 to time 2 is .71, and genetic factors account for 85 percent of it. Changes in the adolescents' nonshared environment constitute the greatest influence on the small changes in social responsibility between earlier and later adolescence.

GLOBAL SELF-WORTH   As explained in Chapter 5, we measured global self-worth differently from all other assessments of adjustment. We used a single measure from a single source—the adolescents themselves. For other variables we used several measures from several sources. Our approach to measurement of self-worth, in all likelihood, reduced the reliability of our assessment. This will lead to over-estimates of nonshared environment at the expense of the shared environment or genetic factors or both. For these reasons, findings for global self-worth are not strictly comparable to findings from the other six domains of adjustment we measured.

Despite these caveats, the model-fitting results for global self-worth, in contrast to the other domains of adjustment we have just examined, provide an intriguing new pattern (see Figure 8.14). The stability of global self-worth is .34. The stability might be weakened or attenuated by the relative unreliability of our measure. Genetic factors are the preponderant influence on this modest stability. In sharp contrast, changes in nonshared factors account for 79 percent of the change in self-esteem of these adolescents.

With respect to Figure 8.1, these analyses help us recognize that the substantial nonshared influences apparent at time 1 and again at time 2 are almost all involved in change of global self-worth from earlier to later adolescence.

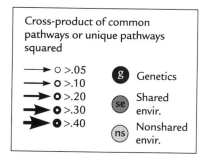

Factors accounting for stability from time 1 to time 2

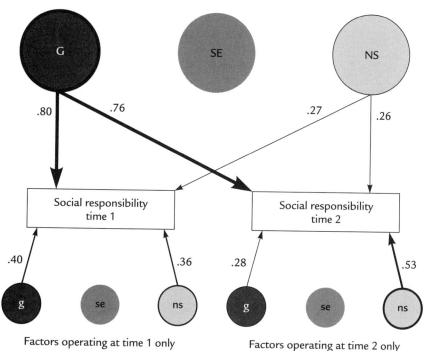

Factors accounting for change from time 1 to time 2

Figure 8.13 Balloon diagram showing results of model testing for factors influencing stability and change in social responsibility

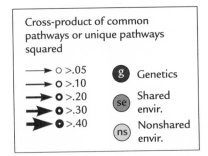

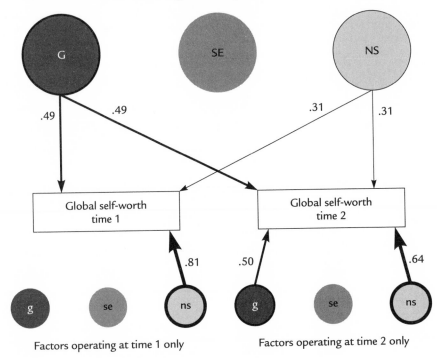

Figure 8.14 Balloon diagram showing results of model testing for factors influencing stability and change in global self-worth

Where to Look for Genetic and Environmental Influences on Adolescent Development

Table 8.1 summarizes our data on genetic and environmental influences on development from early to late adolescence in seven broad domains of adjustment. We can draw three provisional conclusions about influences on development in this period. First, genetic factors are, by far, the most important influences on stability in these domains of adjustment across adolescence. Second, shared environmental factors also play a role in stability, though not as big a role as genetics. Further, changes in shared environments play virtually no role in individual differences in changes in any of the areas of adjustment we have studied. Third, genetic factors are the strongest influence on change from earlier to later adolescence. Fourth, nonshared factors are also important in developmental change but have virtually no role in stability. Let us consider each of these major findings in turn.

Genetic Influences on Stability

The most striking example of the genetic role in continuity across the two time periods is that of cognitive agency, a measure that assesses school

Table 8.1 Genetic and environmental influences on development from earlier to later adolescence

Domain of adjustment	Preponderant influence on stability	Preponderant influence on change
Antisocial behavior	Genetic	Genetic and nonshared
Depressive symptoms	Genetic	Genetic and nonshared
Cognitive agency	Genetic	Genetic and nonshared
Sociability	Genetic and shared	Genetic
Autonomy	Genetic and shared	Genetic
Social responsibility	Genetic	Nonshared
Global self-worth	Genetic	Genetic and nonshared

achievement as indexed by grades as well as child and parent ratings of involvement and success in schoolwork. Our findings here are quite similar to those reported for measures of children's mental abilities (probably strongly correlated with cognitive agency as we define it) in children observed from ages one through nine. Across this span genetic factors accounted for age-to-age continuity, and both genetic and nonshared factors accounted for change from age to age (Cherny and Cardon, 1994). Interestingly, earlier in development—from fourteen to twenty months— shared environment plays a much more prominent role in both stability and change in mental abilities (Plomin et al., 1993). However, in studies of personality change and stability in young adults (McGue, Bacon, and Lykken, 1993) as well as in studies of mental abilities in late life (Plomin, Pedersen, Lichtenstein, and McClearn, 1994), genetic factors are the prevailing determinants of stability across time.

There are three major mechanisms by which genes can account for the stability of adjustment across time. First, genes may be continuously influential—in a biological sense—across a span of development. That is, genes may continue to regulate the neural underpinning of those sensory and motor response styles that, in turn, influence various domains of adjustment. Through this continuous influence on neurobiology they maintain stability of adjustment across time.

Second, genes may have an impact early in development on one or more neurobiological mechanisms. The consequences of these genetic effects may be long-lasting because the genetically influenced neurobiological capacities are stable. This is probably the case with phenylketonuria, for example. High levels of phenylalanine in the blood, the result of one aberrant recessive gene, lead to impaired brain development in young children. The effects of this impairment persist, because the brain cannot recover from this early insult. The active effects of phenylalanine accumulation probably diminish over time since dietary treatment of older children with PKU is ineffective. A reduction in phenylalanine comes too late: the early brain damage endures. In this sense the gene exerts its biological and irreversible influence early: genetic effects seem to persist because their neurobiological consequences do.

Third, genes may influence individual differences in the capacities or talents of children early in life. These differences, in turn, influence which developmental niche adolescents choose for themselves. Environmental factors may then perpetuate genetically initiated membership in different developmental niches (Scarr and McCartney, 1983). One example of this mechanism may be formal and informal ability-tracking in school systems. Genes may have a strong influence on academic competence in the early years. The school environment responds to these differences, so this hypothesis goes, by using them as a basis for assigning children to different classes or enriched educational environments. These classroom environments may, however, "take over" from genetic influences and maintain individual differences long after the genes have ceased to have a discernible biological effect. This would reflect a long-lasting version of developmental evocative gene-environment correlation, discussed in Chapter 4. In effect, environmental mechanisms would splint children into able and deficient lines of development long after any genetic effects had ceased operating.

With respect to cognitive agency, or to any of the seven areas of adjustment we studied, we simply cannot tell from the data which of these three mechanisms accounts for the remarkable role of genetic factors in the stability of individual differences from early to mid-adolescence. There are two reasons for underscoring the possible environmental mechanisms that might "take over" from genetic stratification. First, these mechanisms suggest that environmental factors might mediate genetic effects on stability as well as on change. Second, from a practical point of view, they might suggest ways of offsetting environmental factors that maintain children in the lowest strata of cognitive achievement. The prominence of formal and informal ability groupings in school systems might be just such a mechanism. Children are regularly grouped, either within classes or between classes, according to their intellectual abilities, and these groupings are regularly associated with enhanced academic achievement, self-esteem, and diminished delinquency (Jenkins, 1995; Kelly, 1975; Slavin, 1987; Sorensen and Maureen, 1986).

Yet the scant genetic data available to us suggest that as conspicuous as this practice is in many schools, it is unlikely to serve as a major splint for genetically initiated differences. First, in counties whose school systems assiduously avoid ability groupings there are still major stratifications of children in ability, usually related to the occupational status and educational backgrounds of their parents (Undheim and Nordvik, 1992). Since data suggest that shared environment plays little role in age-to-age stability in this domain, the apparent influences of social factors on stability are likely to reflect genetic influences. Indeed, the more school policies stress equal access for all to the same educational opportunities, the more genetic factors play a role in individual differences in achievement (Heath et al., 1985). Second, in cultures or historical epochs in which intense social stratification, extending into school systems, is rigidly practiced, there is little evidence for genetic influence on individual difference in cognitive ability. In these societies environmental factors can have no role in maintaining genetic differences since the very expression of genetic influences is precluded (Abdel-Rahmin et al., 1990; Heath et al., 1985).

Shared Environmental Influences on Stability

Across all seven domains of adjustment, shared environment accounts for a relatively small amount of either stability or change. For both sociability and autonomy, however, shared environment has a substantial association with stability. There are three possible mechanisms to account for these findings.

First, some aspect of the sibling's shared environment has a similar influence on autonomy or sociability, and such shared environmental factors are relatively constant over time. To be detectable, the variation in these specific shared environmental factors must be important for most of the families in our sample. The shared mechanism might be a social quality of the family: perhaps its level of engagement as a group in social and community activity. For example, some data have associated maternal depression with reduced sociability in very young children (Stein et al., 1991). A variant in this mechanism might be that the shared environment at an earlier time period does not remain constant but leads to or influences the appearance of a successor shared environment. For example, depressed mothers may be less effective than nondepressed mothers in supporting their children in early peer play. Thus, later in development, the child's reduced skills and opportunities for peer interaction might also diminish sociability. This would be seen, in our model testing, as shared environment's accounting for stability over time, since the specific shared environments at times 1 and 2 are so highly correlated.

A second possible mechanism is that shared environments are continuous across time within families but that different shared environments are important for different families. In this case we could not hope to detect these shared environments by systematic study of any sample of families; there is no quantitative procedure that could detect these effects.

A third possible explanation is that sociability or autonomy remains constant across time for nongenetic reasons and, as a consequence, influences some aspect of the environment shared by siblings. That is, the average or combined levels of sociability in a pair of siblings have an effect on family, friends, neighbors, or other aspects of the environment they share. There is some research evidence to support this influence of sociability on a child's social environment (Gifford and Gallagher, 1985; Levitt et al., 1993).

Again, we need more research to weight these three possibilities. The data from our study do provide us with one conclusion: according to the results of analytic models we used, and assuming the validity of their assumptions, it is unlikely that changes in shared environment play any important role in change in children across the adolescent period—even in sociability and autonomy, despite the fact that analysis at two single points in time suggested it might. This rules out a broad range of possible environmental changes during this period and helps focus reformulated theories of development. For example, the following factors are unlikely to have a place in a validated theory of development in adolescence: changes in family economic circumstances, in the mental health of either parent, in the quality of the parental marriage, or in the social networks of the family unit. In other words, changes in these and similar factors cannot influence development in the domains we have studied independent of genetic changes. This conclusion simply restates in terms of developmental change the results of numerous previous studies: shared environment is not an important influence on psychological development in most domains.

Genetic Influences on Change

At the beginning of this chapter we discussed the three mechanisms that might explain findings of a genetic influence on change: intracellular mechanisms, changes in environmental circumstances, and a slow accrual of measurable effects of constant genetic influence. Data from our study cannot weight the likelihood of these three possibilities. But it is worth noting that these effects are pervasive across all domains of adjustment except social responsibility.

Other investigations are now recognizing the importance of genetic change as a major influence on changes in adjustment across time. A previous study, conducted over a much longer period than our own, also found evidence for genetic change in antisocial behavior (Lyons et al., 1995). In early childhood prominent effects of genetic change have been noted for behavioral inhibition, the level of positive or warm feelings, and the development of language competence (Plomin et al., 1993). Substantial influence of genetic change on changes in cognitive abilities has been noted at five separate time points across the age span of one to nine years; indeed, for four of these time points, genetic change was the preponderant influence on change in cognitive abilities (Cherny and Cardon, 1994).

These findings have two important implications for environmental theories of adolescent development. First, studies that have only examined the relationship between environmental change and changes in child and adolescent adjustment, without controlling for genetic factors, may have unintentionally produced misleading results. As will be shown in the next chapter, genetic change can masquerade as environmental change. Second, as noted, genetic change is not necessarily a slow process but may, according to other studies, occur in as short a time as six months (Plomin et al., 1993). As we reformulate our concepts of development, these relatively rapid and substantial effects are a new place to look for the role of the social environment in adolescent development.

Nonshared Influences on Change

Our results point to two important findings about the nonshared environment. The first of these two findings concerns a comparison of nonshared factors with shared and genetic factors. Next to genetic factors, nonshared factors play the most important role in developmental change from earlier to later adolescence. They are far more important than shared factors, which have very little role. This finding replicates those in previous studies of genetic and environmental influence on change (Cherny and Cardon, 1994; Lyons et al., 1995; Neiderhiser and McGuire, 1994; Plomin et al., 1993). The second finding concerns a comparison between the stable and the changing components of nonshared influences observed at a single point of time in earlier or later adolescence. We found that almost all these nonshared influences at any one point in time were likely not to have been operating at the other point in time. That is, the most substantial component of nonshared influences is unstable. This same unstable feature of the nonshared environment was found in all previous studies (Cherny and Cardon, 1994; Lyons et al., 1995; Neiderhiser and McGuire, 1994; Plomin et al., 1993).

There are three possible explanations for this instability of the nonshared environment and its influences on change in adjustment in adolescence. First, it may be attributable to variations in the environment that are salient for all families and differ for siblings in the same family. In Chapter 6, for example, we discussed differences between siblings in their ability to comprehend social frames and reach accurate judgments of the intent of others. We also discussed differential treatment of siblings by their parents. These circumstances tend, however, to be quite stable from earlier to later adolescence. Thus, the only components of circumstances like these that are likely to be influential are those that change over time. These components are small, and thus it is unlikely that they play an important role.

A second possible source of changing nonshared influence is evanescent events that have a strong impact at a single point in time, that are different for sibs in the same family, and whose effects fade fairly rapidly. For example, the accumulation of stressful life events and circumstances may be nonshared and evanescent. Indeed, the correlation from time 1 to time 2 of a score for cumulated, stressful life events for our study is only .21. This is in sharp contrast, for example, to parental warmth and conflict, which show correlations from .60 to .66 from time 1 to time 2.

A third source of nonshared influence, consistent with these findings, is major, life-changing events that may happen at different times or have different impacts for siblings in the same family. For example, one sibling may develop a strong relationship with an influential teacher, or witness a tragic event, or be sexually molested by a relative. The impact of these events on the adolescent could cancel out the effects of other nonshared factors such as differential parental treatment (even if such factors remain constant). The effects of these life-changing events could show up, in our analyses, as new nonshared influences later in adolescence. Recently researchers have shown a strong interest in singular life events—such as illness in the family and rape—that may alter the course of life in adolescence and early adulthood (Cohen et al., 1994; Frazier and Schauben, 1994; Sack et al., 1995; Zall, 1994). Moreover, there are some data on processes such as disruptions in life routines and supportive relationships that may expose adolescents to the full force of these events (Farhood et al., 1993). There is also new interest in social processes, such as resilient and inquisitive families, that may protect children from severe adversity such as severe economic privation (Jarrett, 1995).

It is possible that complex and shifting balances between life-changing and life-stabilizing events constitute an important component of the changing nonshared environment in early adolescence. It seems unlikely, however, that the impact of major life-changing events can account for a major portion of these substantial nonshared influences on change in adjustment. The emerging data in genetic and environmental influences on change suggest that nonshared factors are important engines of change across broad spans of development from early childhood on, but it seems unlikely, except in the most dramatic or tragic of lives, that life-changing events would occur repeatedly throughout the early life course. We return to a more complete consideration of nonshared influences in Chapter 13, after we have reviewed more data on such factors in the chapters that follow.

Chapter 9

ANTITHESIS II: INFLUENCES ON STABILITY
AND CHANGE IN ADOLESCENTS' FAMILIES

In Chapters 6 and 7 we stated our thesis about the nonshared environment. In support of the thesis we presented a review of other research on the topic, some additional research to help us understand how these unique environments may exert their influences, and data showing very strong associations between differential parenting and differential sibling relationships, on the one hand, and adolescent adjustment, on the other. Chapter 8 appeared to present a challenge to our thesis and perhaps to other environmental theories about adolescent development as well. First, we showed that for virtually all domains of adolescent adjustment genetics played a major role in accounting for individual differences. This was particularly true for antisocial behavior, cognitive agency, and social responsibility. Genetic influences were also important for depression, sociability, and autonomous functioning. They seemed less important for the adolescent's self-worth, although by time 2 genetic factors accounted for nearly 50 percent of the differences among adolescents on this measure. Second, we showed that genetics played a preponderant role not only in the stability of these domains of adjustment across time but also, in many cases, in their change. In other words, these data suggested that genetic factors play a surprisingly important role in the two fundamental processes of development: those that maintain differences among individuals across time and those that influence differences in how individuals change across time.

In this chapter we present data that appear to challenge further conventional environmental theories of adolescent development. We show that genetic factors not only play a major role in individual differences in adolescent adjustment, but also have a substantial effect on most but not all subsystems within the family. We examine the following four subsystems, in particular: marital, father-child, mother-child, and sibling. First we review how the adolescent's genes might influence qualities of relationships in the family subsystem. Second, we present data from our entire sample at time 1 and from the reduced sample that was eligible at time 2. In addition to offering a glimpse at developmental processes, these data provide two cross-sectional pictures in order to note which findings appear consistent across time and which do not. Third, we present the longitudinal analyses that we employed to explore the role of genetic and environmental factors in producing both stability and change.

The Concept of Genetic and Environmental Influence on the Family

Analytic Principles

The same analytic principles for detecting genetic and environmental influences on measures of adjustment hold for detecting these influences on measures of the adolescent's social environment. From a computational point of view, we simply treat any measure of the adolescents' social environment as if it were a characteristic or assessment of adjustment of the adolescents themselves. These analyses provide estimates of the genetic influences on the relationships in social environments. They also provide estimates of the shared and nonshared environmental influences on such relationships. The results of these analyses are not straightforward, however, because in all cases we are talking about the adolescents' genes and the adolescents' shared and nonshared environments. How are we to understand these circumstances as influences on parents' or on siblings' behavior toward the adolescent?

Adolescents' Genes and Family Relationships

In Chapter 4 we reviewed how genetic factors might influence the quality of relationships in social systems. Our focus was on parental and sibling relationships. To recapitulate briefly, we identified two basic mechanisms. First, heritable characteristics of adolescents might influence or shape how they behave toward significant others, which, in turn, might influence the response of others. Like other researchers in this field (Plomin et al., 1977; Scarr and McCartney, 1983), we distinguished two forms this influence might take. Evocative mechanisms result from a heritable quality of the adolescents, such as temperament. This quality challenges or elicits responses; little choice, initiative, or agency is attributable to the adolescents themselves. Active mechanisms reflect greater initiative on the adolescent's part; for example, genetic influences on the adolescent's personality or values may, in turn, influence interpersonal choices and strategies of the adolescent, which then shape the nature of his or her relationships. In fact, the differences between evocative and active genetic influence on relationships are not sharp, and each may be thought of as a facet of the same process.

A second mechanism by which genes and relationships are linked is called a "passive" mechanism. Here, the mediating link between an adolescent's genes and the relationship quality is biological and not psychosocial. Adolescents share exactly 50 percent of their genes with each parent and approximately 50 percent with each of their full siblings. Ordinarily, passive genetic influences on family systems are thought to affect only relationships involving blood relatives, since it is only with these relatives that adolescents share genes. Indeed, as Chapter 4 pointed out, adoption designs can distinguish passive from evocative and active genetic effects because adopted children and adoptive parents do not share any genes. In the next chapter we will consider ways in which passive mechanisms may extend beyond the family. Here we focus primarily on the family itself.

Genetic influence on family systems is not necessarily restricted to subsystems in which the adolescent is a direct and continuous participant, such as the sibling subsystem or the parent-child subsystem. It is possible that genetic influences can extend more remotely, for example, to the marital relationship. Irritability in children, for instance, might be associated with marital conflict for genetic reasons. Irritable children could evoke marital conflict in their parents, or, by a passive mechanism, shared genes between the child and the parents may influence irritable behavior in the child and negative patterns of interaction between the parents. In our study we assess these remote effects in a restricted way, focusing on marital conflict. A general measure of marital conflict cannot be included since, by definition, it is the same for each sibling. However, we use a measure of the level of marital conflict about a specific child; this can vary between siblings and thus lends itself to analysis of remote genetic and environmental influences. A broader investigation of remote effects is possible in adoption designs as well as in twin and sibling designs of older siblings who have married and had their own children.

The Shared Environment and Family Relationships

As in analyses of measures of the sibling's adjustment, the shared environmental influences on measures of the adolescent's social environment refer to experiences shared by the adolescent siblings. These shared experiences, however, are now influencing not siblings' own adjustment but measures of relationships in their families. The shared environment refers to a complete, anonymous set of environmental factors that influence the behavior of others toward the siblings and which they share by virtue of being siblings. Like genetic influences, these shared environmental influences may have remote effects by influencing relationships in which the adolescents are not direct participants. These shared environmental influences, both immediate and remote, fall into three broad classes.

First are factors that jointly influence the siblings themselves as well as others in the family, including any common factors in their neighborhood, their culture, or their religion that might lead them to be treated similarly by their parents or siblings. Such factors are likely to vary from family to family in an ordinary research sample and are unrelated to siblings' genetic similarity.

A second broad class of shared variables influencing the family environment consists of the nongenetic effects of the similarity between siblings. These include any of the siblings' nonheritable characteristics that would evoke or alter the behavior of others toward them. As reviewed in Chapter 3, most studies have found very low correlations on any measures of siblings' characteristics that are not attributable to shared genes. It is for this reason that genetically unrelated siblings growing up in the same family are often no more alike than children matched at random.

The best possible estimate of correlations of sibling characteristics, independent of genetics, is an estimate of shared environment in a twin-sibling design of the kind used here. Our own findings on this matter suggest a potential and substantial role for these nongenetic similarities between siblings and, as a consequence, are at variance with prevailing trends in other studies. Figure 9.1 recasts the data first displayed in Figure 8.1. It shows our own best estimates of the balance between nongenetic shared and nongenetic nonshared environmental influences on adolescent adjustment across seven broad domains of adjustment. In accordance with previous findings, shared environmental factors play only a trivial role in depression, cognitive agency, social responsibility, and global self-worth, and have a modest role in antisocial behavior. Surprisingly, as

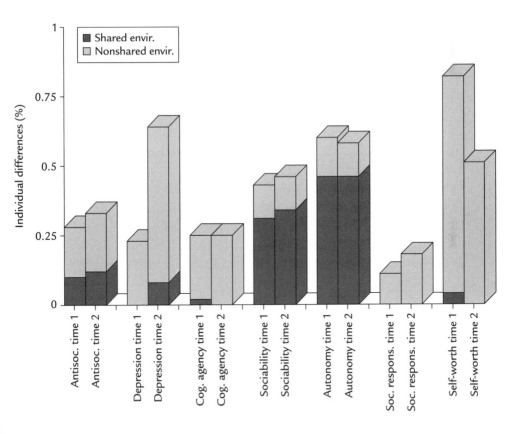

Figure 9.1 Shared and nonshared components of adjustment measures

noted, the shared environment is the prevailing environmental influence on sociability and autonomy. Thus, this second mechanism of shared influence—effects of correlated characteristics of siblings—on family relationships remains plausible.

A third mechanism of shared environmental influence on family relationships arises from stable characteristics of family members that shape their relationships with adolescent children in ways that are consistent across siblings. These characteristics fall into two related subclasses. First are features of the parent such as depression or paranoid attitudes. These may account for variation among parents in different families in their relationships with their children. They may also account for differences between parents in their relationships to their children but within any family lead to consistency of a particular parent's response, attitude, and interaction among all children. It is interesting that where these characteristics of parents are influenced by genetic factors they will show up in our genetic analyses of family relationships partially as shared environmental effects and partially as genetic effects, depending on the extent to which parents and children share the relevant genes.

A second subclass of these stable characteristics consists of perceptual biases of parents, siblings, or other family members. These are most important when measures of relationships rely heavily on the self-report of family members. Data presented in the previous chapter, for example, suggest that parents value equity and fairness in their treatment of their children. Thus, they report treating them quite similarly. A bias or value of this kind will be manifest as a shared environmental influence on measures of parenting that rely, at least in part, on parents' self-reports. There is a corresponding bias possible in siblings' perceptions. If siblings perceive their relationship as reciprocal, whether or not this can be confirmed from other sources of information about their relationship, this too will show up as a shared environmental influence on sibling behavior.

The Nonshared Environment and Family Relationships

Nonshared environmental influences, in our analyses, refer to nongenetic influences on parenting and sib behavior that are different for siblings in the same family. The meaning of the word "different" is quite precise here. It refers to the fact that the relationships of one sibling in a family *cannot be predicted* from the relationships of the other sibling. Inaccuracies or low reliability in the measurements themselves may contribute to this unpredictability in our analyses. We have kept this source of non-shared influence to a minimum by using highly reliable measures. Like shared environmental influences, substantive sources of nonshared environmental influences fall into three broad categories.

The first category consists of factors that jointly influence differences between siblings as well as differences in how they are treated by others in the family. Examples and circumstances described extensively in Chapter 6 can be recast here to show how nonshared factors associated with the adolescents' siblings might shape relationships in their families. Consider again the example of parental bereavement and its potential relationship to schizophrenia (Summers and Walsh, 1980; Walsh, 1978). When a child is born into the family just after the death of a significant adult (in the example we are considering it was a grandparent), both the child and the parents' relationship with the child may be profoundly affected. In contrast to an older child in the family, or to a younger one as yet unborn, the newborn may be experienced as—in part—a restitution for the recent loss and thus bound more closely to parents' feelings, expectations, and disappointments. The death itself is a factor that resides neither in the parent nor in the child but influences both in the forging of a distinctive parent-child relationship.

The second class of nonshared influences on family relationships consists of those arising from differences between siblings that are not attributable to genetic causes. Many nongenetic factors—inside the family and without—may make siblings differ from one another: different sets of peer groups, classroom experiences, intimate heterosexual or homosexual experiences, and stressful life events. These differences, in turn, can influence such sibling characteristics as mood states, personality, and interpersonal skills, all of which may have a secondary impact on the siblings' relationship within the family.

A third class of nonshared influences on family relationships are stable characteristics of family members that lead them to behave differently or report differences in their behavior toward the two siblings. In Chapter 7 we encountered one example of a factor of this kind: the tendency of the siblings themselves to perceive differences in how they are treated by their parents. This perceptual "bias" in adolescents will show up as a nonshared influence on those measures of parent-child relationships that are heavily weighted by the adolescents' self-reports.

Genetic and Environmental Influences on Relationships in Families at Time 1 and Time 2

The statistical models applied here are the same as those we used in the previous chapters. Before presenting their results, we provide a glimpse, in Figure 9.2, of exemplars of the basic computations for these models: the correlations between siblings within each type. We show the data for mother's and father's conflict/negativity toward the adolescent sibling. Here, and for all other data reported in this and subsequent chapters, we

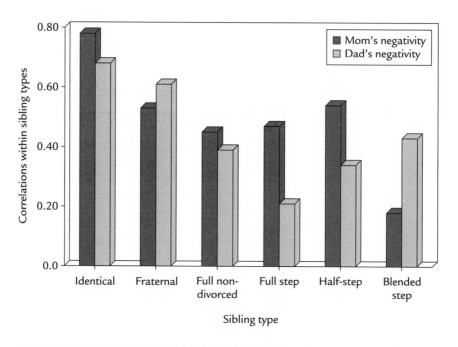

Figure 9.2 Intraclass correlations for estimating the role of genetic and environmental influences on measures of mother's and father's negativity

use scores composed of parent, child, and observer reports combined. In Figure 9.2 we can read correlations in the same way we would if the primary data were measures of the adolescents' adjustment. For example, both mother's and father's data show substantial correlations for all sibling types; most notable are substantial correlations for the genetically unrelated blended sibs. This suggests that shared environmental influences are important for negativity in both mother-adolescent and father-adolescent subsystems. Part of the shared environmental "influence" here is almost certainly the parents' tendency to see themselves as consistent in their behavior toward both adolescent sibs. For this and other environmental measures, however, there is a clear indication of genetic influences as well. Particularly for mother's data there is a clear "genetic cascade," with the exception of data for half-sibs, who show greater correlations than those derived from genetic expectations. The pattern of father's correlations also suggests some genetic influence: substantial discrepancies between identical twins and the three sets of full sibs, fraternal twins, and full sibs in nondivorced families and in step-families. But both the half-sibs and the blended sibs are higher than genetic expectations.

Figure 9.3 presents the full array of analyses for all 5 measures of mothering employed in this study using the full sample at time 1 and the sample eligible for testing at time 2. Recall that the time 1 sample consisted of 720 families, minus 12 with uncertain zygosity, whereas the time 2 sample contains 91 percent of those time 1 families eligible to participate (two children still living at home): 395. In all analyses reported in this book, we checked to see if the subsample of time 1 families that we would see again at time 2 yielded results different from the full sample of 708 time 1 families in comparisons at time 1. In this and other cases there was no difference in results. Hence these results can, with great caution, be read as indexes of genetic influences on family relationships in earlier and later adolescence. Figure 9.4 presents similar data for fathering.

The overall patterns in these findings are quite interesting. First, as anticipated from the pattern of correlations in Figure 9.3, genetic factors in the adolescent have a substantial influence on mother's negativity. This strong effect remains at time 2. Indeed, the influence of genetic factors in the adolescent on mother's negativity is about equal in magnitude to the influence of these genetic factors on the adolescent's own sociability

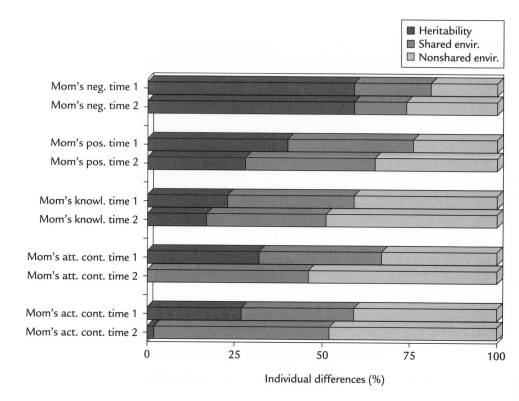

Figure 9.3 Heritability and environmentality of mothering

and exceeds the genetic effects on the adolescent's autonomy, global self-worth, and depression at time 2. In other words, this more indirect effect of adolescent genetic factors on mother's behavior toward the adolescent siblings would be a relatively strong effect if it were some assessment of adjustment of the adolescents themselves. Again, as anticipated, the genetic effects on father's negativity are substantial. The pattern and magnitude of the findings for fathers are quite similar to those for mothers.

Examining the results from both parents, we note two other major features. First, the genetic effects on measures of positivity and monitoring/control are, overall, somewhat less than those for negativity. Second, there are some intriguing changes from time 1 to time 2. Particularly notable are the declines in the genetic influences on mothers' attempts to control and success in controlling their adolescents' behavior from time 1 to time 2. There is a corresponding increase in genetic influence on fathers' knowledge of and attempts to control their adolescents' behavior from time 1 to time 2. It is tempting to speculate that as children get older, fathers replace mothers in responding to the same genetically influenced behavior that requires monitoring and control in the adolescent. That is, genetic factors influence some limit-breaking behavior in adolescents that is responded to with mothers' attempts at and success in control early in adolescence and by monitoring and attempts to control by fathers later in adolescence.

This might be an intriguing first report of a gene or set of genes associated with one relationship earlier in development, here the mother-adolescent relationship, and a different one later, in this case the father-child relationship. Unfortunately, the data do not bear this out. The relation-

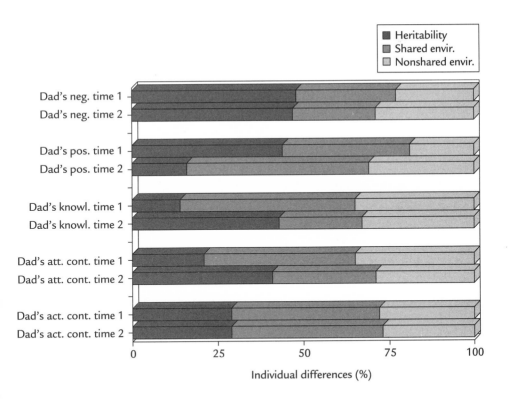

Figure 9.4 Heritability and environmentality of fathering

ship between mother's attempts and successes at control at time 1 has weak or trivial associations with father's knowledge at time 2. (The correlation coefficients are $r = .05$ and $.17$, respectively.) Mother's attempts and success at control are somewhat more highly correlated with father's attempts at control at time 2 ($r = .22$, $p < .0001$ and $.24$, $p < .0001$). As we will describe in Chapter 12, however, these associations between mother's earlier control and father's later control are due to the environment shared by siblings across time and are not at all attributable to genetic factors that influence both mother's and father's response. Thus the genetic factors that influence mothers to both attempt and succeed at control at time 1 are different factors from those that influence fathers to monitor and try to control their adolescents at time 2.

One other change is notable: genetic factors in the adolescent have less influence on father's positivity at time 2 than they did at time 1.

Figure 9.5 presents data on sibling positivity and negativity. Here the results are quite different. Genetic influence on how siblings behave toward each other is small. The preponderant influence here is shared environment, reflecting the high degree of reciprocity in sibling relationships in adolescence, at least when the siblings are the same gender, identical in age, as in twins, or separated by less than three years, as in the rest of the sample. That is, if one sibling acts in a positive manner toward the other, the other reciprocates, and the same is true for negativity. The reciprocity, these findings tell us, is entirely unrelated to the degree of genetic similarity between the siblings. At first glance a finding like this seems surprising in view of the common lore that twins, particularly identical twins, are united by special bonds of closeness and understanding. Indeed, clinical reports suggest that a special closeness among twins makes it difficult for them to separate as they mature (Adelman and Siemon, 1986). Quantitative research tends to support these observations. Twins, particularly MZ twins, do report greater feelings of closeness for each other (Paluszny, 1977; Tambs et al., 1985) than do nontwin siblings and show more severe grief reactions when one of them dies (Segal et al., 1995). The comparisons reported here, however, refer to *reciprocity* rather than to *absolute levels of closeness*. Genetic analyses neglect mean differences among the six different groups of siblings. As mentioned, they focus on how predictable the measures of one sibling are from the measures of another.

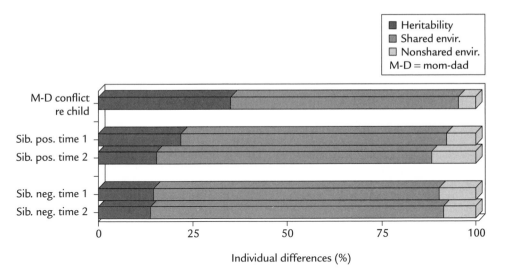

Figure 9.5 Heritability of sibling relationships and marital conflict measures

When we compare average levels of closeness across the six groups of siblings, we find that the data confirm previous reports. MZ twins, and to a lesser extent DZ twins, are more positive toward each other than are other sibling pairs. MZ twins are also less negative than most other groups of siblings. An interesting exception is the group of unrelated siblings in step-families. These siblings are even less negative than are MZ twins, but they are also among the lowest in positivity, suggesting that—compared with other sibling groups—they are disengaged from each other. When we combine both sibling positivity and sibling negativity in a single analysis, however, the overwhelming differences among sibling pairs are within each of the six groups rather than among the six groups; only 11 percent of the differences among all sibling pairs can be accounted for by sibling type. Therefore, we can conclude that differences in the shared environment among families within all six types of siblings are by far the strongest influence on both sibling positivity and sibling negativity.

We turn finally toward our first attempt to examine remote genetic effects. Here we examine a measure of the degree of conflict each parent reports about circumstances relating to each child. The measure pools the mother's report and the father's report on these conflicts. We did not obtain reports from the children and did not code this aspect of marital strife in our videotapes. Figure 9.6 reports the simple correlations for each sibling subgroup for this measure. It is apparent that all correlations are quite high, suggesting that shared environmental factors are the most influential factors. In other words, parents report a good deal of consistency in their fights with both siblings. Nonetheless, there is also a distinct genetic cascade—although a shallow one—implying some remote genetic influence here. Returning to Figure 9.5, we can see results from model testing that bear out these expectations. There is a notable remote genetic effect, but the preponderant influence is shared environmental. Again, we suspect that parents are biased in favor of seeing themselves as equally harmonious (or contentious) about both of their teenage children and that this bias is a major contributor to the shared environmental ef-

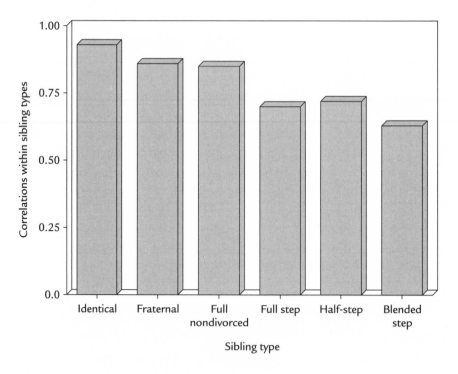

Figure 9.6 Intraclass correlations among sibling groups on mother-father conflict regarding children

fects here. Biased or not, it is notable that this remote genetic influence is equal to or greater in magnitude than the more direct effects of adolescents' genes on parent-child relationships in twelve of twenty analyses shown in Figures 9.3 and 9.4.

Genetic and Environmental Influences on Stability and Change in Family Relationships

Concepts

Chapter 8 provided an extensive introduction to the logic and analyses of genetic and environmental factors that influence stability and change across development. When we use measures of relationships in adolescents' families, we use the same reasoning as when we employ measures of adolescent adjustment. Just as in the case of individual differences, both genetic and environmental factors may influence stability and also change. The reasons or mechanisms that might account for these influences are parallel to those described in detail in the previous chapter. They need to be recast, however, when our analyses focus on relationships as opposed to measures of adjustment.

GENETIC INFLUENCES   In considering genetic influences on the stability of family relationships, we must repeat a point emphasized earlier in the chapter: our design does not permit us to distinguish whose genes are influencing our analyses. First, the adolescents' genes could be the only influence at work; in that case heritable characteristics of the adolescents are evoking parental and sibling responses, or the adolescents are actively shaping these relationships in response to genetic influences on their own perceptions and behaviors. Second, the parents' genes could be the sole influence. Because adolescents share exactly 50 percent of their parents' genes, their genes serve only as a surrogate for their parents': we know something about the parents' genes by examining a group of individuals who have many of the same genes themselves. Or, third, the results could reflect the influence of both parents' and adolescents' genes.

Whether the parents' genes or the adolescents' genes are at work here, the same three mechanisms could account for their stability as account for stability in measures of adolescent adjustment. First, genetic factors could be continuously active across the time period studied (in either parents, adolescents, or both). Second, neurobiological factors, initially influenced by genetic differences among parents or adolescents, may continue to be influential even if the genes themselves have "turned off." Third, the combined effect of many factors nested within environmental niches may perpetuate stable relationships across time. For example, in early childhood genetic factors may influence a child to be unruly, disrespectful of others, and quick to anger. The child develops a reputation as a "bad actor" at home, in the neighborhood, at school, and among peers. Parents and others might respond to the child as a "bad actor," and such responses might in turn perpetuate unruly behavior, long after the genetic influences cease to operate.

The mechanisms for genetic change are more elusive when we consider genetic influence on relationships rather than on individual adolescent adjustment. Indeed, the simplest approach is to consider that such mechanisms might arise from evocative or active gene-environment correlations. The influence of genetic factors changes in adolescents and, *as a consequence,* their parents and siblings treat them differently. To be precise, because of change in genetic influence in the adolescents, the treatment they receive from parents and sibs at time 2 is not predictable from treatment they received from these same family members at time 1.

Genetic influences on change may also be due to passive mechanisms, but these are much harder to conceptualize. Passive gene-environment correlations are possible only because parents and children share 50 percent of their genes. It is very plausible, however, that the genes that influence family relationships in either parent change their effects in different ways and in different amounts at different times during development.

ENVIRONMENTAL INFLUENCES Earlier in this chapter we reviewed three mechanisms that could account for shared environmental influences on family relationships and three that could be responsible for nonshared influences. There is no evidence collected prior to our study to help us determine whether all these mechanisms are equally likely to lead to stability or change across time. From a logical point of view, however, we would expect generation-based rating biases, the bias toward similarity in parents and that toward differences in adolescents, not to change much over three years. Data presented in Figures 7.1, 7.2, and 7.3 would appear to confirm this. There seems to be little change from time 1 to time 2, for example, in the parents' tendency to rate their behavior as consistent across children, in contrast to their rating of the quality of their children's peer groups, which shows much less consistency across their two children at time 1 and at time 2.

As mentioned, shared and nonshared influences that operate on all members of the family, in addition to the effects of nongenetic similarities and nongenetic differences between siblings on their families, might influence parental and sibling behavior toward the adolescents. It is plausible that changes in either or both of these domains could account for the role of shared environment or nonshared environment in changes in the quality of relationships in the parent-child and sibling subsystems of the family.

Genetic and Environmental Influences on Stability and Change in Parenting, Sibling Relationships, and Marital Conflict

When we presented data on stability and change of measures of adolescent adjustment in the last chapter, we had only seven measures to present. A detailed reading of each of seven "balloon diagrams" helped highlight the patterns of findings for each of them. Here, we have twelve separate measures of parenting and sibling relationships. Moreover, these measures are naturally grouped in pairs: mother's and father's conflict/negativity; mother's and father's warmth/support; and so on. Overall patterns in the data, then, are best represented schematically.

As a first step, we have represented in Figure 9.7 the data for change and stability in mother's conflict/negativity by using a balloon diagram according to the conventions of Chapter 8. The stability coefficient is high, .66. Note in Figure 9.7 that genetic factors account for most of this stability (.69 × .67 = .46; .46/.69 = 67%). Almost all the remaining stability is accounted for by shared environment. The coefficients shown in Figure 9.7 won't add up to the stability or instability coefficients because we have not included paths that were below statistical significance. Figure

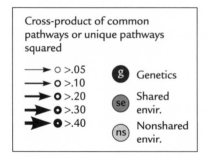

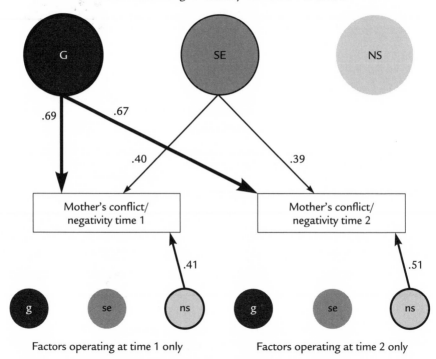

Figure 9.7   Balloon diagram showing results of model testing for factors influencing stability and change in mother's conflict and negativity

9.7 also shows that the main contribution to the coefficient of change (1 − .69 = .31) is nonshared environment; it accounts for 68 percent of the instability or change ($.41^2 + .51^2 = .21; .21/.31 = 68\%$).

Figure 9.8 displays these same data as a stacked bar graph. Consider the top-most bar, which represents genetic and environmental influence on stability and change in maternal conflict/negativity. The vertical arrow represents the balance of stability and instability from time 1 to time 2. If we were to draw an imaginary line down to the horizontal axis of the graph, it would intersect at .69, the stability coefficient. Thus, everything to the left of the arrow constitutes the relative weights of genetic, shared, and nonshared environmental influences on the stability of mother's con-

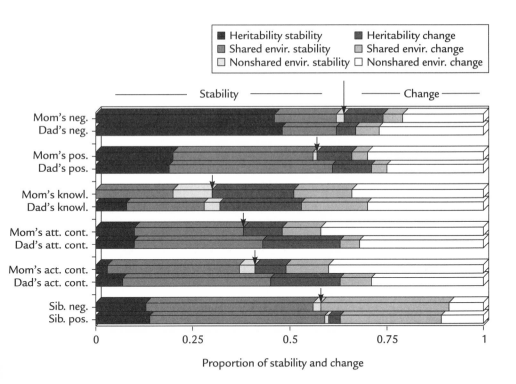

Figure 9.8 The proportion of stability and change in measures of parenting and sibling relationships accounted for by genetic, shared, and nonshared environmental factors

flict/negativity; everything to the right constitutes the relative weights of these three classes of variables on change in mother's conflict/negativity.

Now let us go to the balloon diagram in Figure 9.7 from the top-most stacked bar in Figure 9.8. We can compute the heritability of stability by multiplying the two paths from the common G in Figure 9.8 (.69 × .67 = .46). As indicated in the legend, the black portion of the top-most bar represents the heritability of stability in mother's conflict/negativity. An imaginary line down toward the horizontal axis from the right border of this black zone of the top-most bar will intersect with .46. Comparable computations can be done to estimate the components of influence for change. Here we square the path coefficients, paired within domain, that is, the two components for unique heritability, for unique shared environment, and for unique nonshared environment. Then we obtain an average of these two squares. Recall that these estimates of classes of variables acting uniquely at time 1 or at time 2 are entered on the arrows from the smaller "balloons" underneath the variable being analyzed. The only significant pair of unique influences for mother's conflict/negativity is for nonshared influences where the influences are entirely on change ($.41^2 +$ $.51^2)/ 2 = .21$. The lightest gray zone at the far right of the top-most bar represents these influences, and an imaginary line down to the horizontal axis will intersect with $1 - .21 = .79$.

We can now review the findings on stability and change across all measures of family relationships. First, let us examine parental conflict/ negativity ("neg." in Figure 9.8) and warmth/support ("pos."). Figure 9.8 tells us that these measures are quite stable across time. The coefficient for stability for mother's conflict/negativity is .66, and for father's it is .62. A comparison of these values with the stability of measures of adolescent adjustment is instructive. Figure 9.9 presents coefficients of stability for all seven measures of adolescent adjustment and all twelve measures of family relationships. Mother's, father's, and sibling's positivity and negativity are all highly stable across three years and are comparable to the high stability of adolescent antisocial behavior, depression, sociability, autonomy, and social responsibility. The measures of control strategies for mothers and fathers are lower and are comparable to the stability of adolescent self-worth.

These data convey a picture of considerable stability across time in

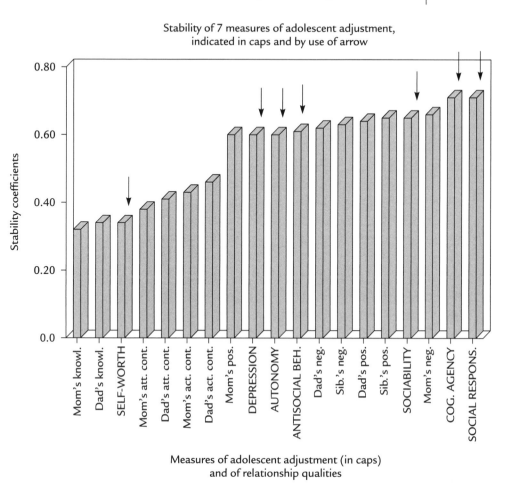

Figure 9.9 Coefficients of stability for all measures of family relationships and all measures of adolescent adjustment

both the adolescents and their families except in the area of control strategies. As noted in Chapter 2, longitudinal studies are most effective in delineating causal sequences among variables during periods of development when there is at least a moderate amount of change so that we can assess the relationship of changes among variables. Figure 9.9 makes it clear that we are studying a period characterized more by stability than by change in both adolescent adjustment and family relationships. Thus, we have a relatively narrow window through which to peer at change processes.

Let us return to Figure 9.8. It is striking, as in the case of so many measures of adolescent adjustment, that most of the stability of both the mother's conflict/negativity (which we saw in the balloons of Figure 9.8) and the father's conflict/negativity is influenced by genetic factors. For fathers, 72 percent of this stability is attributable to genetic factors. In contrast with most of our measures of individual adjustment, however, a very small (and statistically insignificant) portion of change is influenced by genetic factors. In fact, the most important source of change is in factors associated with differences between the siblings—the nonshared components. For fathers, these anonymous nonshared factors account for 71 percent of change in conflict/negativity across three years. Another way of thinking about these findings requires a comparison of Figure 9.8 with Figures 9.3 and 9.4. The latter two show that the shared and nonshared environments contribute about equally to differences among families in parent conflict/negativity. According to Figure 9.8, most of these shared environmental factors are, in effect, stable across three years, whereas most of the nonshared factors change. We examined the pattern of unique variances at times 1 and 2 for nonshared factors, the primary component of change. There was an even balance between factors operating at time 1 but no longer operating at time 2, and the reverse: factors operating at time 2 but not at time 1.

A somewhat different pattern emerges for parental positivity. The shared environment accounts for a much greater portion of the stability across time, and genetic factors account for comparatively little: only 35 percent for mothers and 31 percent for fathers. As with parental conflict/negativity, however, nonshared factors account for most of the change. Again, there was an even balance of unique influences at times 1 and 2. As was the case for parental negativity and parental positivity, the unique nonshared factors at times 1 and 2 were about equal.

A third pattern of results emerges for control strategies, knowledge about or surveillance of children, attempts to control and actual success in control ("att" means "attempted control" and "act" means "actual control"). Here the primary role of genetic factors is to influence change across time rather than to influence stability, with the shared environment playing an almost exclusive role in influencing stability, and nonshared factors, once again, playing the dominant role in change. This distinctive role for genetic factors in change rather than stability in parental monitoring and control is part of an increasingly intriguing pattern of results relating to this aspect of parenting in adolescence; we provide more complete analysis and interpretation of these findings in Chapter 12 as part of our synthesis. Again, in this domain, the unique nonshared factors at times 1 and 2 were about equal. This was, however, not the case for genetic factors. For mom's attempted and successful control, some genetic factors operating at time 1 did not operate at time 2, but no new ones came to take their place. For dad's knowledge and attempts there was only a small unique effect at time 1; by time 2 a substantial amount of new genetic influence was operative.

Yet a fourth pattern characterizes sibling positivity and negativity. These relationship qualities are very stable across time; they show approximately the same stability as positivity and negativity in parent-child relationships. In contrast to the other three patterns of findings summarized above, however, shared environment plays the dominant role in both stability and change across the three-year period of study. The unique shared environmental influences were about equal between times 1 and 2.

To summarize, nonshared factors play the preponderant role in influencing change in family relationships across time. Genetic factors play a more modest role, particularly for control strategies. Only for siblings do shared environmental factors play a role in change. In almost all instances of change, shared and nonshared environmental factors that are influential in early adolescence lose their importance by mid-adolescence, when they are replaced by other, uncorrelated nonshared environmental factors. The picture looks different for genetic factors. In many instances, early influences seem to fade away without being replaced by new genetic influences, or new genetic influences arise without earlier ones disappearing.

Summary

Perhaps the most important implication of the data presented in this chapter is that genetic influences are notable, and in some cases quite strong, on measures of family processes. These findings move us one step further to a severe challenge of our thesis. If genetic factors influence measures of the family environment *and* measures of adjustment, then they might account for many associations between the two. In other words, genetic factors might be important in relationships that have, hitherto, been thought to reflect only psychosocial processes. We examine the role of genetic factors in explaining these associations in the next chapter. But before we move on to this crucial analysis of overlapping factors, we need to pause, briefly, to see what we have learned in this analysis.

The first notable lesson is that genetic factors play a substantial role in differences among families in parent-child relationships and marital conflict about children. Second, in contrast to genetic influences on most domains of the adolescent's own adjustment, the shared environment also plays a notable role in influencing all family systems. Although parents' desire to appear consistent with their children must be one influence on this effect of the shared environment, other explanations are also plausible. Factors such as community, social class, and subcultural identity may be factors common to the siblings and may also influence parenting practices. Our data make clear that these shared influences, whatever they are, remain very stable across a three-year span from early to late adolescence.

Third, the pattern of genetic and environmental influences for sibling relationships is quite different from genetic influences on parent-child relationships and on the marital relationships of the parents. Shared environment plays the main role in explaining differences among sibships in their warmth and support as well as in their conflict and negativity at earlier and later adolescence. Indeed, the role of shared environment in sibling relationships is nearly twice the size of its role in parent-child relationships and marital conflict. The contrast between parent-child and sibling relationships will become a major theme later in this book. Here we can already draw some inferences of importance for understanding family systems and thus begin to pave the way for the third section: the synthesis.

Recall that there are three possibilities for explaining the important role of the shared environment in family relations. First, factors external to the family but common to its members—such as a neighborhood or social class—might be the primary influence on the relationships. If factors of this kind are operating, they must affect siblings primarily, since shared environmental factors have such a conspicuous role in their relationships. Thus, *adolescent-specific* shared environments such as schools and peers are likely sources.

It is also possible to think of parental relationships with the siblings as an "external" source as well. Many examples from previous research connect the level of parent-child conflict, for example, to conflict between siblings (Brody and Stoneman, 1994; Dunn and Kendrick, 1982; Dunn and Munn, 1987; Volling and Belsky, 1992). Indeed, data from our own study show a strong link between the quality of parent-child relationships and sibling relationships. For example, the negativity in dad's relationship with his children at time 1 showed a significant correlation with both the positivity of sibling relationships ($r = -.26$) and the negativity of such relationships ($r = .40$). Dad's positivity with his children was also correlated with sibling positivity ($r = .34$). A similar pattern of findings was noted for mother's relationships with her children.

A second possibility is that sibling similarities may be jointly and equally influencing the tenor of the sibling relationships. This must, of course, be equally true across all six sibling groups for shared environment to have such an overwhelming influence on their relationships. This alternative seems less likely than the first, since there are few measures in which adolescent siblings are similar for nongenetic reasons. In our data autonomy and sociability are examples, but with the exception of delinquency and heavy drinking there are few others in the literature. The phenotypic correlations between autonomy and sociability, on the one hand, and sibling warmth and support, on the other, are modest but significant statistically ($r = .23$ and .30, respectively, at time 1, and .28 and .24 at time 2). Antisocial behavior also shows some sibling similarity independent of genetic influences. For example, the correlation between blended sibs is .27 at time 1. Moreover, the phenotypic correlations between antisocial behavior and both sibling warmth and sibling conflict are considerable at time 1 ($r = -.37$ and .45, respectively) and at time 2 ($r = -.38$ and .42).

A third possibility, consistent with the second, is that sibling systems are highly reciprocal systems where aggressive or affectionate behavior in one sib is highly likely to lead to reciprocation from the other. This would be a third explanation for the enormous role of shared environmental influences on the quality of sibling relationships. This reciprocity might override any genetic influences that might play a major role in these relationships and that play such a central role in parent-child relationships.

Indeed, all three of these proposed mechanisms might work in tandem. In Chapter 13 we return to this quality of sibling relationships as a central feature of our synthesis, exploring genetic and family relationships data that etch an intriguing portrait of life in the family of the adolescent.

Chapter 10

ANTITHESIS III: LINKING FAMILY RELATIONSHIPS AND ADOLESCENT DEVELOPMENT

After discussing the logic and design that underlie our study, we presented three major lines of evidence. First, we presented data on the associations between the characteristics of family subsystems and assessments of seven domains of adolescent adjustment. We used a special feature of our design to clarify the effects of subsystems in which the adolescents were direct participants, such as their relationships with each parent. We distinguished these from subsystems in which the adolescents were not members, such as the relationships of the adolescents' siblings with father and with mother. In our analyses of family relationships and adolescent adjustment, we found substantial associations between measures of both the adolescent-parent relationship and the adolescent-sibling relationship, on the one hand, and both psychological symptoms and positive adjustment, on the other. At times these associations accounted for more than 50 percent of the variation in our measures of adjustment, an astonishing level of association for psychosocial data. Added to these were more indirect effects of the parent-sibling subsystems and the marital subsystem, so that the total variance accounted for was unprecedented in previous studies. Data of this kind would, in most developmental studies, implicate family relationships as major influences on adolescent development, or imply that adolescent characteristics are major provocations of family processes. In either case, psychosocial mechanisms would seem to be central in these linkages.

In a second line of analyses, however, we obtained equally strong findings on the influence of genetic factors on the very same measures of adjustment. For the majority of these measures we found that genetic fac-

269

tors also accounted for well over 50 percent of differences among the adolescents and that the magnitude of genetic effects was comparable in data at both time 1 and time 2. Moreover, we found that genetic factors were the preponderant influence on *change* in three domains of adjustment and were important influences in the remaining four. The apposition of these genetic findings and those on family factors in adolescent adjustment points to a familiar paradox: previous studies by geneticists have shown substantial genetic influence on adolescent adjustment, whereas previous studies by psychosocial researchers have suggested the importance of family relationships in adolescent adjustment. Although our study is the first to combine fully both perspectives, the data we obtained merely confirm the Solomonic quandary: how can both sets of data be right?

We have suggested that a third line of evidence might resolve this paradox. In the last chapter, we noted that genetic factors influence the same family relationships that are themselves distinctly associated with adolescent adjustment. Parent-child subsystems that are strongly associated with antisocial behavior, cognitive agency, or social responsibility also show particularly notable genetic influence. In fact, about half of all differences among families in parent-adolescent conflict can be explained by genetic differences among the adolescents. Genetic influences are also notable in the marital subsystem and, to a lesser extent, in the sibling subsystem. We wondered whether these genetic influences were just a curiosity, or if they might be important in resolving one of the major paradoxes not only of our study but of all current studies in psychological development.

In this chapter we take a second look at the three previous lines of evidence we have presented—phenotypic associations, genetic influences on adjustment, and genetic influences on family subsystems—in an attempt to resolve the paradox. We begin with a recognition that, in our own study, measurements of family process account for about half of the variation in adolescent adjustment, and genetic factors for the other half. We have summarized this finding, using data on parent conflict/negativity and antisocial behavior, in Figure 10.1. The top two bars are the simple associations between mother's negativity and antisocial behavior and father's negativity and antisocial behavior. We presented these data, graphically,

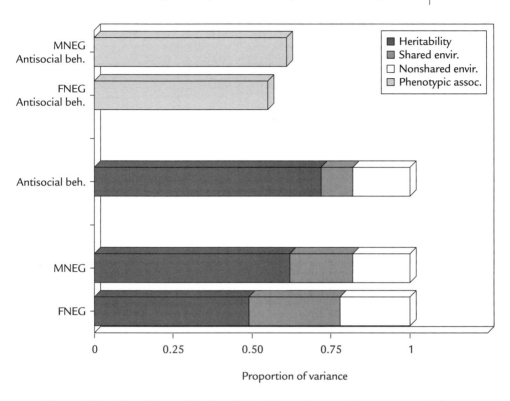

Figure 10.1 The effects of family relationships and genetic factors on variations in adolescent adjustment

in Figures 7.7 and 7.8. Findings of this kind are well known to family researchers both in the United States and abroad. The third bar is a finding familiar to behavior geneticists: it shows the substantial heritability of antisocial behavior in our sample. The figure also illustrates an intriguing third feature of the data: more than 50 percent of differences among families in mother and father conflict/negativity are, like antisocial behavior itself, influenced by genetic factors.

We will use data similar to those exemplified in Figure 10.1 to try to understand how both family processes and genetic influences can be powerfully implicated in adolescent adjustment. Could it be that social processes, such as parental conflict/negativity, account for about half the differences in adjustment among adolescents, and that genetic factors account for the other half? If this is the case, do social processes operate directly on the developing psychobiological mechanisms of adolescent development, and do genetic factors also influence these mechanisms, separately and independently? Or perhaps an entirely different scenario is more plausible: perhaps genetic and social factors are deeply intertwined so that neither operates independent of the other. In this case, the two sets of factors together might explain one half of the picture with the other half still to be explored.

Unfortunately, we cannot answer these questions as fully as we might like. As powerful as the data are, they raise as many questions as they answer. Nonetheless, we can examine three more focused questions that illumine the issues laid out by our overarching concerns about development. We address the first of these more focused questions in this chapter and the second two in the chapter that follows.

First, are the genetic influences on measures of adjustment the same as those that influence measures of family subsystems? The answer to this question is important; if it is "yes," then we know that there is some connection between the large phenotypic associations between family process and adjustment reported in Chapter 7 and the even larger heritabilities for these same measures described in Chapter 8. That is, the phenomena described in each chapter do not refer to entirely distinct or separate sets of influences on adolescent adjustment; rather, these findings reflect some connections between the two sets of findings. Answers to our simple question about overlap will not, however, shed much light on the specific mechanisms that may account for this connection.

If we find that the genetic factors that influence family relationships overlap with those that influence adjustment, the next logical question is how much the association between family process and adjustment might be attributable to this overlap. If the overlap in genetic influence accounts for the majority of these associations, then we can move a step closer to resolving our paradox. Genetic factors can then be understood to account, in some way, for both the high heritabilities of adjustment and the high associations of family process with adjustment. In other words, the high phenotypic associations and high heritabilities shown in Figure 10.1 account for the *same half* of the variance in antisocial behavior. We could say that genetic and environmental mechanisms do not operate independently; they do not each account for about half the variation in adolescent adjustment. Both are intertwined, but how they are intertwined will still not be clear at this point in our analysis.

The analyses we use to address this question have another important yield: they also estimate how much overlap there is between environmental influences on family processes and environmental influences on adjustment. Although these are equally important data, they play a marginal role in resolving the major paradox of this book. We will attend to these findings but will use them only as ancillary support for a theory of development.

A second question is merely an extension of the first: Do our findings on the overlapping influences of genetic factors—influences on family processes and on adjustment measures—remain relatively the same over time? For example, it is possible that there is an overlap of genetic influences on parental conflict/negativity and antisocial behavior in early adolescence, but that this overlap disappears by mid-adolescence even though the magnitude of genetic influence on antisocial behavior and on parental conflict and negativity remains the same. To extend this conjecture, it is entirely conceivable that while the magnitude of genetic influence remains the same across time, new genetic factors arise in development to replace old ones. The new ones may no longer show the same overlap between their influence on social processes and their influence on adjustment. This is, of course, precisely the case for antisocial behavior, where we have shown that almost a third of the genetic factors operating in early adolescence no longer influence antisocial behavior at mid-adolescence; they are entirely replaced by new factors that were not operating earlier.

If these newer genetic influences dramatically attenuate the role of genetic overlap, we might conclude that in early adolescence genes and environment may account for the *same* half of variation in adolescent adjustment, but that by mid-adolescence they account for *different* halves. If we discover that this is the case, we will have to grin and bear it; a developmental theory to match findings of this kind would be complex, conditional, and frustrating. As will be shown, comparing these overlaps at time 1 and time 2, particularly in relationship to other findings, permits us to draw inferences about some mechanisms that might account for genetic and environmental influences on development.

A third question picks up this theme of change and prompts us to examine it more directly. Although our interest in individual differences among adolescents at a single point in time remains very keen, we are even more interested in factors accounting for change from one time period to another. In their approach to estimating influences on developmental change, most researchers focus on the circumstances that exist immediately prior to change. They ask whether differences among individuals in these circumstances accurately predict the extent of change in measures of adjustment. For example, in a psychosocial study we might ask whether differences among adolescents in conflictual relationships with mother at time 1 accurately predict differences in their *change* in antisocial behavior from time 1 to time 2. Because our design is genetically informed, we can rephrase this question: Do differences in measures of any family subsystem at time 1 predict change in adolescent adjustment independent of genetic differences in the adolescents at time 1? In Figure 9.10 we showed that for most measures of adjustment and family process there is substantial stability across the time of our study, more than we expected from theories that depict adolescence as a period of change and unpredictability. This stability narrows the window for observing family process related to change. Nonetheless, if genetic factors are the major predictors of adolescent change in these analyses, this would confirm their role as a major engine of development during this period of maturation.

Overlapping Genetic and Environmental Influences

Concepts

GENETIC INFLUENCES   We began our discussion of overlapping genetic influences in Chapter 4. We continue it in greater detail here. Figure 10.2 is a revision of Figure 4.7. We have redrawn this figure to emphasize that among the various explanations for the overlap of genetic influence there is a fundamental distinction between passive mechanisms, on the one hand, and a number of different active/evocative mechanisms, on the other. These two categories of mechanisms provide alternate explanations for any finding indicating genetic overlap: either one could have the preponderant role, or both could be important.

Passive mechanisms can be solely responsible for an observed overlap of

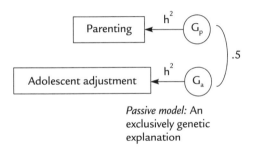

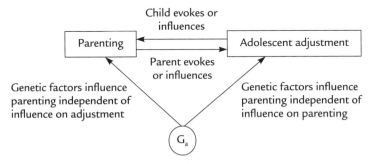

Figure 10.2 Schematic contrast between passive model and evocative/active model

genetic influence on a measure of family process and a measure of adolescent adjustment. A passive mechanism is possible only if three conditions are met: (1) the measures of adjustment show genetic influence; (2) the measure of family process, let us say, parenting, also shows genetic influence; and (3) the genetic factors that influence the measure of adjustment are the same factors that influence parenting. A correlation between parenting and adjustment could then be explained by the 50 percent overlap in genes between a parent and a child. This mechanism can be considered a "null hypothesis." Passive mechanisms imply that no social influences—from parent to child or vice versa—play any role in the observed associations between parenting and adolescent adjustment. If this were true, all correlations between environment and adjustment—now the bedrock of developmental theory—would be epiphenomena only. That is, if only passive genetic mechanisms were important, these observed correlations would not reflect any significant role for family relationships in adolescent development, either as influences or as consequences of adolescent adjustment. Current theories of adolescent development, almost all of which center on adolescents' social relationships, would have to be completely overhauled.

Given the potential consequences for current theory, are passive mechanisms even plausible, let alone likely explanations for phenotypic associations? One of the three components of passive mechanisms, the 50 percent genetic overlap between parents and children, is no longer a matter of dispute. But what about the other components of this argument? For example, is it plausible to argue that parents' genes might influence how they treat their children and that these genetic factors may, if transmitted to their children, influence measures of adolescent adjustment? Strikingly, recent research makes this a very plausible argument. There is now increasing evidence that a substantial proportion of variation among parents in their treatment of their children is influenced by the parents' genes (Kendler, 1996; Rowe et al., 1992). Further, longitudinal research is now demonstrating that adolescents with behavior and related disorders grow up to have seriously disturbed relationships with others, including their children (Brooks, 1981; Elder, 1986; Kandel and Davies, 1986; Kandel et al., 1986).

Details of our findings do permit us to gauge very roughly whether passive mechanisms are the only ones at play. But the extent to which passive mechanisms contribute to overlapping genetic influences on family processes and on adolescent adjustment can only be determined, with certainty, in an adoption design. As noted in Chapter 4, by using such designs, researchers can compare the correlation of a parental behavior with child adjustment for birth mothers rearing their own children with a comparable correlation for adoptive parents rearing their adopted children. Since birth mothers and adoptive mothers can be assumed to be equally vulnerable to evocative and active influences from heritable characteristics of their children, a greater correlation for birth mothers and their children strongly suggests passive mechanisms.

In addition to passive mechanisms, a broad array of other mechanisms might account in whole or in part for observed overlap of genetic influences on family process and adolescent adjustment. In all these mechanisms, psychological and social processes are critical in some way to connecting family processes and adolescent adjustment. The bottom portion of Figure 10.2 illustrates some of the possible paths of influence. In these models genetic factors may influence child adjustment independent of their influence on parenting, or the reverse might occur, or some combination of both may be the case. If the genetic effects on child adjustment are primary, then the genetic overlap could be explained by a secondary influence of heritable child characteristics on the parent. These effects are secondary in that they must follow genetic influences in the sense that the presence in the adolescent's genome and potential activity of specific genes can be safely assumed to be uninfluenced by either parenting or adolescent adjustment. As will be discussed later, however, the possibility that the child's adjustment or the parent's relationship with the child might influence genetic expression in behavioral differences cannot be so easily ruled out. For the moment we will set aside these more complex influences.

Figure 10.3 pulls apart two subcategories of the evocative/active set of mechanisms. The top portion of the figure shows a circumstance in which the adolescent's genes directly influence adolescent adjustment; the parents' relationships with the child are a consequence of this genetically influenced adjustment. "Directly influence" is a relative and statistical term. It means only that the influence on adolescent adjustment is not contingent on a prior influence of genetic factors on parenting. The pathway indicated by the simple arrow from G_a to adolescent adjustment is hardly simple and almost certainly involves hundreds of specifiable molecular, neurobiological, and related processes.

The bottom portion of Figure 10.3 shows another type of evocative/active mechanism. In this circumstance, genetic factors influence parenting independent of any influence on child adjustment. The levels of child adjustment are then secondary to genetically influenced variations in parenting. The path from G_a, the adolescent's genes, to parenting must be even more extended. Molecular and neurobiological steps, analogous to those of the child-effects model at the top of the figure certainly must be required here as well. But since we are considering only the adolescents'

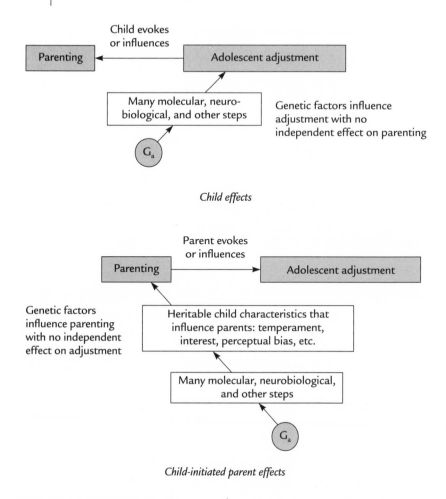

Figure 10.3 Two possible evocative/active models accounting for overlapping genetic influences on family processes and adolescent adjustment

genes, we need to posit a mechanism for linking these molecular and neurobiological events to parental behavior. One possibility is that the adolescents' genes influence their temperaments or their ways of perceiving the social world and its cues or their interests and values. These genetic influences may occur very early in life, perhaps long before adolescence. In turn, these heritable characteristics of the adolescents influence parental responses and shape long-term patterns of relationships with both parents. Ultimately, as a secondary effect, these patterns influence adolescent adjustment.

The child-effects model and the child-initiated parent-effects model may both be important. We cannot unravel these models here, but in Chapter 12 we weigh their plausibility and outline research approaches for delineating the role of each of these mechanisms.

ENVIRONMENTAL INFLUENCES   The analysis we use in this chapter can also estimate the extent of overlapping environmental influences. But first let us return to Figure 10.1. Note that both mother's negativity and father's negativity are equally determined by genetic influences and environmental influences, with a preponderance of shared environment in the latter category. Similarly, nongenetic, environmental influences have a modest effect on antisocial behavior as well. Here nonshared factors slightly outweigh shared environmental factors. We can now ask several important questions: Do any of the environmental influences on mother's or father's negativity overlap with the environmental influences on antisocial behavior? Do any of the nonshared influences on mother's or father's negativity overlap with those on antisocial behavior? If we do find an overlap of environmental influences, is it attributable to shared or nonshared influences? And, if we find shared or nonshared influences, how can we explain observations of this genre?

For shared environmental influences, there are three possibilities. First, consistent negativity, directed equally at both siblings by each parent, may cause antisocial behavior in the adolescents. The first and easiest way to think of this is that parents' consistent style of parenting—say, irritable, contentious, punitive, and aggressive—has an influence on their adolescent children and is as likely to influence one child as the other. A second explanation is that antisocial behavior—particularly joint, imitative, or equally frequent antisocial acts of children—induces punitive and aggressive parenting from parents. In this instance, either child is equally likely to elicit this behavior, and, over time, both children make more or less equal contributions to this effect on parents. A third possibility is that some external factor operates to make parents consistently negative to both children and that same factor also makes it equally likely that both children will commit antisocial acts. For example, a family may be part of a migrating group in which economic privation, prejudice of native dwellers in a new country, and physical danger may make fathers irritable and combative while encouraging aggressive behavior in the children. All three of these mechanisms might be combined to explain overlapping shared environmental influences. Further, the balance of each of these might differ from family to family.

These same three possibilities hold for overlapping nonshared influences. First, father's differential treatment of his two children leads one of them to antisocial behavior but protects the other from the same fate. A second possibility is that differential aggressiveness toward father by children who themselves differ in antisocial tendencies elicits differential parenting. Third, it is remotely possible that some outside factor influences the behavior of the father toward the child and also influences the antisocial behavior of the child, although parenting and child behavior are not themselves causally linked. As with shared environmental influences, the balance of mechanisms need not be the same in each family. Moreover, it may not be a single external circumstance that acts on both father and adolescent but a combination of disruptive and provocative circumstances.

ANALYTIC PRINCIPLES   We use the same approach to analyzing the components of correlations here as we used in previous chapters to examine two other questions: the influences on stability and change in measures of adjustment and on measures of family process. The correlations we study here all examine the association between a measure of family process and a measure of adolescent adjustment. Again, the central statistic in this analysis is the cross-correlation. We correlate the parenting or sibling behavior directed toward the adolescent with the measure of adjustment of the sibling. We compute this for each of the six groups of siblings in our study and then compare these cross-correlations across the six groups. (See Chapters 8 and 9 for a more detailed description of the logic of these analyses.) We will take readers step by step through the inferences that can be drawn by simple inspection of these correlations so that those unfamiliar with structural-equations modeling can do for themselves the logical operations that are roughly comparable to those performed by our more complex model-testing procedures.

Figure 10.4 shows the cross-correlations for both mother's and father's negativity and adolescent's antisocial behavior. This figure should be compared with Figure 10.1. Note again that in 10.1 antisocial behavior and mother's and father's negativity are strongly influenced by genetic factors. Figure 10.4 provides information on the extent to which these influences are the same or overlapping genetic effects. In Figure 10.4 the slope of an imaginary line resting on top of the bars is definitely downward, moving from siblings with more genetic similarity to those with less. This suggests that genetic influences are important in the correlation of antisocial behavior with both mother's and father's negativity and antisocial behavior. That is, it is likely that there will be substantial overlap of genetic influences on these two measures. For mother's negativity,

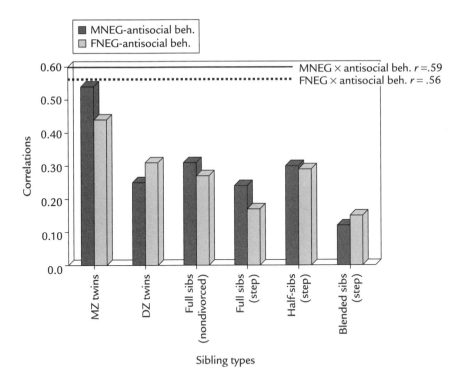

Figure 10.4 Cross-correlations between mother's and father's negativity toward the adolescent and the sibling's antisocial behavior

MZ twins show much higher correlations than DZ twins. The latter are roughly comparable to the full sibs in nondivorced families and in step-families. Half-sibs show a somewhat higher correlation than would be predicted by genetic expectations alone, but the blended siblings, sharing no genes, show the lowest correlation of all six groups.

The phenotypic correlation between mother's negativity and adolescent's antisocial behavior is .59. We can roughly estimate the importance of genetic factors by comparing the correlations for MZ twins and DZ twins ($.54 - .25 = .29$) and then doubling the difference ($2 \times .29 = .58$). Alternatively, we can compare the correlations of full and blended sibs in step-families ($.24 - .12 = .12$) and double the differences ($2 \times .12 = .24$). Since both approaches, using twins and using sibs in step-families, are rational ways to estimate genetic influence, the results from these comparisons could be pooled. A truer value for heritability might be somewhere in between, approximately $(.58 + .24)/2 = .41$. We can call this very approximate figure (and the more precise figure to be estimated below) the "heritability of association" between mother's negativity and adolescent's antisocial behavior. The figure would account for roughly 69 percent (.41 divided by the phenotypic correlation of .59) of the phenotypic association; that is, about two-thirds of the observed association between maternal negativity and antisocial behavior is due to overlapping genetic influences. Our model-testing procedures integrate these two comparisons with all other comparisons that are possible among the six groups in our study.

As Figure 10.1 indicates, the heritability of both mother's negativity and antisocial behavior exceeds our estimated heritability of association. We can crudely estimate this "leftover" heritability by subtracting the heritability of association from the heritability of mother's negativity ($.58 - .41 = .16$). Thus there is a small amount of genetic influence on mother's negativity that is unique and does not affect antisocial behavior. We can perform the same subtraction to estimate crudely the unique or nonoverlapping heritability of antisocial behavior ($.67 - .41 = .26$). One way to think of these unique genetic influences on mother's negativity and on antisocial behavior is that they contribute to the lack of association, or disjunction, of the two variables. That is, these variables fail to correlate perfectly in part because some of the genetic influences on one variable are different from some of the genetic influences on the second variable. Indeed, we can refer to the sum of these unique influences as the "heritability of disjunction," which can be precisely estimated using our model-testing procedure.

Figure 10.4 also contains important information about overlapping and unique environmental influences. Note that the correlations between unrelated siblings is modest (.12 for mother's negativity and .15 for father's). These can be used as direct estimates of the shared environmentality of association or contribution of the overlapping shared environmental influences on mother's and father's negativity and antisocial behavior. As shown, overlapping genetic influences already account for .41 of the .58 association between mother's negativity and antisocial behavior. If we add .12 from shared environmental factors, this leaves a total, crudely estimated, of about .05 (.58 − .41 − .12) for nonshared factors. This number corresponds to the differences between the phenotypic correlation and the MZ correlation (.59 − .54 = .05), which is one way of crudely estimating the contribution of overlapping nonshared factors—the nonshared environmentality of association—to the phenotypic association.

Figure 10.5 shows the results of the model-testing procedure. We expect these findings to be roughly in line with our crude estimates based on simple inspection of the cross-correlations, but there will, of course, be some differences because the model testing takes into account comparisons among all the groups. The heritability of association is indicated, again, by the product of the path coefficients from the large "G" balloon (.72 × .58 = .42); the shared environmentality of association by the product of path coefficients from the large "SE" balloon (.35 × .28 = .10); and the nonshared environmentality of the association by the product of the path coefficients from the large "NS" balloon (.28 × .23 = .06). These more precise estimates come very close to those from our direct inspection of the correlation coefficients in Figure 10.4 (direct inspection: .41, .12, .05; model testing: .42, .10, .06). Importantly, the model-testing results suggest that overlap of genetic influence is by far the most significant reason for the association between mother's negativity and adolescent's antisocial behavior: .42/.59 = 71%.

Unique genetic factors are also the strongest influence on disjunction

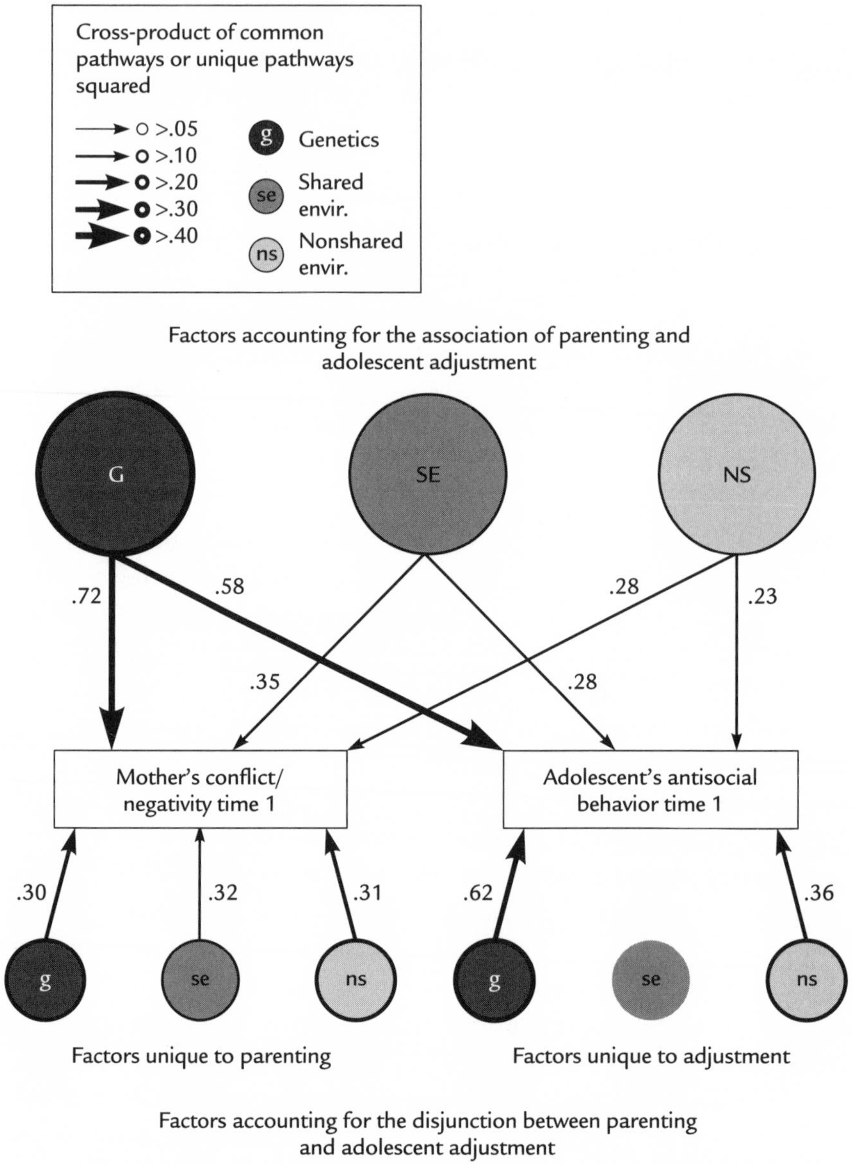
Cross-product of common pathways or unique pathways squared
>.05
>.10
>.20
>.30
>.40
g Genetics
se Shared envir.
ns Nonshared envir.
Factors accounting for the association of parenting and adolescent adjustment
G
SE
NS
.72
.58
.28
.23
.35
.28
Mother's conflict/ negativity time 1
Adolescent's antisocial behavior time 1
.30
.32
.31
.62
.36
g
se
ns
g
se
ns
Factors unique to parenting
Factors unique to adjustment
Factors accounting for the disjunction between parenting and adolescent adjustment

Figure 10.5 Balloon diagram showing components of the association between mother's negativity and adolescent's antisocial behavior

between the two variables. The same pattern of results holds for father's negativity and antisocial behavior. We can reasonably estimate the contribution of genetic, shared, and nonshared factors to the disjunction in two steps. First, we square each of the path coefficients. Then for each of the components of the disjunction, we add these squares from the paired relevant paths and divide by 2. For example, we add the squared coefficient from the little "g" attached to negativity and the little "g" attached to antisocial behavior for the heritability of disjunction, with similar procedures for "se" and "ns." For example, in Figure 10.5 we begin the computation of the heritability of disjunction by squaring .30 and .62 (.09 + .38 = .47). Then we divide this number by 2 to reach .235.

Overlap in Measures of Family Systems and Adolescent Adjustment

Figure 10.6 summarizes many analyses identical to the ones we have just illustrated with an informal inspection of cross-correlations and with model testing. (Details are in Appendix B.) We present the results in this section and in the next one using the bar-graph format. The very top bar represents the analysis of mother's negativity and antisocial behavior just reviewed in detail. The arrow marks the juncture of our representation of the association and disjunction. An imaginary line drawn from the arrow to the horizontal axis would intersect at .59, the phenotypic correlation. Everything to the left of the arrow represents the relative contributions of the heritability and the shared and nonshared environmentality of the association between the two variables. Everything to the right represents the same components of the disjunction. Note that we have grouped the bar graphs so that each dimension of mother's parenting is shown with all measures of adolescent adjustment where the phenotypic association was .2 or greater. Also, we have indicated with a minus sign (−) where the correlation between parenting and the adolescent adjustment domain was negative. For example, the *higher* the mother's negativity, the *lower* the adolescent's scores on social responsibility.

In order to anticipate and grasp the findings in the next four figures, which present data analyzed from the entire sample at time 1, readers may wish to consult Figures 7.7–7.10, 8.1, 9.3, and 9.4. These graphs report the phenotypic associations between measures of family subsystems and

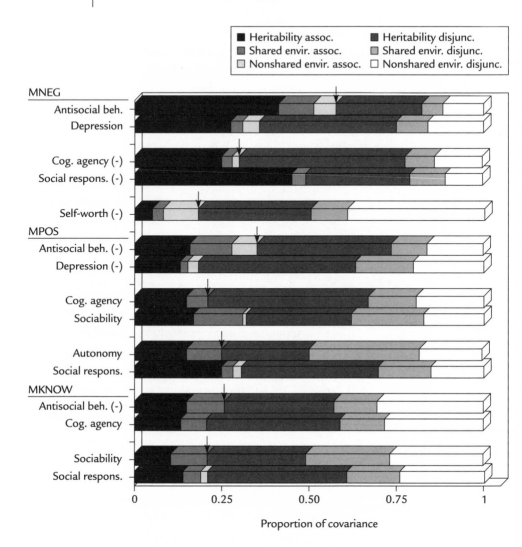

Figure 10.6 Analysis of the association between mother's parenting and all seven domains of adolescent adjustment

adjustment as well as genetic and shared and nonshared environmental influences on both measures of adjustment and measures of family process. These influences set certain constraints on the extent to which overlapping influences can affect the association between two different variables. For example, the very high heritability of social responsibility and the negligible role of shared environment for that variable make it likely that genetic factors will play an important role in association with a family-process variable that is also highly heritable, such as maternal negativity. Nonshared factors are the only other determinants of social responsibility, and their overlap with nonshared influences on family process is possible.

Mother's Parenting

NEGATIVITY   Mother's negativity was associated with five of the seven domains of adolescent adjustment we measured: positively with antisocial behavior and depression, and negatively with cognitive agency, social responsibility, and global self-worth (see Figure 7.7).

Antisocial Behavior and Depressive Symptoms.   Figure 10.6 clearly indicates that genetic factors play the preponderant role in both association and disjunction. Overlapping shared environmental factors play a modest role in association, as do nonshared factors in disjunction. We have already shown, in Figures 8.1, 9.3, and 10.1, how these findings fit with our analyses of the heritability and environmentality of both antisocial behavior and mother's negativity. Since both show such high heritabilities, we would expect genetic overlap as well as genetic nonoverlap to play big roles in both association and disjunction.

Although the association between mother's negativity and depression is less (.35) than that between mother's negativity and antisocial behavior, the same pattern of findings on overlapping and nonoverlapping factors prevails. For depression, overlapping shared environmental factors, although shown on the graph by a very narrow segment, fall far short of statistical significance. Indeed, we would expect virtually no role for overlapping shared environment because, as Figure 8.1 showed, there was no significant influence of shared environment on depression at time 1; in effect there were no influences to overlap with any other variables.

Cognitive Agency and Social Responsibility.   The correlations of cognitive agency and social responsibility with negativity are −.31 and −.48, respectively. In this case genetic factors play an even more important role in both association and disjunction than they do for antisocial behavior and depression. Again, this was to be expected given that cognitive agency and social responsibility are the most heritable of the domains of adolescent adjustment. Nonshared factors are the only substantial environmental influence on cognitive agency and social responsibility at time 1, but they are clearly involved only in disjunction, not association.

Although all four of these findings follow logically from our analyses of both adjustment and mother's parenting, they must come as a surprise to both geneticists and environmentalists alike. For four of the five areas of adjustment associated with mother's negativity, environmental factors acting alone account for almost none of the observed association. Currently no theory of the development of psychopathology or competence in adolescence could have predicted these results or can encompass them now that they are so plainly evident.

Self-Worth. Table 8.1 indicated that findings for global self-worth and its relationship to all measures of family relationships will be different from other measures. At time 1, Table 8.1 showed that global self-worth has very low heritability; most of the individual differences were due to nonshared environment. The correlation with maternal negativity just made our cutoff at $-.20$. It was conceivable that all the genetic influence on global self-worth might overlap with mother's negativity. In that case most of this small correlation might also be due to overlapping genetic influences. As Figure 10.6 shows, however, this is not the case. This is one of the very few correlations that can be attributed to overlapping environmental variables, in this case, nonshared factors.

POSITIVITY A review of Table 9.3 reveals that mother's positivity shows lower heritability and more influence of shared environment than does her negativity. This circumstance, by itself, makes it more likely that overlapping shared environment may play a more prominent part than negativity in association between mother's positivity and those areas of adjustment that also show substantial influence of shared environment. As Figure 8.1 indicated, the most conspicuous of these variables are sociability and autonomy.

Antisocial Behavior and Depression. The phenotypic correlations for antisocial behavior and depression are smaller than the corresponding correlations for mother's negativity (−.32 and −.20, respectively). For antisocial behavior, overlapping genetic factors are the most important explanations of the association, .16/.32 = 50%. Shared environment is more important than nonshared environment in determining the correlation between mother's positivity and antisocial behavior. The small inverse association between depression and mother's positivity is almost entirely attributable to overlapping genetic influences.

Cognitive Agency and Sociability. The phenotypic correlations for cognitive agency and sociability with maternal positivity are again modest: .24 and .32. Once more, overlap in genetic factors is the greatest influence on the association, particularly for cognitive agency, where it accounts for 58 percent of the association. With reference to Figure 8.1 it is clear that, especially for sociability, genetic and shared environmental factors play the preponderant role in individual differences. It is not surprising, then, that they play the preponderant role in both association and disjunction.

Autonomy and Social Responsibility. The phenotypic associations with autonomy and social responsibility are .25 and .35, respectively. Overlapping genetic factors play the predominant role in both associations. For social responsibility, genetic factors are also mainly responsible for the disjunction with mother's positivity. As can be anticipated from Figure 8.1, shared environment plays a more important role than nonshared environment for both association and disjunction of autonomy with mother's positivity.

MOTHER'S KNOWLEDGE OF CHILD'S ACTIVITIES As Figure 9.3 indicated, mother's knowledge about her child's activities has only a modest heritability. The larger environmental influence is equally divided between shared and nonshared influences. This makes it less likely that overlapping genetic factors will play such a major role in the association of mother's knowledge with adolescent adjustment. In other words, there are many shared and nonshared influences on mother's knowledge of her child's activities that could overlap with analogous influences on the adolescent's adjustment and thus account for all or most of any observed phenotypic associations.

Against this backdrop it is surprising that, in all four associations significant at .20 or above, overlapping genetic factors once again play the most significant roles and nonshared factors play almost none. The modest level of the phenotypic associations, all in the lower or mid-20s, should not obscure the dramatic and unanticipated findings here. Even when genetic influences are small on family process, they seem crucial for almost any associations they have with child adjustment. In contrast, even when nonshared influences are large on both measures of family process and measures of adjustment—in this case for both mother's knowledge and cognitive agency and depression—they account for only trivial amounts of association.

Father's Parenting

In considering whether the same pattern of findings we discovered for mother's parenting holds for father's as well, it is again useful to anticipate the possibilities by referring to data already presented. In this case, Figure 9.4 provides some clues. These data suggest that we might see a replication. Although there are subtle differences in the distribution of genetic and environmental influences on father's parenting in comparison with mother's, the basic pattern is the same. For example, at time 1, there are substantial genetic influences on father's negativity and positivity and, as with mother's, far less on father's knowledge. Moreover, there are very high correlations between all measures of mother's and father's parenting. That is, in those families in which mothers treat their children negatively, so do fathers. As a result, mother's and father's scores on this measure are highly correlated.

Nonetheless, neither the similarity of distribution of types of influences—genetic and environmental—nor the high correlations assure us that the same genetic or environmental factors are influencing mother's and father's parenting. Indeed, as will be shown in Chapter 12, even though mother's and father's positive parenting is very highly correlated, there is no significant genetic overlap between the two. The same is true for all three measures of parental control: knowledge, attempted control, and actual control. Thus, it is crucial that we scrutinize the associations of father's parenting as carefully as we did mother's because they may provide fresh data confirming or disconfirming the pattern of findings for mother's parenting. In other words, if we see the same pattern of findings for fathers, we can regard them as a partially independent replication of the pattern of findings just reviewed for mothers. Figure 10.7 provides these data.

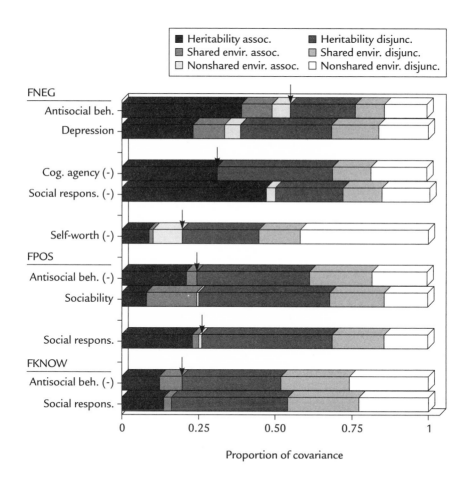

Figure 10.7 Analysis of the association between father's parenting and all seven domains of adolescent adjustment

NEGATIVITY   Despite a number of factors that might have led to substantial differences in the results for fathers in comparison with mothers, the actual findings are remarkably the same. Genetic factors play a preponderant and almost exclusive role in all associations with the exception of self-worth. Nonshared factors, when they have an important role, contribute mainly to disjunction, not to association. Again, the exception is self-worth, where they are the most significant influence on a very modest phenotypic association.

POSITIVITY   Only three of the seven domains correlated significantly with father's positivity. The shared environmental overlap with sociability shows about the same influence on association as it did for mother's, but the influence of genetic overlap is somewhat less. By contrast, the influence of genetic overlap is much more important for antisocial behavior for fathers than for mothers, accounting for nearly all of an association of −.26. The findings for social responsibility parallel those for mothers.

FATHER'S KNOWLEDGE OF CHILD'S ACTIVITIES   Once again, the phenotypic findings for father's knowledge of his adolescent's activities are modest. Father's knowledge correlated with only two of the seven domains of adjustment we measured. The findings emphasize a widespread pattern, however. As Figure 9.4 showed, the genetic influence on father's knowledge is even less than it is on mother's knowledge. As a corollary, the role of shared environment is even more important. We might have expected the modest influence of genetic factors on father's knowledge to tilt the equation away from the importance of overlapping genetic factors (since there is so much less genetic influence to provide an opportunity for overlap). In fact, the reverse seems to be true. *All* the genetic influence on father's knowledge is involved in the overlap with antisocial behavior and with social responsibility. Once again, nonshared factors play no role in the association but are important factors in the disjunction.

Sibling Relationships

Figure 9.5 showed that the distribution of influences on sibling interactions is strikingly different from that on parenting. For positivity and negativity, at times 1 and 2, the influence of genetic factors is small and virtually all the environmental influences are shared. This may well be due to the high degree of reciprocity in the relationships of adolescent siblings who are the same gender and close in age. These findings point to the high likelihood that shared environmental factors will overlap to account for associations between sibling relationships and adolescent adjustment. This is especially true for adjustment domains such as sociability and autonomy that are also strongly influenced by the shared environment. Figure 10.8 shows the results for measures of sibling relationships.

NEGATIVITY   Three domains of adjustment showed an association, at .20 or higher, with sibling negativity: antisocial behavior (.45); depressive symptoms (.23); and social responsibility (−.33). Each shows a somewhat different pattern of findings.

Antisocial Behavior and Depressive Symptoms.   For both of these measures of adjustment, the overlapping shared environment is the preponderant influence on the association; genetic, shared, and nonshared factors contribute equally to disjunction. Genetic overlap accounts for a smaller portion of the association with antisocial behavior than with depressive symptoms; all the genetic factors influencing sibling negativity also influence antisocial behavior. The importance of shared environment, or any form of environmental influence, on these associations is unusual and greater than any other association we will consider at time 1. Of the twenty-seven associations of .20 or above examined thus far, only three others have shown environmental factors as the most significant influence: mother's negativity with self-worth, father's negativity with self-worth, and father's positivity with sociability.

Social Responsibility.   Although overlapping shared environmental factors account for some of the association with social responsibility, genetic factors are clearly the most significant.

POSITIVITY   Six of the seven domains of adjustment correlated with sibling positivity, some at substantial levels: antisocial behavior (−.37); depressive symptoms (−.28); cognitive agency (.25); sociability (.30); autonomy (.23); and social responsibility (.37). Since, as Figure 9.5 in-

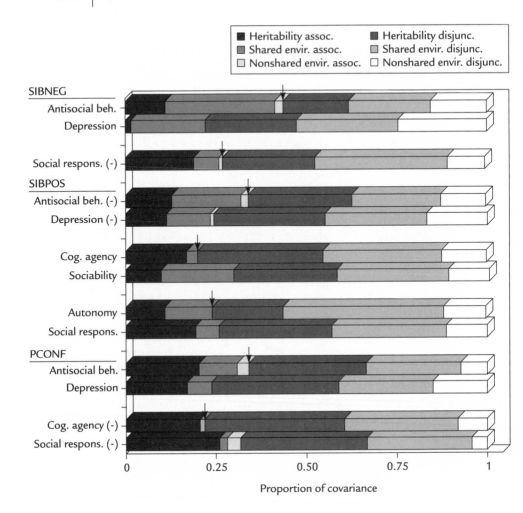

Figure 10.8   Analysis of the association between sibling interaction and all seven domains of adolescent adjustment

dicated, sibling positivity is more than twice as heritable as sibling negativity, we might anticipate that genetic overlap will play a more substantial role in these observed, phenotypic associations. Moreover, as might be expected, sibling positivity correlates inversely with sibling negativity (−.45), suggesting that some factors—genetic and/or environmental— may have overlapping influences. Since the correlation is *only* −.45, however, these variables are more distinct than similar; we can be quantitatively precise by saying they share only 20 percent of their variance ($-.45^2$ $\times$ 100). There are undoubtedly many sib pairs in our sample who are high in positivity and low in negativity and vice versa. Thus, an analysis of the associations of sibling positivity with measures of adjustment does not simply duplicate, in reverse, the data we have already examined for sibling negativity.

Antisocial Behavior and Depressive Symptoms. Although, as anticipated, genetic factors do play a notable role in the phenotypic associations, they are outweighed by shared environmental factors—similar to our findings for the association between sibling negativity and antisocial behavior and depressive symptoms. For positivity, however, genetic factors are important for depressive symptoms, whereas they played no role in the association with negativity.

Cognitive Agency and Sociability. For cognitive agency the results are quite similar to those for negativity and social responsibility. Some of these findings are attributable to the very high heritabilities of both cognitive agency and social responsibility. Table 8.1 showed these heritabilities to exceed 75 percent for cognitive agency and social responsibility. Sociability, however, showed a preponderance of overlapping shared environmental factors. This is consistent with the less prominent role of genetic factors and the more prominent role of shared environmental factors in individual differences in sociability.

Autonomy and Social Responsibility. Autonomy also shows a preponderance of overlapping shared environmental effects; again this follows, in part, from the large role these factors play in individual differences in autonomy. Social responsibility shows the reverse: a preponderance of overlapping genetic influences.

Summary: Contrasts between Parents and Siblings

In this section we have explored the role of family relationships and genetics in adolescent development, asking whether each of these influences accounts for one-half the differences in adjustment among adolescents, or whether they account for the same half. We will return to this question, with special attention to developmental mechanisms, in the next chapter, but here we can give a brief summary of the implications of our findings thus far.

The data for siblings in our study are strikingly different from the data for parenting. Of the twenty-five associations between parenting and adjustment, only one shows a preponderant influence of overlapping shared environment on observed associations. Two more, both involving self-worth, show overlapping nonshared environment as the most important influence but—for reasons we review later—these last two findings should be weighted less than most of the others. For the nine associations involving sibling measures, however, six show the preponderance of shared environmental overlap. Our genetic tools have permitted us to draw attention to differences among family systems. The associations between the mother-child and father-child systems and adolescent adjustment are due, primarily, to the overlap of genetic influences on family and adjustment measures. The association between sibling relationships and adolescent adjustment is attributable, in very large measure, to overlapping shared environmental influences. In broad strokes, parenting and genetic factors account, more or less, for the same half of differences among the adolescents, whereas genetic factors and siblings account for different halves. That is, in six of the associations between sibling relationships and adjustment, genetic overlap counts for very little.

Variations from Earlier to Later Adolescence

A Possible Basis for Change in Overlap

The findings thus far have been striking. Might they change from earlier to later adolescence? Figures 10.9 and 10.10 suggest that they might. Figure 10.9 summarizes data presented in Chapter 8 and shows that for at least three of our measures of adjustment there are substantial changes in genetic influences across time. These are sociability, antisocial behavior, and autonomy. It is entirely possible that, for these three measures, the new genetic influences in later adolescence do not overlap with those genetic factors that influence parenting. Thus we might expect an attenua-

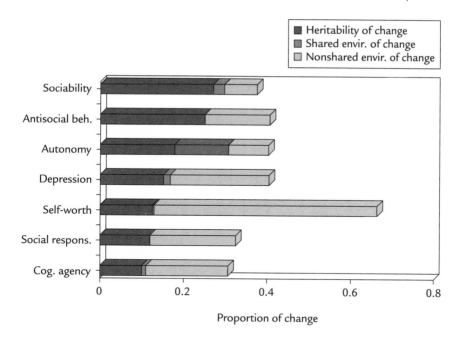

Figure 10.9 Components of change in the seven domains of adolescent adjustment

tion of the magnitude of genetic overlap between family measures and adjustment. By contrast, the reverse argument is equally plausible: new genetic influences on adjustment might overlap more with those factors influencing family systems.

In Figure 10.9, the overall length of the sidebars equals $1 - r$ for any particular variable where r is the stability coefficient, the correlation of the time 1 measurement of a variable with the time 2 measurement. The genetic component reflects the contribution of genetic factors to change over time; we have referred to this as the "heritability of change." Figure 10.9 presents the variables in order of the magnitude of genetic change. For antisocial behavior, for example, 24 percent of genetic factors operating at time 1 do not operate at time 2, and 26 percent of those operating at time 2 do not operate at time 1. The average of these two figures is exactly equal to the length of the densely shaded portion of the sidebar for antisocial behavior.

Figure 10.9 also shows the contribution of environmental factors to change; they are mostly nonshared factors. Thus, it is possible that by later adolescence these new nonshared factors will show more overlap with nonshared factors that influence parenting. The most extreme example of change in environmental factors concerns global self-worth. Here 66 percent of the nonshared factors influencing global self-worth at time 1 do not operate at time 2, and 41 percent of the nonshared factors that are influential at time 2 are new and were not influential at time 1.

There is one more reason to expect major changes in genetic and environmental overlap from earlier to later adolescence: the genetic and environmental influences on changes in parenting and in sibling relationships also change. This could lead to a decrease or increase in overlap with genetic and environmental factors influencing adjustment. Figure 10.10 summarizes critical data from Chapter 9. It shows the amount of change in genetic influences in our measures of family systems. Again, the length of the sidebar reflects the magnitude of overall change in the measure

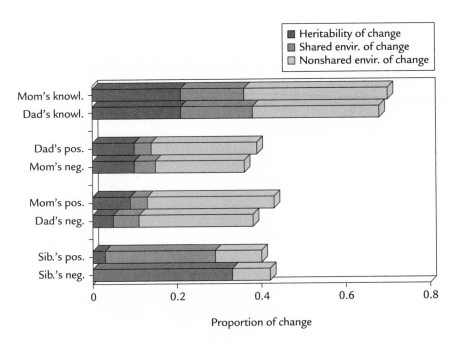

Figure 10.10 Components of change in the eight domains of family subsystems

$(1 - r,$ the correlation or stability from time 1 to time 2). Data for mom's and dad's knowledge show not only the most change but also the largest role for genetic factors in that change. Indeed, the magnitude of genetic change here nearly equals that for the adjustment variables of sociability and antisocial behavior.

Changes in Genetic and Environmental Overlap

MOTHER'S PARENTING   Figure 10.11 provides data on genetic and environmental overlap between mother's negativity and all the domains of adjustment that correlate with it at the level of 0.20 or greater. We can compare it with Figure 10.6 to detect major changes in the distribution of overlapping influences from time 1 to time 2. The changes in the overlap of shared and nonshared environment are quite small. They exceed $+.07$ in only two instances: mother's positivity with cognitive agency and with social responsibility, both of which decrease .11 from time 1 to time 2.

All but one of the changes in genetic overlap are more substantial at time 2 than at time 1, and all but one are in the positive direction. These changes are summarized in Figure 10.12. Four changes are modest: mother's negativity with self-worth, mother's positivity with cognitive agency, mother's positivity with social responsibility, and mother's knowledge with social responsibility.

FATHER'S PARENTING   The important comparisons in father's parenting are between findings already presented from time 1 observations in Figure 10.7 and findings for time 2 contained in Figure 10.13. For father's parenting, the changes in overlap of shared and nonshared environment do not exceed .05 with two exceptions: shared environmental overlap increases by .07 and nonshared increases by .08 for father's negativity and depression. Some of the changes in genetic overlap are somewhat larger and are illustrated in Figure 10.14. The patterns of these findings on genetic overlap are different from those for mothers, as there are a number of very small changes that actually show a decline as well as changes that show an increase. The changes in genetic overlap for father's knowledge and social responsibility and antisocial behavior parallel those for mother's.

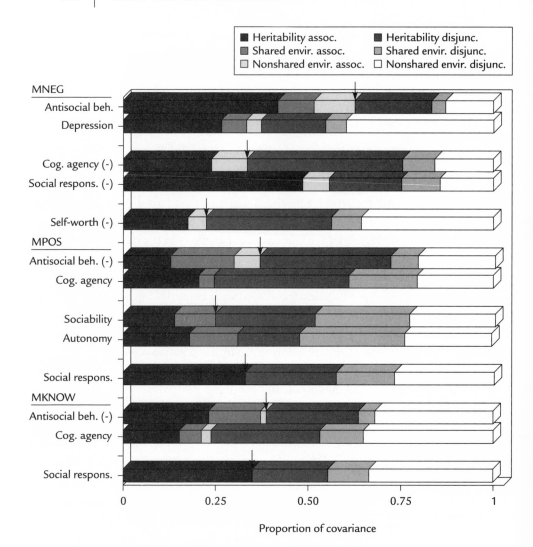

Figure 10.11 Analysis of the association between mother's parenting and all seven domains of adolescent adjustment at time 2

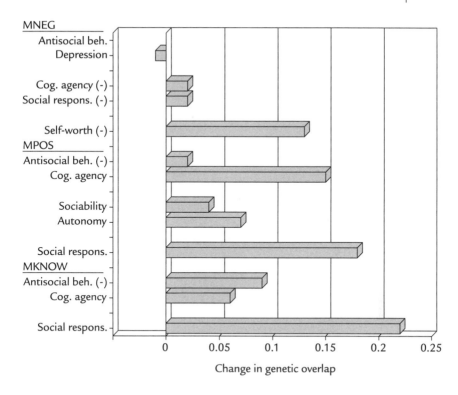

Figure 10.12 Summary of changes in genetic overlap for mothering

As with mother's positivity and negativity, for father's positivity and negativity we find a striking *lack of change* in overlap with sociability and antisocial behavior despite very substantial genetic change in both of these adjustment measures.

SIBLING RELATIONSHIPS The comparison between time 1 and time 2 assessments of genetic and environmental overlap for sibling data is shown in Figures 10.8 and 10.15. Figure 10.16 summarizes the changes in overlap between times 1 and 2. Although the changes in genetic and shared environmental overlap are small they reveal an intriguing pattern: with just one exception, all the shared environmental overlap *increases* from earlier to later adolescence and all the genetic overlap *decreases*. A

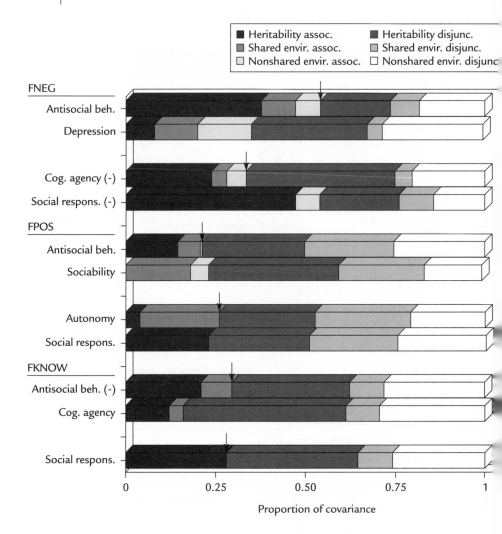

Figure 10.13 Analysis of the association between father's parenting and all seven domains of adolescent adjustment at time 2

pattern of this kind is very unlikely to be due to chance alone. Changes in the nonshared environment for sibling relationships are negligible.

Summary: Contrasts between Sibling and Parent Subsystems Become More Marked in Later Adolescence

SHARED ENVIRONMENTAL INFLUENCES   The most important role for environmental overlap in our findings was in sibling relationships. Fourteen of the nineteen analyses of associations between sibling relationships and adolescent adjustment at times 1 and 2 showed a preponderance of environmental influences, and the remaining four showed a preponderance of genetic influences. Of these fourteen associations, all were for shared environment. Moreover, of the nine comparisons between time 1 and time 2, six showed a small to modest increase in the importance of shared environment; most of these were for sibling positivity with a correspondingly small decrease in genetic influence. In Chapter 13 we return

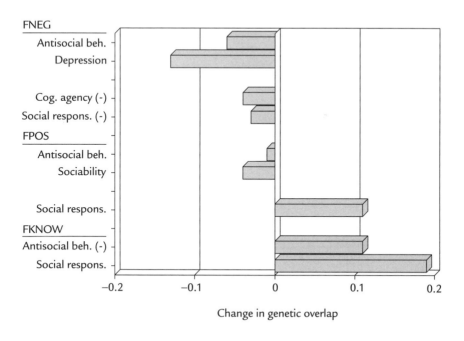

Figure 10.14   Summary of changes in genetic overlap for fathering

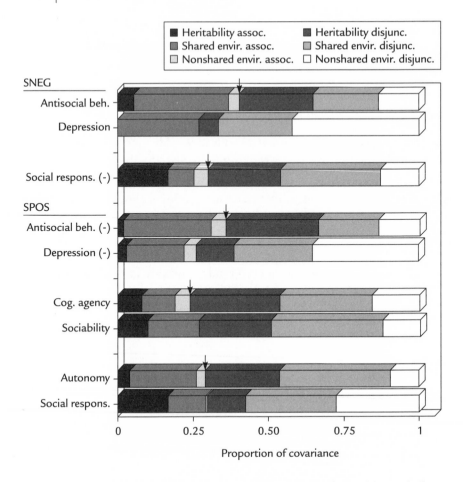

Figure 10.15   Analysis of the association between sibling relationships and all seven domains of adolescent adjustment at time 2

to the sharp contrasts between the sibling and parent-child findings when we discuss the implications of our research for a refreshed understanding of family dynamics.

Shared environmental factors were the strongest influence in five of the fifty-two associations between parental measures and adolescent adjustment we examined. Four of the five concerned sociability and autonomy, the only measures of adjustment for which shared factors are the most significant environmental influence (see Figure 8.1). Indeed, between mother's knowledge, father's positivity, and sibling positivity at time 1 and father's and sibling's positivity at time 2, we have probably identified, in this study, the major nongenetic correlates of these two important domains of adjustment.

These shared environmental characteristics of the family could operate jointly on both autonomy and sociability. That is, they could explain why siblings in certain families score high in both or low in both. They could also explain the disjunction between the two: why siblings might be high in one domain and low in another. At time 1 and at time 2 autonomy and sociability are positively correlated, at .47 and .53. This means that in

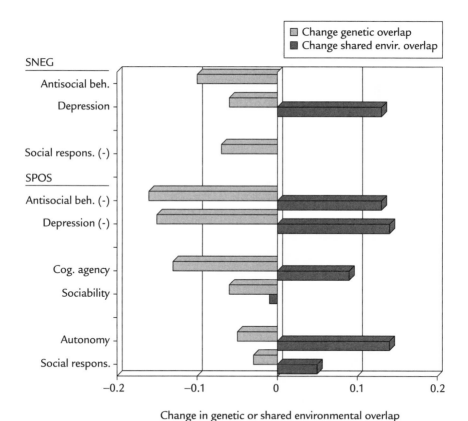

Figure 10.16 Summary of changes in genetic and shared environmental overlap for siblings

most families siblings are either high on both or low on both, but for a large number of families this is not the case. Anonymous shared environment accounts for more than a third of this overlap between autonomy and sociability: 34 percent at time 1 and 41 percent at time 2. At time 1, shared environment accounted more for the disjunction of autonomy and sociability than for their overlap, but by time 2 these two measures of adjustment had more overlapping shared environmental influences than unique shared environmental influences.

THE SILENCE OF THE NONSHARED FAMILY ENVIRONMENT   Among the most striking findings of this chapter is that the nonshared environment seems to have a negligible influence on adjustment. At times 1 and 2 nonshared factors failed to explain *any* appreciable overlap between almost all measures of family process and adjustment. Three exceptions are worth noting.

The first two are the inverse relationships between mother's and father's negativity and the adolescent's global self-worth at time 1, a finding not repeated at time 2. It is interesting to compare these findings with our phenotypic analysis in Chapter 7. There we found that the specific negativity directed at the adolescent by both father and mother accounted for about 10 percent of the variance in global self-worth at both time 1 and time 2. Our analyses here confirm that between one-third and one-half of this observed, phenotypic association is unrelated to genetically influenced differences between the adolescent siblings in self-worth. By time 2, however, genetic differences between children are the most significant influences on the association between parental negativity and global self-worth (see Figures 10.11 and 10.13). As Figure 8.1 indicates, anonymous nonshared factors still account for 50 percent of the differences among adolescents at time 2 in global self-worth. Regrettably, we have identified none of these time 2 factors in this study.

Among the exceptionally rich array of analyses made possible by our study design, one more association for nonshared environmental overlap is worth noting: that between father's negativity and adolescent depression. Here, nonshared overlap is the preponderant influence on the association between father's negativity and depression at time 2, although shared environmental and genetic factors are also important (see Figure

10.13). Interestingly, in our phenotypic analysis in Chapter 7 we had found that father's negativity had a substantial association with depression at times 1 and 2. The time 1 correlation now turns out to be attributable, for the most part, to genetic factors, but the time 2 association, in some measure, is attributable to nonshared factors. Even here, however, we can claim only limited success in identifying a nongenetic, nonshared correlate of adolescent adjustment. By time 2, anonymous nonshared factors are the preponderant influence on adolescent depression, accounting for well over 50 percent of the variation among adolescents (see Table 8.1). Differential negativity by father toward his children accounts for only a small fraction of these effects, just over 2 percent, which remain, for the most part, anonymous.

These three scattered findings are a very disappointing yield for a study designed to ferret out nongenetic, nonshared effects. The disappointment is enhanced when we recognize that global self-worth, a measure that provides 66 percent of our findings on nongenetic, nonshared correlates, is assessed by only a single scale given only to the adolescent. While the *concept* of self-esteem has been central in thinking about adolescent development in the family context, its *measurement* here is very circumscribed and dependent on a single measuring instrument and a single source and hence may contain considerable error.

CHANGES IN GENETIC OVERLAP   Our disappointment in this small yield of findings concerning the nonshared environment is well compensated by a cornucopia of entirely unanticipated findings on genetic factors. That genetic factors are important in adolescent development is no surprise. What is a stunning surprise is that genetic factors are (1) the most important contributors to the association between parenting and adolescent adjustment; and (2) the major engines of change in the associations between parent-child relationships and adjustment. Of the fifty-two associations between parental measures and adjustment that we examined in this chapter, forty-four indicated that genetic influences were more important than nonshared and shared environmental influences combined. In exactly half of these forty-four associations, shared and nonshared environmental influences combined were minute (less than .08). When comparing our findings from earlier and later adolescence, we observed no increase in environmental overlap exceeding .08 and no decrease exceeding .11. All other instances of change of greater magnitude than these, from time 1 to time 2, in the association of any of our measures of family subsystems and adolescent adjustment were attributable to changes in genetic influences. In sum, the impressive role of overlapping genetic factors in accounting for the association between parenting and adjustment stayed roughly the same for fathers and increased for mothers.

Chapter 11

SYNTHESIS I: GENETIC INFLUENCES
ON CHANGE IN FAMILY RELATIONSHIPS
AND ADOLESCENT DEVELOPMENT

According to our thesis, nonshared experiences of siblings hold a key to understanding how the environment, including the social milieu, exerts its influence on psychological development. Indeed, data we presented in Chapter 7 appeared to give very strong support to this thesis. We reported strong associations between qualities of the parent-child and sibling relationships, on the one hand, and levels of adjustment in all seven domains we studied, on the other. The evidence seemed to provide emphatic support for the importance of the social environment that was unique to each sibling.

As our antithesis suggests, however, this may all be a mirage. Genetic factors are important for individual differences in the same seven domains of adjustment. They are important not only for variations in adolescents across different families, but also for differences between adolescents in the same family. Moreover, genetic factors also influence all the domains of family relationships we have studied: marital, father-child, mother-child, and sibling. At the conclusion of the last chapter matters looked even worse for our thesis. The associations between family relationships and adolescent adjustment—associations that seemed like such a promising support for our environmental thesis—were attributable mostly to genetic factors. The only modest exception was for sibling relationships, whose links with adjustment were attributable not just to genetic factors but also to shared environmental factors. This was cold comfort, however, to our thesis. These shared environmental influences were clearly only a small fraction of the total set of influences on the seven domains of adjustment we studied. Even worse, our thesis never argued for shared envi-

309

ronmental effects because previous genetic studies had indicated that they were, with few exceptions, not major influences. Moreover, across all four family subsystems examined in our study, we could find virtually no role for nonshared factors in accounting for phenotypic associations between measures of family subsystems and measures of adjustment. Thus, the case appeared quite hopeless for the nonshared-environmental thesis.

We could leave matters just where they are. Other environmentalist theses, such as those related to some severe mental disorders, seem to have been buried by data that did not support them. Oddly, many psychosocial researchers seem to have willingly joined genetic researchers at what may have been an unnecessary funeral. Our thesis concerning the influence of marital, parent-child, and sibling relationships on adolescent development might have been the next candidate for entombment had we not begun this book with a very different conception of reasoning in the behavioral sciences. Data don't bury theories, especially not data from genetically informed designs juxtaposed with data from psychosocial studies. Close examination of the logical principles underlying behavioral science research, such as that undertaken in Chapters 2, 3, and 4, underscores that genetic data can sharpen and clarify psychosocial theories—not annihilate them.

Regrettably, the genetic data we have presented thus far can do little sharpening or clarifying; indeed, they have only muddied the waters. Our findings have indicated that psychosocial theories of adolescent development do not go far enough in explaining cause and effect. Many of the leading psychosocial correlates of adolescent development—marital, parent-child, and sibling relationships—are linked to adolescent adjustment by genetic and not purely psychosocial mechanisms. But the nature of this link is a mystery that clouds previous psychosocial theories without providing a clear alternative explanation.

In this chapter we take the first step toward providing an alternative theory that fits the data from our study but that cannot be proven by them. We estimate the relative contribution of two scenarios. In one scenario, heritable characteristics of the offspring may influence family relationships, and these relationships may, in turn, influence the ontogeny of adolescent adjustment. This is the parent-effects model discussed at several points thus far. If supportable, it represents a true synthesis: genetic factors and family factors are both important in adolescent development, but not as influences independent of each other. Indeed, the parent-effects model argues for the tightest possible interdependence between the two: genetic factors cannot be expressed in behavior without the mediation of the social environment, and the social environment, as assessed by a broad range of measures, has little effect independent of genetic influences.

A second scenario is that heritable characteristics of the child directly influence his or her adjustment with little or no direct effect on family subsystems. According to this scenario, however, the family does respond to these genetically influenced variations in adjustment. Thus the impact of genetic factors on the family system during adolescence is indirect or mediated through the adolescent's adjustment. In this scenario, the family has little continuing role in modifying adjustment.

Outline of a Synthesis

We have reached a critical turning point. We have brought the analyses of the data in our study as far as we had anticipated when we began our research project, and as far as conventional analytic models will take us. We have said what can be said with security. Ending our work here would merely reinforce the dilemma now facing the field of developmental studies: equally compelling data argue for the primacy of the social environment as argue for the primacy of genetic influences. The data just presented in Chapter 10 only heighten this paradox, providing firm evidence for the genetic perspective. We are left wondering if decades of psychosocial research on adolescent development, and perhaps other phases of development, has simply been wrong. And if so, how can we account for the effectiveness of a range of programs and interventions that have flowed from this research?

There is no accepted or well-developed approach to resolving this paradox. A simple resolution of the dilemma at this point, based on the data from our study and their analyses thus far, would have to favor developmental models in which genetic factors played the central role. Indeed, we have found that genetic factors play a major role in individual differences in adjustment and in change in adjustment across time. Genetic factors also influence all our measures of four crucial family subsystems: marital, mother-child, father-child, and sibling. Most important, we have found that the substantial associations between measures of family subsystems—particularly the parent-child subsystems—and measures of adjustment are due more to genetic factors with overlapping influences on family and adjustment than to overlapping environmental factors. Notably, there are exceptions to this overall pattern of high genetic overlap.

But there is good reason not to stop here. Indeed, the most fascinating aspect of our study begins now. We return to the data to look at critical details that suggest a more integrative synthesis than any presented thus far, one in which neither genetic factors nor environmental factors are central. More than likely these two sets of influences are intertwined in ways we never imagined when we began our study. Our synthesis takes place in four steps, with a chapter devoted to each.

First, we look carefully at the subtle though informative sequences of genetic effects on family dynamics and adolescent adjustment. We ask whether genetic effects have their impact on family process first, and, if so, whether there is evidence that these family processes mediate genetic influences on development. If this is the case, then we have evidence favoring what we have called the parent-effects model (see Figure 10.3). We have redrawn this model with some small changes in Figure 11.1, which

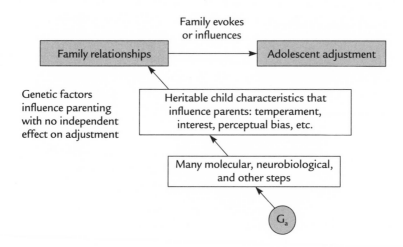

Figure 11.1 Child-initiated family-effects model

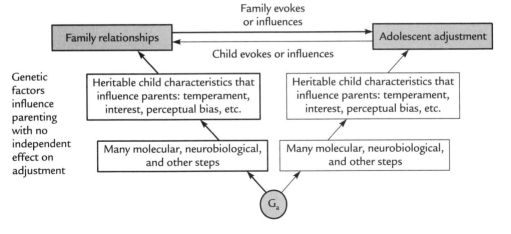

Figure 11.2 Alternative sequences of effects for evocative models

depicts heritable characteristics of the child that influence family relationships instead of just parent-child relationships, as in Figure 10.3. Here we allow for other effects of heritable characteristics on the family system, namely, effects on sibling and marital subsystems. The child-initiated family-effects model shown in Figure 11.3 postulates a particularly close intertwining of genetic and social factors: genetic factors require social processes in order to be expressed in behavior, but social processes do not act independent of genetic variation. For example, heritable characteristics of the child shape the level of parental hostility toward that child. Parental hostility then intensifies the child characteristics, making antisocial behavior more likely.

This chapter provides a schematic summary of the data on this crucial sequence of genetic influence. We estimate the extent to which genetic factors influence parenting first and adolescent adjustment second; this sequence is shown in the lefthand portion of Figure 11.2. An alternate possibility, which we can detect with our analyses, is that genetic influence on adjustment precedes genetic influence on family processes, a sequence shown in the righthand portion of Figure 11.2.

As a second step in our synthesis, we explore in Chapter 12 the capacity of family processes to transmit genetic information. This is another way of exploring the plausibility of the family-effects model pictured in Figures 11.1 and 11.2. The concept of "transmission" implies that there is already good reason to believe that family processes may serve as a mediating step in the expression of genetic influence in levels of adjustment. On the basis of this premise, we explore two closely related issues.

First, we examine the genetic overlap among all our measures of adjustment, asking if these measures all reflect some underlying, monolithic genetic influence, or whether unique genetic factors are important influences on adolescent adjustment. For example, at time 1, antisocial behavior is inversely correlated with sociability ($-.22$). About half of this correlation is attributable to overlapping genetic influences on antisocial behavior and sociability, that is, the same genetic factors that influence adolescent antisocial behavior also influence reduced sociability. Each has important and unique genetic influences that are not shared with the other, however. Indeed, the major reason that antisocial behavior and sociability have such a low correlation ($-.22$ instead of $-.60$ or $-.80$) is that each has genetic influences that are distinct from the other's. The coefficient of disjunction ($1 - r$) is .78; 66 percent of this coefficient is accounted for by distinctive genetic influences on antisocial behavior and sociability. To put the matter another way, genetic influences account for two-thirds of the lack of correlation between antisocial behavior and sociability. Or, in yet other words, only 13 percent of the genetic influence on antisocial behavior overlaps with that of sociability. Similarly, only 23 percent of the genetic influence on sociability overlaps with that of antisocial behavior. So far as these two domains of adjustment are concerned, there is no single monolithic genetic influence underlying both.

If this is a typical pattern, then family process—as a mediator—will have to be responsive to or reflect these distinctive genetic influences. If it is to be a mediator, it cannot be an obtuse responder to a broad variety of distinctive genetic influences. Rather, it must also be responsive to many *different* genetic influences. We test this idea in Chapter 12. If family process has this distinctive responsivity, it might well constitute a relationship code capable of transmitting the broad array of distinctive influences on adolescent adjustment. Figure 11.3 shows the two possible findings in these analyses. The upper portion of the figure illustrates how distinctive genetic influences might have a differential impact on two different family processes. For example, one set of genetic factors might influence mother's warmth and support toward her adolescent children; another set

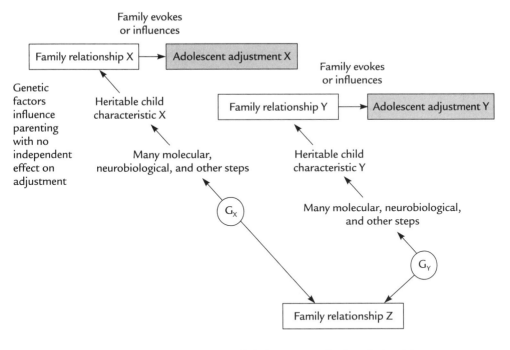

Figure 11.3 Distinctive genetic influences on family relationships

of factors might influence her conflict and negativity. This would allow, to continue the example, for the genetic influences on adolescent autonomy to be "transmitted" by mother's warmth and the genetic influences on depression to be "transmitted," distinctively, by mother's conflict and negativity. The lower portion of Figure 11.3 illustrates schematically an "obtuse" family process, one that cannot respond differentially to two distinctive genetic influences. This process can "transmit" only genetic influences common to two or more domains of adjustment but cannot transmit the critically important distinctive genetic influences.

In a third step toward synthesis, we re-examine in detail what we have learned about the shared and the nonshared environments. Data about the role of environmental factors have spoken in a soft voice so far. First, the specifiable effects of the social environment, independent of genetic influences, have been small. Second, the main target of our study, nonshared influences, remains elusive. However, there have been intriguing leads on the shared and nonshared environments that we have set aside in previous chapters. For example, we have reported the surprising influence of the shared environment on autonomy and sociability as well as on the role of sibling relationships in adolescent adjustment, a role that seems somewhat different in our analyses from the role of parent-child relationships. Perhaps most important, we must return to the nonshared environment. Indeed, our results have confirmed the major implications of previous findings: this aspect of the environment is the most important influence on adolescent adjustment. Yet it is both elusive—we have not been able to specify just what it is—and unstable across time. Can we come to grips with what these influences might be?

The fourth step in our synthesis looks to the future. We cannot prove, with the data we have in hand, the synthetic proposal we make here. Our synthesis is the result of inferences arising from a study that was designed with simpler aims. Indeed, we did not design the study to investigate the central hypothesis that has arisen from our findings, the hypothesis of the relationship code. But the synthesis itself has little value unless there are existing methods to test it. Chapters 2 and 4 reviewed the designs available to developmental research. We emphasized the logic of inference inherent in each design and formulated a review of these designs as a set of logical principles. The most crucial components of a model of development are specific causal links. We argued in Chapters 2 and 4 that cross-sectional and longitudinal studies provide some clues to such links. But theory-testing experiments are the strongest evidence for causal sequences in development. As noted, the most frequently used experiments in the developmental field are treatment studies for people whose development has led to serious maladjustment, and prevention studies to block the development of maladjustment. These true experiments serve two major functions for our synthesis. First, they remind us that some psychosocial theories of development have strong backing and that our new, genetically informed theories of development must account for the positive outcomes of these experiments. Second, they serve as a template for research designs of comparable strength in genetics. Genetically informed designs rarely use experiments, and none have been used to delineate the behavioral mechanisms involved in gene expression. In Chapter 12 we outline an experiment, based on a preventive-intervention model, that would fill this conspicuous gap.

Special Challenges in Studying Sequences of Events

Researchers study development across time in order to understand the causes of change in human behavior and to learn which factors lead to high levels of adjustment and which lead to poor adjustment. Even more important, they try to determine which factors lead to *change* in adjustment across time. They hope to gain enough information about adolescent development to be able to promote high levels of adjustment and prevent serious problems in children, adolescents, and adults. As pointed out in Chapters 2 and 4, it is difficult to define causes of development. For example, many developmental researchers recognize serious limitations inherent in longitudinal studies, though such studies often provide the best information available on influences on development. This is the case with our current study. As noted, a stronger case for cause-and-effect relationships can be made with true experiments than with longitudinal studies if the experiments are a reasonable model of sequences as they actually occur in development. Indeed, Chapter 14 will specify tests for our synthesis by laying out a program of true experiments in this field. To carry out this program, we must extract as much information as possible from longitudinal studies. Only then can we construct good hypotheses that will—in turn—serve as the basis for true experiments.

Although measuring developmental change presents serious problems in most studies, it is a particularly daunting task in this one. Six features of our study provide especially difficult challenges.

First, we are not studying a single independent or potentially causal variable; rather, we are studying four separate family subsystems: mother-child, father-child, marital, and sibling. Further, we have assessed several different qualities of these relationships. In addition, our study included stressful life events as well as qualities of the adolescents' peer groups, teachers, and best friends. Second, we are studying seven large domains of adjustment: antisocial behavior, depression, cognitive agency, sociability, autonomy, social responsibility, and self-worth. Third, we are searching for both environmental and genetic mechanisms of causality. Fourth, we are examining two broad classes of environmental variables: shared and nonshared. Fifth, we focus not only on anonymous genetic and environmental variables, but also on specific environmental variables that describe the family subsystems. Thus, the picture of causal links between environmental factors and development, derived from this complex mélange of data in our study, threatens to be complex indeed.

A sixth difficulty deserves special mention. Despite the fabled *Sturm und Drung* of adolescence, we have not observed extraordinary changes in the adolescents, vis-à-vis one another, on most of our measures. A quick review of Figure 9.10 reveals that the stability of almost all measures is substantial across three years. It is only when we turn to parental measures of monitoring and control that we see major changes, with the coefficient of stability falling below .50. The low stability of our measure of self-esteem, among this set of measures, is most likely due to our use of only one measure and one informant, the adolescent. It probably does not reflect instability or change in the underlying psychological process. In order to detect processes relevant to change in our sample of stable adolescents and their families, we must pay close attention to the data.

Resolving Interpretive Dilemmas with a Cross-Lag Analysis

To illustrate our approach to longitudinal analyses, let us return to findings first reported in Figures 10.6, 10.7, 10.11, and 10.13. These findings, summarized in Figure 11.4, concern mother's and father's knowledge or surveillance of their child and the child's antisocial behavior. These are not the most dramatic findings we have presented, but Figure 11.4 is informative nonetheless. It tells us that the phenotypic correlations between parental knowledge and antisocial behavior, an inverse relationship, are very modest at time 1 but nearly double at time 2 (note that the arrows shift to the right for both moms and dads).

Figure 11.4 also tells us that the genetic components of the correlation increase along with the phenotypic correlation, and that the increase for each of these genetic components is proportionate to or greater than the increase in the phenotypic correlation. Since we know that there is a substantial change in the genetic influence on antisocial behavior from time

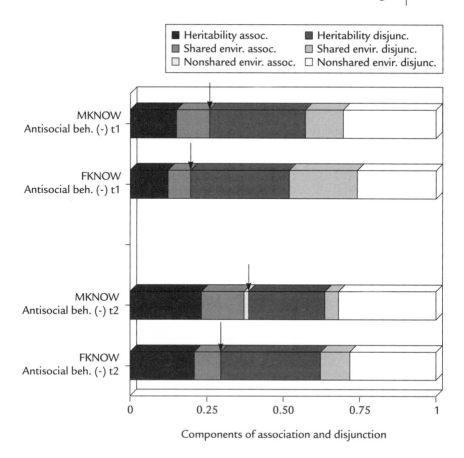

Figure 11.4 Genetic and environmental components of the association between parental monitoring and antisocial behavior

1 to time 2 (see Figures 8.5 and 8.6), it is tempting to interpret these modest shifts in phenotypic and genetic correlations as the parents' reactions to the genetic changes in antisocial behavior in their offspring. Perhaps the genetic influences that awaken in late adolescence produce a more malignant form of antisocial behavior than is active in early adolescence, and, as a consequence, both parents reduce their monitoring; they "back off." According to this view, later in adolescence parents of the most troublesome offspring reduce their surveillance, whereas parents with children who show much less antisocial behavior maintain their vigi-

lance to be sure their adolescents don't slip up. Quite possibly the parents of a very troublesome child turn their attention to the less troublesome sibling to prevent that child from following his or her wayward sibling. We have referred to this possibility as "parental salvage." This sequence is an instance of an evocative genetic-environmental correlation of the child-effect variety. We have illustrated this model for parental surveillance and antisocial behavior in Figure 11.5. This figure is a variant of Figure 11.2, and it provides a specific example for the righthand portion of 11.2.

This image of parental retreat is troubling, leading us to search for a more benign explanation for findings of this sort. According to one alternative theory, illustrated in Figure 11.6, it is possible that genetic factors influence a precursor to antisocial behavior. This precursor does not, itself, lead to the same serious maladjustment that antisocial behavior does. It is detectable by parents, however, galvanizing them into action. The more a child evinces this genetically influenced trait, the more the parent monitors the child. Thus, there is a genetic correlation between the magnitude of the troubling trait and the magnitude of parental surveillance. If the process stopped there, then the association between parental monitoring, on the one hand, and antisocial behavior that develops from the precursor, on the other, should be positive, not negative. This would indicate that parental monitoring is not effective: it is initiated by the child's evincing a troublesome precursor. Despite the parents' increased monitor-

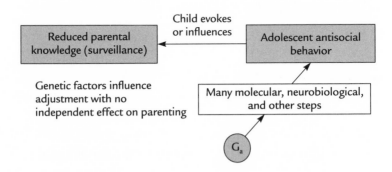

Figure 11.5 Child-effects model of parental salvage

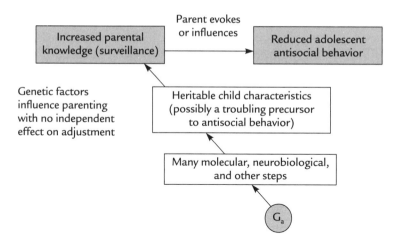

Figure 11.6 Parent-effects model of parental monitoring and antisocial behavior

ing of the troublesome child, the antisocial behavior develops anyway. A crucial second step in this sequence is necessary to fit our findings. The parents' monitoring behavior successfully *reduces* the child's potential for antisocial behavior. Indeed, the children most effectively monitored earlier in development are those with the least probability of developing antisocial behavior as they traverse adolescence. The hypothesis further states that this protective effect of monitoring, while it may begin before early adolescence, becomes most conspicuous in late adolescence. Hence we see the enhanced genetic overlap between genetic influences and parental monitoring at time 2 in comparison with time 1. In sum, this is a theory of "parental protection" in sharp contrast to a theory of parental retreat or, with reference to the better-behaved sib, a theory of parental salvage.

We need an analytic approach that helps distinguish between two such diametrically opposed explanations of the data. Ideally we would perform an experiment. If the child-effect model is correct, then any treatment that successfully reduces antisocial behavior should enhance parental monitoring; after our experiment, the parent would switch from exclusive surveillance of the well-behaved child and provide an equal degree of monitoring to the effectively treated child. Alternately, we could demonstrate a parent-effect model by helping parents improve their monitoring. We could take two mutually compatible strategies. First, we might instruct parents of children who show little of the heritable precursor behavior that monitoring is important because significant environmental influences may enhance antisocial behavior. They would be encouraged to improve their monitoring to protect the children from a variety of shared and nonshared environmental risk factors for antisocial behavior. Let us suppose that, subsequent to this intervention, the children in the parent-treated group do show reduced antisocial behavior in comparison with an untreated control group. We have then demonstrated part of this model: the effectiveness of monitoring in protecting a child from antisocial behavior. If we effectively treated only the group of parents with nontroublesome children, we would sharply *reduce* the genetic correlation of antisocial behavior and parental monitoring. This is because we are eliminating the effectiveness of the heritable "signal" inherent in the child's troublesome behavior. In an untreated group it is these children who elicit effective monitoring. Hence, the genetic correlation: children with heritable troubling behavior elicit surveillance and children without this behavior do not. In a treated group these troublesome children and their nontroublesome counterparts receive enhanced surveillance. Thus, the "signal" is overridden by the effective treatment. In effect, the treatment provided to parents of the nontroublesome children provides a "signal" equal in strength and effect to that provided by the troublesome children.

Second, we could also enhance the ability of parents to detect early signs of trouble and improve their monitoring. This should actually *increase* the genetic correlation of antisocial behavior and parental monitoring. We would be enhancing a natural process by helping parents detect early signs of trouble more accurately than they would under ordinary circumstances while enhancing their reaction to what they observe.

Before undertaking the enormous expense entailed in experimental interventions of this kind, we must perform the most powerful analyses possible on the data currently available to us, in order to provide plausible scenarios. We have designed analytic procedures for this study that will help us weight the likelihood, for example, of parental salvage or parental protection as mechanisms accounting for data of the kind summarized in Figure 11.4. Our analytic strategy focuses on sequence: do genetically influenced components of antisocial behavior (or other measures of adjustment) in earlier adolescence *precede* changes in parental monitoring (or other measures of family subsystems) later in adolescence? Perhaps the reverse case is true: changes in genetic influences on parenting precede changes in antisocial behavior.

Analytic Principles

A *cross-lagged correlation* is one way to perform an "experiment" on longitudinal data. It is an effort to repeat statistically the main steps of an actual experiment. Figure 11.7 shows the basic elements of a cross-lagged analytic model. The four central measurements are those of family subsystem characteristics at time 1 and again at time 2, as well as measurements of adjustment at times 1 and 2. We would like to know whether, for example, a family characteristic at time 1 precedes a change in adjustment between time 1 and time 2, represented by arrow X in Figure 11.7. Alternatively, we would like to know if the level of adjustment at time 1 precedes change in the family subsystems between times 1 and 2, represented by arrow Y.

However, we cannot simply correlate family subsystems at time 1 with adjustment at time 2. A relationship of that kind is contaminated by other relationships that are not relevant to establishing a sequence. Consider the top half of Figure 11.7. The relationship depicted by arrow X will have a higher value the higher the contemporaneous or cross-sectional relationship is between family process and adolescent adjustment at time 1 and also at time 2. This is particularly so if the stability of the measure of family relationships and of adjustment is high. A conservative test of any hypothesized sequence involves restricting the analytic model shown in Figure 11.7. Understanding the genetic and environmental components of the relationship between time 1 and time 2 requires additional steps. Schematically, we describe our analytic model here in three steps, although the computational procedure accomplishes all three simultaneously. A more detailed explanation of the model is published elsewhere (Neiderhiser, Reiss, Hetherington, and Plomin, 1999).

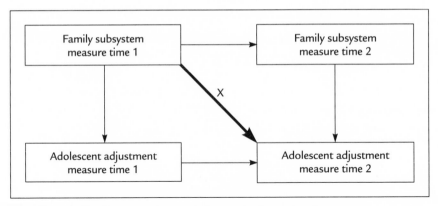

Family system characteristic *precedes* change in adjustment

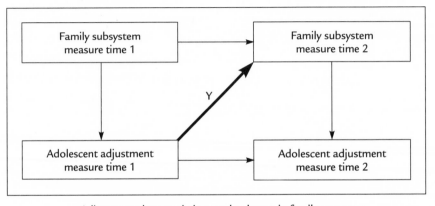

Adjustment characteristic *precedes* change in family system

Figure 11.7   Basic phenotypic cross-lagged correlation

STEP 1: PARTIALING OUT CONFOUNDING RELATIONSHIPS FROM THE CROSS-LAGGED ASSOCIATION   The first step involves two components. First, we want to separate the family measure at time 1 from two correlates that may confound our estimates of the relationship between family process at time 1 and change in adolescent adjustment between time 1 and time 2 (see Figure 11.8, part A).

The first correlate is the stability of the family measure. If we remove this by *partialing* it from our analyses, we are left with family process as measured at time 1, but this does not continue or predict family process at time 2. (The term "partialing" refers to a statistical procedure for estimating and then removing the effects of a confounding variable from an analysis.) Removing this confounding role of the stability of family process fixes our analysis on the unique aspects of time 1 family processes. A second unwanted correlate is the relationship between family process and adjustment at time 1. It is impossible to know, from our data, which of

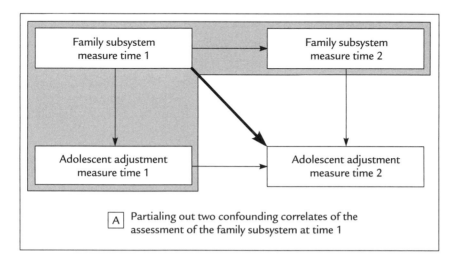

A   Partialing out two confounding correlates of the assessment of the family subsystem at time 1

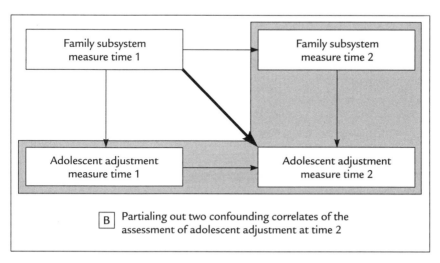

B   Partialing out two confounding correlates of the assessment of adolescent adjustment at time 2

Figure 11.8   Partialing out confounding relationships from a cross-lagged association

these—family process or adjustment—precedes the other. This indeterminate relationship could influence our estimate of the relationship of time 1 family processes to changes in adjustment from time 1 to time 2. Thus, we try to eliminate this contemporaneous association between family process and adjustment at time 1 by a statistical partialing procedure.

Second, we want to separate our measure of adjustment from two confounding correlates (see Figure 11.8, part B). Since we are interested only in the effects of time 1 family process on change in adjustment, we want to eliminate the influence of time 2 family process on time 2 adjustment. We are also interested only in change in adjustment across two time periods; hence, we partial out the effects of time 1 adjustment on time 2 adjustment. The resulting cross-lag will be small, particularly if the confounded relationships we have partialed out are large, but it will be a conservative estimate of the sequence of events we are tracking here.

STEP 2: ESTIMATING GENETIC AND ENVIRONMENTAL COMPONENTS OF THE CROSS-LAGGED CORRELATIONS Using strategies that are analogous to those in our more simple bivariate analyses from Chapters 8, 9, and 10, it is possible to construct latent variables that serve as estimates of the components both of the relationships that are partialed and of the remaining cross-lagged correlation. Figure 11.9 illustrates this very schematically. Regrettably, even this "simplified" diagram looks a bit like the remnants of a balloon sale after a tornado has struck. A bit of patience, however, will help readers recognize the basic logic of the cross-lag model in this tangle.

First, examine the balloons in sector A of the diagram. Then go back to Figure 11.8 and observe the phenotypic partialing in part A of that figure. Sector A in Figure 11.9 is the genetic expansion or detailing of this more familiar phenotypic strategy. That is, sector A of this tangle accomplishes the same partialing as do the statistical procedures in part A of Figure 11.8. The latent variables in sector A of Figure 11.9, the three balloons labeled "G," "NS," and "SE," are estimates of the genetic and environmental components of the partialed-out variables that confound assessment of time 1 family-process measures. For reasons we have already stated, these variables must be partialed from our analyses but are of little interest in themselves. The same is true of the genetic and environmental compo-

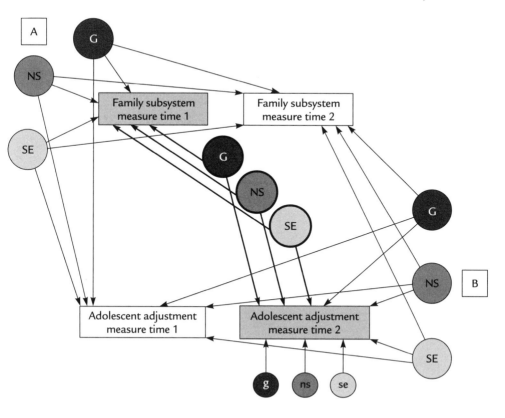

Figure 11.9 Model estimating genetic and environmental components of the cross-lagged association from family process to adolescent adjustment (G = genetics; NS = nonshared environment; SE = shared environment)

nents in sector B of this diagram. These are equivalent to partialing variables that affect the assessment of changes in adjustment from time 1 to time 2. The contemporaneous assessments at time 2 are also partialed here. Thus, the tangle of balloons in sector B of Figure 11.9 is, in effect, the genetic detailing of the phenotypic partialing shown in part B of Figure 11.8.

Now, we come to the point of all this: the genetic and environmental components of the cross-lagged correlation. These three components are represented by the three balloons with bold borders and bold arrows in the very center of Figure 11.9. Added together, they estimate the extent to which variation in time 1 family process precedes changes in adolescent adjustment. The genetic component, the primary interest of this chapter, tells us the extent to which this sequence is attributable to genetic mechanisms. Let us return to our example of parental knowledge (surveillance) and antisocial behavior. Suppose that the value of "G" is high (it is estimated by the product of the path coefficients on each of the two arrows leading away from "G"), and that "G" is the predominant component of this cross-lagged correlation. That is, the coefficients on the arrows leading away from "NS" and "SE" are very small. The following two statements are plausible interpretations of this finding:

1. Differences among families in parental monitoring unique to time 1 precede changes in adolescent antisocial behavior from time 1 to time 2 *for genetic reasons*.
2. The relationship between genetic factors and family process that is unique to time 1 precedes changes in adolescent antisocial behavior from time 1 to time 2.

The following is not a plausible interpretation of these findings:

The sequence "genetic influence on parenting at time 1 precedes change in adolescent antisocial behavior from time 1 to time 2" is less likely than the opposite sequence, "genetic influence on antisocial behavior at time 1 precedes change in parental monitoring from time 1 to time 2."

Unfortunately, there is no good method for comparing the two sequences enclosed in quotation marks in the last paragraph. We have not been able to design a single model to compare rigorously these two sequences; nor are there acceptable techniques for statistically comparing two different models. Nonetheless, we can address this crucial comparison, lying at the heart of our effort at synthesis, by taking our analytic efforts one step further.

STEP 3: REVERSING THE CROSS-LAG: EXAMINING THE IMPACT OF ADJUSTMENT ON FAMILY PROCESS The third step simply follows the same path we outlined in step 1 and step 2 above. We begin, however, with the relationship between adolescent adjustment that is unique to time 1 and estimate the extent to which this precedes changes in family process from time 1 to time 2 (see arrow Y in Figure 11.7). We then come to an analytic strategy identical to the one pictured in Figure 11.9. The only change is that the contents of the "gray boxes" are reversed. In the lower left we have "adolescent adjustment measure time 1," and in the upper right, "family subsystem measure time 2." This is illustrated in Figure 11.10. We will again focus on "G" in the diagonal. If the coefficients for the arrows leading away from "G" are very large, and those leading away from "NS" and "SE" are small, the following inferences are plausible:

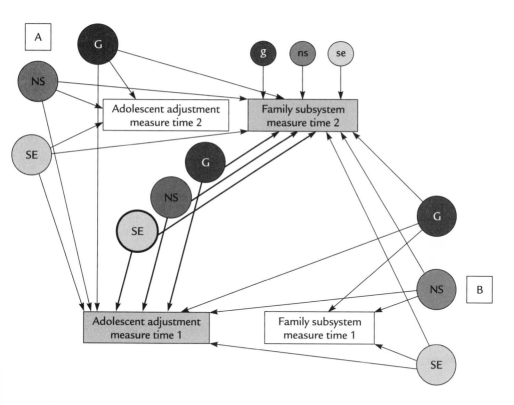

Figure 11.10 Model estimating genetic and environmental components of the cross-lagged association from adolescent adjustment to family process (G = genetics; NS = nonshared environment; SE = shared environment)

1. Differences among families in adolescent antisocial behavior unique to time 1 precede changes in parental monitoring from time 1 to time 2 *for genetic reasons.*
2. The relationship between genetic factors and adolescent antisocial behavior that is unique to time 1 precedes changes in parental monitoring from time 1 to time 2.

Now refer back to Figures 11.2, 11.5, and 11.6. The burden of analysis here is to decide which is more plausible: the family-effects model as shown in the lefthand portion of Figure 11.2, or the child-effects model as shown in the righthand portion of the same figure. Or to describe this aim in more specific terms: Do parents recognize some heritable anlage of antisocial behavior and then respond with increased monitoring, with a consequent reduction in the child's antisocial behavior? Or is it the other way around: do genetic factors influence antisocial behavior directly, without an independent effect on parenting? As a consequence of this genetically influenced antisocial behavior, parents reduce their surveillance, backing off.

There is only one broadly accepted, rigorous method to compare the models depicted in Figures 11.9 and 11.10: testing to see if one model fits the data better than the other, using statistical tests of fit. This statistical comparison, although acceptable, is weak and inconclusive because a test of fit pays equal attention to every component of each model, not just to the crucially informative cross-lags. As a consequence of these limitations, we did not employ this method of comparison. Instead, we used a more direct, though much more intuitive, approach: we examined the partialed cross-lag correlations themselves, comparing arrows X and Y in Figure 11.7 after they have been partialed as illustrated in Figure 11.8. Are there conspicuous differences? If so, they may help us weight the plausibility of the parent-effects model versus the child-effects model.

The Sequence of Family Process and Adolescent Adjustment: The Results of Cross-Lagged Analyses

We present the results of our analyses in two steps. First, we examine the phenotypic, partialed cross-lag correlations. There are two groups of these. The first group consists of the cross-lags between time 1 family process (f1) and time 2 adolescent adjustment (a2). The second group consists of the correlations reflecting the reverse sequence: time 1 adolescent adjustment (a1) and time 2 family process (f2). We know in advance that these correlations will be small because the contemporaneous associations and stabilities, both of which are substantial, will have been removed from these associations by partialing. These phenotypic cross-lags will, as noted, be quite informative. If most of the f1 → a2 correlations are substantial and most of the a1 → f2 associations are minimal, this provides some support for the idea that variation in family process precedes changes in adolescent adjustment. This step will not, however, tell us about the role of genetic factors in this sequence. Thus, our data analysis has a crucial second step, illustrated in Figures 11.9 and 11.10. (Details of these analyses are in Appendix B.)

Family Subsystems and Phenotypic Cross-Lags

MOTHER'S PARENTING   For all our analyses we first examined unpartialed cross-lags. We continued with the analysis only if unpartialed cross-lags equaled or exceeded .20. We then estimated genetic and environmental components of the cross-lag only if the partialed cross-lag equaled or exceeded .15. As Figure 11.11 shows, only eight partialed cross-lag correlations for mother's parenting met our criteria for further analyses. All of these were f1 → a2. None of the a1 → f2 associations exceeded this threshold. The strongest contrasts in direction of effect are for mother's negativity and cognitive agency, mother's positivity and antisocial behavior, mother's positivity and sociability, mother's knowledge and antisocial behavior, and mother's knowledge and cognitive agency. In these instances the f1 → a2 associations are at least twice as large as the a1 → f2 cross-lags.

FATHER'S PARENTING   Figure 11.12 shows the partialed cross-lags for father's parenting. Six of these meet or exceed our threshold of .15, and five of these are f1 → a2. Also, of these five, four show notable differences between f1 → a2 and a1 → f2; the former are at least twice as large as the latter. These cross-lags are father's negativity and antisocial behavior, father's negativity and cognitive agency, father's positivity and antisocial behavior, and father's knowledge and antisocial behavior.

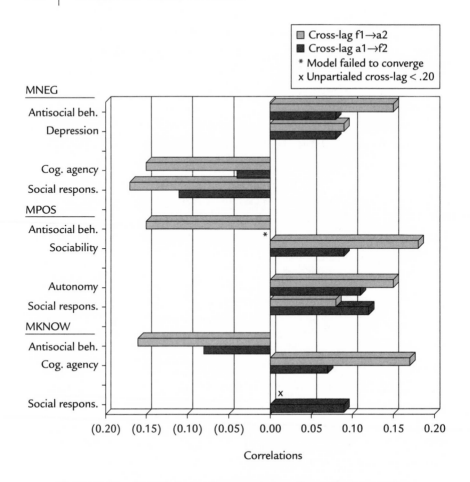

Figure 11.11 Partialed cross-lag associations for mother's parenting

SIBLING RELATIONSHIPS Figure 11.13 shows that four cross-lagged associations between sibling relationships and adjustment meet or exceed our threshold of .15. These are sibling negativity and antisocial behavior, sibling positivity and antisocial behavior, sibling positivity and sociability, and sibling positivity and social responsibility. Of these, three are f1 → a2 associations and two of these show a distinct difference between the f1 → a2 and the a1 → f2 associations: positivity and antisocial behavior and positivity and sociability.

In all, fifteen associations over .15 are of the f1 → a2 type and only three are of the a1 → f2 type. Of the former, eleven show notable differences between f1 → a2 and a1 → f2. In only one of these is a1 → f2 distinctly different from f1 → a2. Thus, the *pattern* of findings—but not a rigorous test of them—suggests that in the majority of sequential relationships we have analyzed there is either not enough change across time to delineate clear sequences of change or time 1 family characteristics *precede* changes in adjustment from time 1 to time 2.

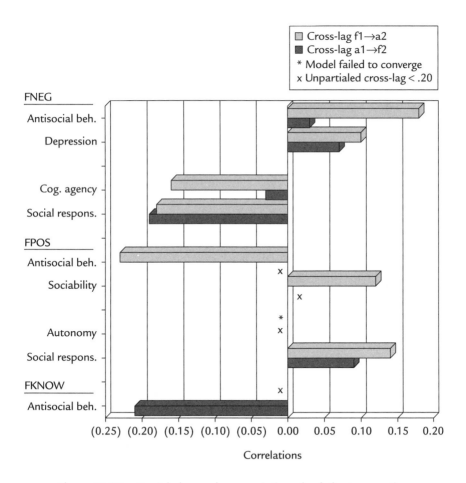

Figure 11.12 Partialed cross-lag associations for father's parenting

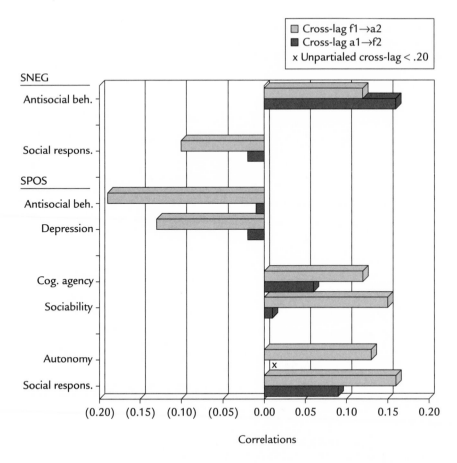

Figure 11.13 Partialed cross-lag associations for sibling relationships

Genetic and Environmental Components of the Cross-Lagged Correlations

Figures 11.14 and 11.15 show all cross-lags in Figures 11.11, 11.12, and 11.13 that equal or exceed .15. In all, eighteen cross-lags are shown, fifteen of the f1 → a2 type and three of the a1 → f2 type. Of these eighteen, fourteen show genetic factors equaling or exceeding shared and nonshared factors combined. These data are revealing. Of particular interest are sequences that can be interpreted as follows:

The sequence "genetic influence on parenting at time 1 precedes change in adolescent antisocial behavior from time 1 to time 2" is more likely than the opposite sequence, "genetic influence on antisocial behavior at time 1 precedes change in parental monitoring from time 1 to time 2."

A statement of this kind seems plausible if a relationship between a family-process variable and an adolescent-adjustment variable meets the following criteria: (a) the partialed cross-lag equaled or exceeded .15; (b) the partialed cross-lags were conspicuously greater for a2 → f2 than for the reverse (although there is no rigorous comparison possible here, we pay special attention to relationships between family process and adjustment where f1 → a2 exceeds a1 → f2 by a factor of two or more);

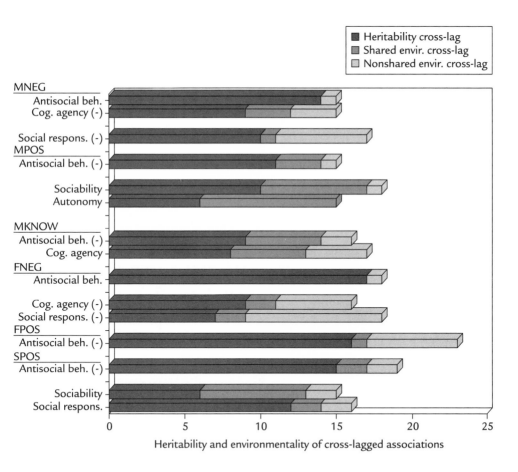

Figure 11.14   Genetic and environmental components of f1→a2 cross-lags at .15 or higher

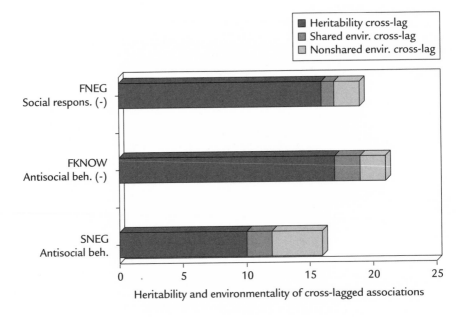

Figure 11.15 Genetic and environmental components of a1→ f2 cross-lags at .15 or higher

(c) the genetic component of the cross-lagged association was larger than the combined nonshared and shared environmental component. Figure 11.16 provides a schematic summary of all eight of these relationships.

In each of these eight sequences the cross-lagged analysis is consistent with the following sequence: (a) genetic factors influence a precursor of sociability, antisocial behavior, or cognitive agency; (b) this precursor evokes or elicits a parental response; (c) the parental response further shapes this precursor so that it emerges as a full-fledged characteristic of adjustment. It is possible, however, to infer that the parental and sibling responses have quite different influences on the ontogeny of different domains of adjustment. For example, the cross-lag correlation between mother's positivity and sociability is positive and its genetic component is predominant. It seems reasonable to conclude that mother's positivity, itself evoked by a precursor to sociability, *enhances* this trait in the child to make a more fully developed sociability more likely and effective. For this reason we have added the word "enhances" over the appropriate arrow in

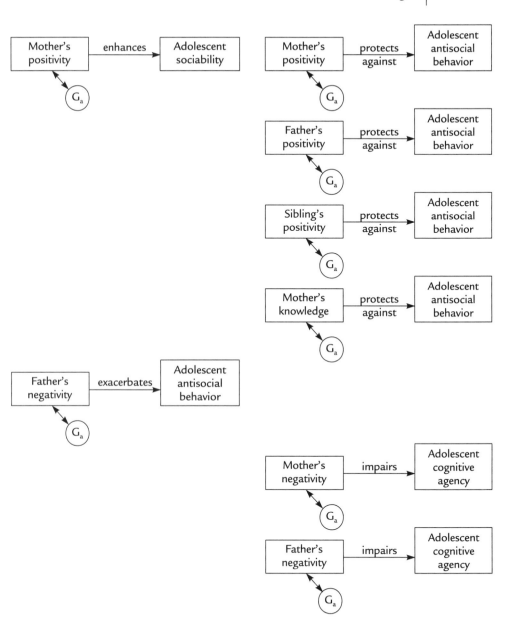

Figure 11.16 Summary of eight f1→ a2 sequences

Figure 11.16. But given that we did not measure the genetically influenced precursor of sociability—the precursor that we assume influences mother's positivity—the term "enhances" is a reasonable conjecture, not a more immediate or conservative inference from our test of statistical models.

The relationship between mother's and father's positivity and antisocial behavior is of a different nature. The salient cross-lag here is negative, meaning that the more maternal or paternal positivity there is at time 1, the less likely the child is to become more antisocial from time 1 to time 2. We cannot be certain whether the parents are responding to some precursor of antisocial behavior, perhaps some deceitfulness, impulsiveness, or aggressiveness. If so, they may be acting to counter this trait with warmth and support and thus suppressing the development of more negative antisocial behavior. It is also possible that they are responding to something positive in their child. Perhaps the child is both a rogue and a charmer. Responding to the child's positive, heritable trait may then have a protective effect, preventing the negative trait from developing into something quite dreadful in the three years between time 1 and time 2. It is also possible that mother and father are responding to different heritable traits in the same child. Indeed, we present evidence in the next chapter showing that there is *no* overlap in the genetic influences on mother's and father's positivity.

It is notable that sibling positivity seems to play the same role as parental positivity. The sibling responds to a heritable characteristic in the target adolescent, and this positive reaction appears to protect the adolescent from antisocial behavior. It is striking that there is virtually no genetic overlap between sibling positivity and either mother's or father's positivity. For that matter, there is no genetic overlap between sibling positivity and *any* parental behavior. Thus, if the family-effects model is supportable, the sibling may be responding to a quality in the adolescent that is quite different from qualities possessed by either mother or father. Indeed, all three family members—mother, father, and sibling—seem to be responding to a heritable, and probably positive, characteristic unique to each of them. Each of these three separate positive stimuli from the adolescent may lead to three distinctive and separate protective systems in the family. Yet a fourth partially independent protective mechanism seems to operate here: mother's knowledge or surveillance, also possibly stimulated by a heritable characteristic in the child, also protects the child from antisocial behavior. We have speculated that mother's surveillance is not a response to a positive feature of her child—although we cannot rule this out. Rather, it seems more parsimonious to assume that mothers would increase their surveillance only of children they were worried about, children showing some early signs of getting into trouble.

The relationship between father's negativity and adolescent's antisocial behavior provides yet another form of parent effect. Here a heritable quality of the child—probably an unpleasant one—evokes a negative, not a positive, response. This negative response most likely *exacerbates* the child's unpleasant trait, making it into a serious problem of maladjustment.

The relationship between mother's and father's negativity and cognitive agency may reflect a fourth mode of parent effect. Here, some heritable attribute of the child elicits negativity from a parent, which in turn *impairs* cognitive agency. It does not seem likely that early signs of intelligence and industry, perhaps the precursors of cognitive agency, would elicit a negative response from most parents. Rather, it is more likely that irritability and aggression in the child are the culprits. Cognitive agency is an innocent bystander: when a parent responds with conflict and punitiveness to an irritable and challenging child, he or she not only exacerbates that child's aggressive and antisocial behavior, but also impairs the child's cognitive agency.

We use one other convention of notation in Figure 11.16. The arrow between genetic influence and characteristics of the social system is double-headed. This expresses our lack of certainty about the pathways between genetic influence and family process. As noted, there was probably a reciprocal influence, extending across years of development, between heritable characteristics in the child and the response of the child's family before we made our first observations in early adolescence. This extended reciprocal influence is summarized by the two-headed arrows.

Figure 11.17 shows a schematic summary of the one finding that meets informal criteria for a genetically influenced impact of a measure of adjustment on measures of parenting: the relationship between antisocial behavior and father's knowledge or surveillance. This is the only instance

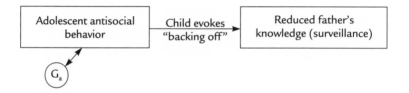

Figure 11.17 Schematic summary of single clear a1→ f2 sequence

in which (a) the partialed cross-lag equaled or exceeded .15; (b) the a2 → f1 cross-lag was at least twice as large as the f1 → a2 cross-lag; and (c) genetic factors exceeded environmental factors in the cross-lag. Indeed, in this instance there was no computable f1 → a2 cross-lag because the unpartialed cross-lag was so low. The data suggest that fathers *back off* in the face of unruly children: they reduce their level of surveillance in response to unruly children—a pattern that is suggested, particularly for time 2, by trends in our phenotypic analyses (see Figure 7.16). These analyses also suggest that, by later adolescence, father may be monitoring the better-behaved sibling.

This pattern comes together more forcefully for dad's negativity and social responsibility. As Figure 11.12 showed, for this association there is a relatively strong a1 → f2 cross-lag along with a slightly smaller f1 → a2 cross-lag. Both met criteria for further analyses: they exceed .15. The f1 → a2 analysis showed that the strongest influences on the association were nonshared factors—a rare positive finding for this type of environmental influence. This finding supports a modest role for differential paternal negativity *preceding* and possibly influencing antisocial behavior, a point to which we return in Chapter 13. More to the point of delineating mechanisms of expression for genetic factors is that the a1 → f2 relationship (see Figure 11.15) is almost entirely influenced by genetic factors. We can understand this as follows: the genetically influenced component of social responsibility influences father's negativity: the more socially responsible child is spared father's irritable and punitive behavior. Along this same vein, the phenotypic analyses in Chapter 7 (see Figure 7.12) suggest that fathers aim their anger, conflict, and punitiveness at the wayward siblings. Here evidence comes together to suggest a salvage process for fathers. For mothers this process is less clear, as none of their a1 → f2 cross-lags are high enough to analyze further.

Summary

We can draw three important conclusions from the data in this chapter. First, the stability of our measures of family systems seriously constrains our sequence analyses. When we compare family measures at time 1 with adjustment at time 2, or vice versa, the correlations are exceedingly low once stability has been partialed out.

Second, of the eighteen analyses that were possible for mothers, fathers, and siblings, fourteen showed genetic factors exceeding the combined effects of shared and nonshared environments on the cross-lagged correlations. This continues a trend from previous chapters. In this case we can say that genetic factors are the most important influences on the sequence in time between parenting and adolescent adjustment.

Third, of the eighteen analyses that were computable, fourteen showed a preponderant influence of $f1 \rightarrow a2$, and six of these were notable instances of genetic influence on the $f1 \rightarrow a2$ relationship. Three relationships showed a preponderant $a1 \rightarrow f2$ relationship, and here genetic factors accounted for virtually the entire association. Thus, the data suggest that a child-initiated parent-effects model may explain much of the association between measures of the family system and adolescent adjustment. More specifically, we have defined five possible modes by which parents may respond to heritable characteristics in their children and participate in transforming them into central patterns of adjustment:

1. Parents may enhance nascent gifts and strengths in their children. This seems plausible in the domain of sociability.
2. Parents may react to a troublesome feature in the child by becoming more protective. They may either react to early forms of difficulty or pay attention to related strengths. This seems to be the case for mother's and father's positivity and its protection against antisocial behavior.
3. Parents may exacerbate nascent difficulties in their children. An example of this may be father's negativity, presumably a response to an irritable characteristic of his child, and the evolution of antisocial behavior.
4. Negative parental response—perhaps to an unpleasant heritable characteristic of the child—may spill over and impair another domain of

development. This seems plausible in the case of both father's and mother's negativity and cognitive agency.

5. Finally, troublesome behavior in the child may coerce a parent into backing off. This may be the case with heritable aspects of antisocial behavior: when these emerge as clear signs of maladjustment during adolescence, they may drive father to reduce his surveillance of the troublesome child.

Chapter 12

SYNTHESIS II:

THE RELATIONSHIP CODE

We presented a thesis underscoring the importance of the nonshared environment, as well as data that appeared to provide it with compelling support. Then we presented a genetic antithesis whose supporting data were so impressive that they appeared to thoroughly deflate the thesis. In the antithesis, we introduced the idea that genetic influences can be important for measures of the environment, just as they are for measures of adjustment. Indeed, the data supported this idea for every measure of the family subsystem we examined in our study. Moreover, the genetic factors that influenced these measures of the family environment were, in large measure, the same genetic factors that influenced adjustment. Thus, genetic factors were the most important component of the correlations we observed between measures of the family milieu and measures of adolescent adjustment.

There are three explanations for this surprising influence on these important associations. A passive model provides the first explanation. Parents and children share exactly 50 percent of the genes responsible for individual differences: those genetic factors that influence measures of adjustment in the child may, in the parents, shape mother's and father's behavior toward their children. A similar explanation is applicable to siblings, as they also share approximately 50 percent of their genes. A child-effects model provides the second explanation: genetically influenced patterns of adjustment in the child elicit or evoke responses from parents, and perhaps siblings and peers as well. A family relations–effects model provides a third explanation. Here, a heritable precursor of a domain of

343

adjustment affects a parent or another member of the family; the response of that family member then influences further development in that domain of adjustment. Data in the last chapter provided preliminary evidence that parental and sibling response to a heritable precursor of adjustment in the child may have several different consequences: it may enhance the precursor if it is positive or protect the child against unfavorable sequelae if it is negative, or the parental behavior may exacerbate a difficult trait, making it more likely that it will seriously impair the child's adjustment.

The family relationships–effects model, if supportable by our data and those from other studies, constitutes the kernel of our synthesis. This model allows for a substantial genetic effect on both family process and adolescent adjustment. It also fits with a great range of psychosocial data that suggest that the social milieu is an important cause of differences among children in their psychological development. Is there any evidence to support this more synthetic perspective over the two other possible explanations, the passive-effects model and the child-effects model?

Supporting data from our study are far from overwhelming. Indeed, we are looking for small but suggestive clues. In the previous chapter we examined, in detail, one relevant sector of the data. We reasoned that if the parent-effects model were true, then there should be evidence that the genetic influence on parenting (and on sibling relationships as well) should occur first in the life of the developing adolescent, to be followed by change in adjustment. Indeed, the pattern of evidence suggested just such a scenario. While these data are also consistent with a complex passive model that cannot be tested in our twin-sibling design, they weaken the child-effects model.

Our analysis of the sequence of influences presented in the last chapter suggests that family relationships might mediate genetic influences. In terms of analytic models available to us, this chapter brought us to the limits of our ability to weight the three possible explanations for the major role that genetic factors play in the observed associations between measures of the social milieu and measures of adolescent adjustment. Is there any other clue that would shed light on the plausibility of the family relationships–effects model? In the last chapter we alluded to some. These clues are relevant only if we assume that the family relationships–effects model is important. They help us imagine how family process might serve as a mediator of genetic influence on development, but they do not, by themselves, give any added weight to this possibility.

The most important clue is that genetic factors, at least in some comparisons of pairs of measures, are quite unique for each domain of adjustment. We provided a telling example in the case of the small inverse relationship between antisocial behavior and sociability. This small inverse correlation means that antisocial children can be either sociable or not sociable, although there is a small likelihood that antisocial children will be less social than their better-behaved peers. The reason for the relative independence of the two domains is that they are influenced, for the most part, by different genetic factors. In this chapter we begin by asking whether this distinctiveness of genetic influence between pairs of adjustment domains is an exception or the rule.

If this finding on specific genetic influences is the rule, it helps sharpen our understanding of what may be required of family process if family process plays a major role in the expression of genetic influence on adolescent adjustment. We cannot conceive of family process as mediating or "transmitting" specific genetic influences on development unless family process itself were responsive to a range of distinctive genetic influences. Thus, the bulk of this chapter is devoted to exploring how distinctive genetic influences are on various aspects of family process. What if we find that these influences are quite distinctive? What would that prove? By itself, such a finding would prove little. Indeed, the only inference possible from these analyses is a null one. Suppose we find that family processes are influenced by a single monolithic genetic influence, or very nearly so. This would suggest that, by itself, family process could not mediate highly distinctive genetic influences such as those responsible for the disjunction between antisocial behavior and sociability. Thus, finding little or no specificity of genetic influence would weaken the relationships–effects model. But even if we do find genetic specificity, we still will not resolve our major interpretive dilemma: the relative contributions of the passive-effects, child-effects, and family-effects models to the main findings of this study. Nonetheless, such a finding, along with the data presented in Chapter 11, would keep the relationships-effects model very much alive as a plausible synthesis.

Common and Specific Genetic Influences in Pairs of Adjustment Domains

This chapter crosses an important but almost imperceptible divide. Up till now, we have tried to extract as much relevant information as possible from the data to advance both a thesis and an antithesis. In the last chapter we drew one further inference from the data: for the most part, genetic influences on family relationships unique to early adolescence precede changes in adjustment from early to late adolescence. It is less plausible that the reverse is true: genetic influences on adjustment, unique to early adolescence, precede changes in family relationships. If we assume that passive mechanisms play little role in the data we collected, and this assumption still needs firm support, then these data argue in favor of a family relationships–effects model, not a child-effects model. The latter may be applicable to a much more restricted set of developmental processes. We delineated one example: evidence suggesting that fathers back off from unruly or difficult children. But according to the data presented in the last chapter, this sequence—genetic influence on adjustment preceding change in family process—is most unusual.

We could end matters right here with the customary scholarly sign off that more work needs to be done. But we feel that our own work is still unfinished. Constructing an informed theory is a central process in developmental studies. In this chapter we go further out on a limb in an attempt to gain some perspective. Of the three alternate explanations for our major findings—the passive-effects model, the child-effects model, and the relationship-effects model—the last is the only one that offers a genuine synthesis of our thesis and antithesis, explaining how there might be an important role for family process in adolescent adjustment as well as strong genetic influences on the same domains of adjustment. The relationship-effects model may also offer a format for resolving major paradoxes engendered by other psychosocial and genetic studies of development in addition to our own. Because it offers the promise of such an important synthesis, we take one more careful look at the data to see how the family relationships–effects model might work if it does have a regnant position in mechanisms of development.

We begin by examining the balance between common and specific genetic influences for all pairs of our measures of adjustment that had a correlation equaling or exceeding .20. Because our study uses a genetically informed design and examines many areas of adjustment, we may regard the seven domains of adjustment that we included in our design as a reasonable sample of a more complete population of successes and failures that are important to the developing adolescent. Our analysis provides vital information about the genetic influences that are important for the adjustment of adolescents who are not selected either for unusual competence or for psychopathology. Results might have been quite different had we measured other domains or focused on more troubled adolescents. If we find that genetic influences on the seven domains of adjustment we studied are diverse, this suggests that any process that mediates genetic influence on adolescent development will have to convey or transmit a broad range of genetic information.

Our search for overlapping and specific genetic influences, in pairs of variables, uses the same analytic techniques that we used earlier to examine associations between pairs of variables. In Chapter 8 we examined the relationship between time 1 and time 2 measures within each domain of adjustment. In Chapter 9 we performed similar analyses for measures of the family milieu. And, in Chapter 10, we used the same analyses to examine the association between family and adjustment measures. Here we look at measures of adjustment whose correlation equals or exceeds .20.

Figure 12.1 returns to our example from Chapter 11, the small and inverse correlation between antisocial behavior and sociability. Recall that a diagram of this kind reflects a comparison of cross-correlations across the six groups of families in our sample. In this instance we have computed cross-correlations for antisocial behavior and sociability. For example, for identical twins we correlated the level of antisocial behavior in one twin with the level of sociability in the other. We computed comparable correlations for the other five groups of families. Once again, our analytic procedures search for a genetic cascade in which identical twins show the highest correlation and other groups show comparably lower correlations in accord with the genetic relatedness of the siblings. In this instance a steep cascade reflects the importance of genetic factors in the correlation or covariation between two variables. The phenotypic correlation here is only −.22; thus, the maximum amount of shared genetic influence possi-

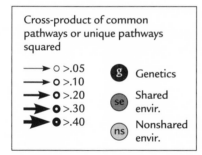

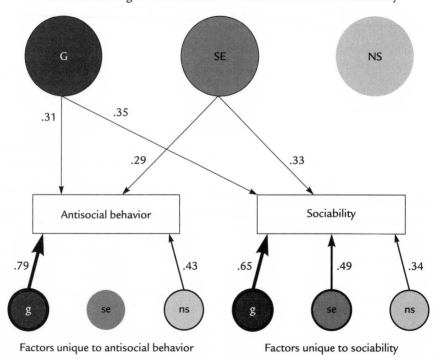

Factors accounting for the association of antisocial behavior and sociability

Factors unique to antisocial behavior   Factors unique to sociability

Factors accounting for the disjunction of antisocial behavior and sociability

Figure 12.1   Genetic and environmental components of the relationship between antisocial behavior and sociability

ble on these two variables is quite small. Examining the unique genetic influences, that is, the genetic influences on one member of this pair that do not overlap with genetic influences on the other member, is even more informative. In this case the unique effects are exceptionally large.

For each variable we can compute the proportion of total genetic influence that is unique. For antisocial behavior this is $.79^2$ ($= .62$): the coefficient on the arrow from the balloon indicating unique genetic variance, divided by the heritability of antisocial behavior, which is the combination of the overlapping genetic variance $.31^2$ ($=.10$) and the unique genetic variance. The result of this computation suggests that 82 percent of the genetic influences on antisocial behavior are distinct from those influencing sociability.

Perhaps even more to the point, we can estimate the proportion of the disjunction between antisocial behavior and sociability that is accounted for by genetic factors. The coefficient of disjunction equals $1 - r$, or .78. This coefficient also equals the sum of mean of the squares of the unique genetic paths, of the unique shared environmental paths, and of the unique nonshared environmental paths. All of these are the upward-pointing paths from the small balloons. The proportion of this coefficient accounted for by genetics is 66 percent. In short, genetic factors are—by far—the most significant influence on the disjunction between these two variables. Any theory concerning mechanisms of gene expression or the mediation of genetic influences on adjustment will have to account for how these distinct genetic influences are transformed into distinct patterns of adjustment.

Figure 12.2 shows a sharply contrasting comparison: the association of antisocial behavior and social responsibility. Here, the phenotypic correlation between them is also inverse, but at $-.69$ it is much higher than that of antisocial behavior and sociability. The coefficients on the paths from the factors common to both variables, the large balloons, should be multiplied to estimate the contribution of the common factor to the phenotypic correlation. The product of the paths from "G" equals $.62 \times .91 = .56$; from "NS" equals $.26 \times .38 = .11$; and the total of the two products equals .67. Recall that our balloon diagrams only show coefficients that are statistically significant at the .05 level. Thus the coefficients shown in the diagram may not always add up perfectly to the relevant phenotypic

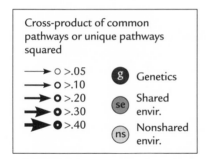

Factors accounting for the association of antisocial behavior and social responsibility

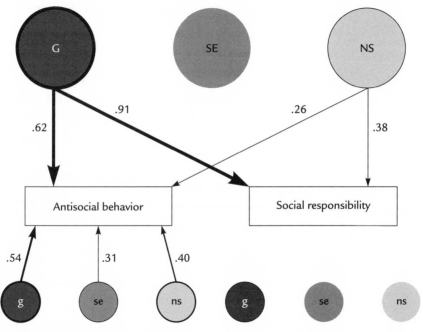

Figure 12.2 Genetic and environmental components of the relationship between antisocial behavior and social responsibility

correlations. Also, we always reverse the sign of inverse correlations for our analyses.

In Figure 12.2 we see that there is no unique genetic influence on social responsibility that is independent from its influence on antisocial behavior, but there is some unique genetic influence on antisocial behavior. Here genetic factors are primarily responsible for the correlation between the two variables. These factors account for .56/(.56 + .11), or 84 percent, of the correlation between the two. Indeed, antisocial behavior and social responsibility seem to have nearly the same genetic roots. It is likely that the factors mediating genetic influence on this pair of variables overlap considerably.

Figure 12.3 summarizes analyses of this kind for all seventeen pairs of adjustment measures that correlated with one another at .20 or greater. Adjustment variables that have lower correlations and are strongly influenced by genetic factors, as all our adjustment factors were, may be presumed to have unique genetic influences. Examine first the bar graphs relating to Figures 12.1 and 12.2 (the fourth and fifth bars down from the top). The magnitude of the phenotypic correlation is represented by the location of the boundary between the white segment (representing the influence of the nonshared environment on the association) and the black segment immediately to its right (representing genetic influences on the disjunction between the two). This boundary is indicated by an arrow for many of these bars. Note the small correlation for antisocial behavior and sociability and the much larger one for antisocial behavior and social responsibility. For the latter, the length of the leftmost dark segment of the bar indicates the sizable genetic overlap between the two. For both of these comparisons it is intriguing to note that genetic factors are the predominant component of the disjunctions. That is, genetic factors are responsible for the fact that, within each pair, the correlation is less than 1.

Figure 12.3 shows an important pattern for almost all these comparisons of pairs. For twelve of the seventeen pairs, the genetic influence on the disjunction between the pair is greater than the genetic influence on the association. For fourteen of the pairs, genetic factors equal or exceed environmental factors in their influence on disjunction. Any theory of genetic expression in adolescence must account for this finding.

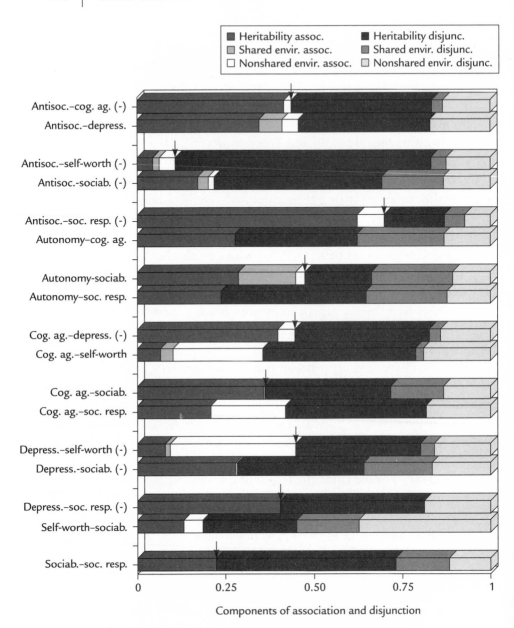

Figure 12.3 Genetic and environmental components of the relationship between all pairs of domains of adjustment where the correlations equaled or exceeded .20

How Genetic Influences Are Expressed

Our hypothesis about a relationship code is an elaboration of the family relationships–effects model. To summarize, it addresses the following findings in our study. First, there are very strong associations between measures of the family environment and all domains of adolescent adjustment. Second, a substantial number of these genetic influences are unique to particular domains of adjustment. Third, there are substantial genetic influences on all the measures of family relationships—marital, parental, and sibling—employed in this study. Fourth, the genetic factors that influence adjustment overlap, to a large extent, with those that influence family process. Fifth, with a few exceptions, there is some evidence that the genetic influences on family relationships that are unique to earlier adolescence precede changes in adolescent adjustment from earlier to later adolescence.

We propose that family relationships are responsive to heritable traits in developing children. These heritable traits might be temperamental features or modes by which children perceive their social world. For example, at several points, we have mentioned irritability and social responsiveness as temperamental styles. Our hypothesis is that these heritable traits in adolescents evoke characteristic responses from parents and other members of the family, including sibs. That is, under ordinary circumstances, there is a reliable association between evolving temperaments and social responsiveness in children, on the one hand, and families' responses to these children, on the other. If we know a child's temperament we can predict, with reasonable accuracy, the responses of his or her family members. For example, aggressive children often elicit counter-aggression in their parents.

But, our hypothesis continues, the family serves as more than a passive responder to the adolescent. The response of family members to heritable characteristics in the child continues to influence that child's behavior. Counter-aggression in a parent, in response to aggression in the child, may only exacerbate the child's aggression. As the work of Patterson and his group has shown, families need to set firm, consistent limits on aggression in children. For a great many families this is difficult to do, and hence outside intervention is required (Patterson, 1982; Patterson,

1985; Patterson, 1989; Patterson and Capaldi, 1991; Patterson, Reid, and Dishion, 1992). To carry our hypothesis one step further, it seems unlikely that heritable traits in children would become manifest as either full-blown successes or full-blown failures without these formative transactions with their families. It is because families play this crucial role, our hypothesis continues, that they are not merely passive responders. They are themselves initiators: they respond to a genetic "signal" from the child—or really an array of signals—and their response is transmitted back to the child. In the case of troublesome behavior, they may exacerbate it. In the case of competent behavior, they may enhance it. Other possible reactions were suggested in our analysis of sequence in the last chapter.

When a family responds with a profile of reactions to a developing child with particular heritable traits, this can be thought of as encoding, very approximately, genetic information. Deciphering this code would be possible only if the genetic influences on relationship qualities were preponderant or exclusive. Then, with enough research data on the specific links between heritable individual dispositions and relationship systems, we could predict genetic influences from relationship data or vice versa. In this ideal case we could readily "decode" family-process data in order to delineate discrete genetic influences. As shown in Chapter 9, however, environmental factors (meaning factors unrelated to genetic factors in the children) play the most significant role in differences among families in their relationships with adolescent children and siblings and in conflicts between husbands and wives about their children. Moreover, genetic factors in parents and siblings—factors not shared with the adolescents—are almost certainly constraining their response to the teens. Thus, we cannot reason backwards to determine which genetic influences, originating in the children, are operative just from a knowledge of the qualities of the relationship systems themselves.

But reasoning in the opposite direction may be central for developmental theory: perhaps separable genetic influences, originating in the child, can be *encoded* in family relationships that directly or indirectly involve the child. At the intracellular level, specific DNA configurations encode equally specific processes of protein synthesis, biochemical sequences that are critical for specific genetic expression in the structure and function of cellular and organ systems. In a somewhat analogous fashion, certain genetically influenced temperaments and patterns of social responsiveness in the child may be encoded by relationship processes that are essential to the full expression of these genetic influences as the child develops. We could follow this encoding process if we had reasonable candidate temperaments and social-response patterns in children and could observe the early impact of these characteristics on others in the family. Genetically informed designs, which we describe in the final chapter of this book, can help decipher which components of this child action and family reaction are attributable to genetic factors.

This sequential process of signaling, encoding, and responding cannot even be inferred from the data presented here. For some of the child signal–family response systems, we have almost certainly arrived on the scene too late, long after these processes have become established. Even more to the point, we do not have good assessments of the genetically influenced progenitors that we are here hypothesizing. Nonetheless, we can conduct analyses that suggest which aspects of family relationships *might* serve to encode genetic information, but we must do so without an understanding of the heritable stimuli involved nor of the consequences of these responses for the further development of the children.

It seems particularly important that we ask whether family relationships *might* be sufficiently discriminating in response to genetic influence to mediate the high level of distinctiveness of genetic influence on adjustment patterns we observed in Figure 12.3. For example, let us say mother's, father's, and sibling's warmth and support toward the adolescent were all influenced by the same genetic factors. These three family subsystems would be redundant in their mediating role of transforming heritable traits into patterns of adjustment. If these family subsystems respond to different genetic influences, however, they reflect a potential repertoire of family responsiveness to a greater diversity of genetic signals from the child. As a consequence, they have a more complete potential for mediating these genetic influences into patterns of adjustment that are themselves genetically distinct. For example, in Chapter 11 the data presented suggested that mothers, in the relatively stable, two-parent families we studied, may have a unique role in responding to heritable progenitors of sociability; we speculated that they respond to this characteristic by enhancing its development. In contrast, if mother's responses of warmth and support are completely redundant with father's and sibling's, there may have been no one in the family to pick up and enhance this genetic potential.

Three caveats are in order. First, this initial delineation of our hypothesis focuses on children in families. Their heritable behavior is regarded as a stream of "signals" to which the family responds. Moreover, we have implied that signaling starts at an early age. Genetic influences on relationships may, however, be important across the life span of both relationships and individuals. For example, according to the data presented in Chapter 8, there is evidence that new genetic factors emerge between earlier and later adolescence. This is particularly pronounced for sociability and antisocial behavior. In the case of sociability, almost half the genetic influence operating in early adolescence disappears by late adolescence, when it is replaced by new genetic influences that were not manifest three years earlier. The same holds for antisocial behavior. We do not know, of course, whether families are sensitive, consciously or below the limn of their awareness, to these changes. Nonetheless, it is plausible to argue that genetic factors, stable or changing, may influence families whether "signals" come from children or from adults.

Second, for technical reasons, our study focused only on two-parent families with at least two adolescent children. It is far beyond the scope of our present inquiry to explore the relevance of our hypothesis to families with one parent or with no parents. There is certainly no reason to believe that these families, with structures different from the ones we have studied, would play either more or less of a role in the mediating process we are positing.

Third, the same logic of encoding applies to environmental factors—both shared and nonshared—as applies to genetic influences. For these factors, too, we could have a family system that is redundant or undifferentiated, responding to a multitude of environmental influences with a single, undifferentiated response. In the undifferentiated case only one or two major environmental factors influence all relationship parameters, or there are a number of such influences but they affect all relationship factors in a relatively undifferentiated way. In the specific case of many different environmental factors, each has specific effects on particular aspects of family relationships. Although we allude briefly to findings relevant to "environmental encoding" in this chapter, we give them less attention than genetic findings. As Chapter 9 has shown, environmental factors are clearly more important than genetic factors in explaining the variations among families in relationship quality. But they have less pertinence for reconciling our thesis and antithesis, a reconciliation that must focus on genetic influences.

In exploring which attributes of family relationships might encode genetic information, we find no guidance in past research. We can only hope that the attributes we have studied, for quite different purposes, will provide leads. Recall that we measured three aspects of relationships in the family system. First, we measured four distinctive subsystems. It is possible that genetic influences on mother's warmth toward her adolescent children might be quite distinct from genetic influences on father's warmth. Since we are focusing on the *children's* genes, we could say, in this instance, that the relationship subsystem in which the child is embedded encodes some information from the child's genome. That is, some genetic factor might influence the ultimate adjustment of the child through its impact on father's warmth, whereas other genetic factors might exert their influence through an impact on mother's warmth. In this instance, four features of parental warmth convey genetic information: high father, low father, high mother, low mother. A profile of these four scores would reflect, according to our hypothesis, reliable but very crude information about the child's genome.

Second, we measured differences in relationship quality within subsystems. For mothers and fathers we measured positivity (warmth/support), negativity (conflict and punitiveness), and monitoring/control. We have broken the last of these into three subcomponents: knowledge or surveillance, attempted control, and actual control. This brings our total to five different qualities for mothers and five for fathers. We could break these down to even more detailed dimensions, utilizing the measurement scales that are components of these broader categories of parenting, but we resist doing so in order to keep our task manageable. For siblings we measured only positivity and negativity. For marriage, only one measure specific to each child was available: marital conflict about the child.

Third, we measured all aspects of the family system, except for marital conflict about the child, earlier and later in adolescence. Thus, relationship developmental time might encode genetic information if there are genetic influences on relationship subsystems that are unique to either earlier or later adolescence.

Findings on the Specificity of Genetic Influences on Family Relationships

In exploring how genes influence family processes, we follow four steps. First, we examine the simple correlations among all our measures of four family subsystems in earlier adolescence: mother-child; father-child; sibling; and marital. These provide some initial clues about the specificity of genetic influences. We know that in early adolescence there is at least some genetic influence on all these measures, and that in later adolescence there is influence on all but two of them. If each of two measures, in a comparison of a pair of measures, shows substantial genetic influence but these measures are uncorrelated with each other, we can assume that genetic influences on each are completely or substantially distinct.

Second, we examine the genetic and environmental overlap between each pair of measures of the environment that show at least modest correlations. The analysis uses the same format as the one just reported for measures of adjustment. Here, we focus not on genetic overlap but on genetic factors that operate as unique influences on each relationship measure, in comparison with the other. We are interested in the heritability of disjunction between each pair of correlated measures of the family. Uniqueness of genetic influence and specificity of genetic influence on the family measure are synonymous. The higher the heritability of disjunction across all correlated pairs of family measures, the more specific the genetic influences on family relationships. The perfect specific case would be for each measure to have an entirely unique set of genetic influences. The perfect undifferentiated or nonspecific case would be for all measures to have the same genetic influence; the genetic overlap between all pairs would be high and the heritability of disjunction would, in each case, be zero.

The third step is to re-examine the overlapping and specific genetic influences in later adolescence, asking if a specific genetic influence increases or decreases over time.

A fourth step has already been performed in Chapter 9 but is relevant here and thus worth reviewing. In Chapter 9 we asked how distinctive the genetic influences on each relationship measure in later adolescence are from those that are influential in earlier adolescence. As noted, relationship development is one of three features of family relationships that can encode genetic influence. Here, for each measure of the family, we compare two estimates of heritability. The first is the heritability of stability, which reflects the nonspecific effect of genetic factors on a relationship measure in both early and late adolescence. The second is the heritability of change, which estimates the genetic influences acting uniquely in both early and late adolescence. If the latter far exceeds the former, we can say that relationship development, over time, also encodes a substantial amount of genetic information.

Major Measures of Family Relationships in Early Adolescence

Figure 12.4 lays out all thirteen of the family variables we measured at time 1. We have displayed them in this fashion so that we may examine, in detail, the relationship between every possible pair of variables, or a total of $((13 \times 13) - 13)/2 = 78$ comparisons. Thus we have seventy-eight boxes in Figure 12.4, each of which represents a comparison between two variables. For example, the first box in the "mom's neg." column and the "dad's neg." row is a comparison between these two measures at time 1. Our search for encoding processes involves examining all the seventy-eight comparisons illustrated in Figure 12.4. A comparison between any two variables, as noted, involves two steps. First we ask whether the two variables are correlated with each other. In the case of the first box, the correlation is quite high: .61. This means, of course, that in a family in which mom shows a lot of negativity toward any particular child, dad is very likely to be negative to that child as well. If two variables show little or no association, we presume that the determinants of those variables must be quite different. Hence, even in this first step there is some evidence for specificity of genetic influence on family measures. Thus, if two variables show little or no correlation but both show substantial genetic influence, we can presume that different genetic factors influence the uncorrelated variables.

But a great many of our measures of family relations show modest or

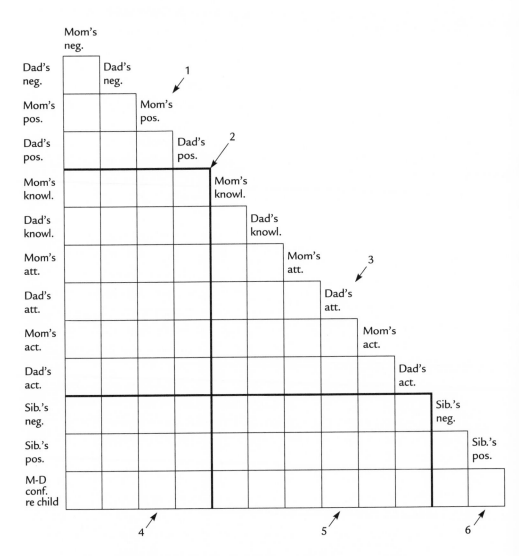

Figure 12.4 Comparison of family-relationship variables

strong associations. Here, determining the specificity of genetic influence requires a second step: we need to know how much of the observed association between any two variables is due to common genetic factors. Theoretically, this percentage could range from 0 to 100. We also need to determine the extent to which genetic factors account for the lack of correlation or uniqueness of each relationship variable.

Before reviewing how this is done, let us return to Figure 12.4. Notice that we have divided the seventy-eight comparisons (or boxes) into six sectors. The sectors are divided by heavy horizontal and vertical lines and are numbered (see bold numbers and arrows adjacent to the sectors). Considering each sector in turn will make it easier to follow these analyses.

Sector 1 focuses on comparisons among measures of mom's and dad's positivity and negativity. For each parent it asks how related positivity and negativity are, and then asks two additional questions comparing moms and dads; for example, is mom's positivity related to dad's positivity? Is their negativity related? Finally, this sector asks whether mom's positivity is related to dad's negativity, and whether mom's negativity is related to dad's positivity.

Sector 2 focuses on the relationships between mom's and dad's negativity and positivity, on the one hand, and their monitoring and control strategies, on the other. Thus, we can examine, within each parent, whether positivity and negativity are related to any of the control strategies. We can also examine these associations across parents. For example, we can ask whether mom's positivity is related to dad's successful, or "actual," control strategies ("dad's act.").

Sector 3 is analogous to sector 1. Here we examine the relationships among three measures of control: parental knowledge of the child's activities ("mom's knowl." and "dad's knowl."), parental attempts at control ("mom's att." and "dad's att."), and parental success in control ("mom's act." and "dad's act."). These relationships are examined within each parent (e.g., "mom's knowl." with "mom's act.") and between parents (e.g., "mom's att." with "dad's att.," or "mom's att." with "dad's act.").

Sector 4 introduces sibling and marital data. It permits all possible comparisons between mom's and dad's positivity and negativity toward the adolescent and siblings' positive and negative relationships with each other. It also shows the relationship between parental positivity and negativity and the parent's own level of conflict about a particular child. For example, the last box in the lower left-hand corner of the chart asks about the association between mom's negativity toward one of her two adolescent children and the level of conflict she has with her husband about that child.

Sector 5 is analogous to sector 4. Here we examine the relationships between parental monitoring and control, on the one hand, and sibling and marital relationships, on the other. Sector 6 is the smallest of the sectors. It examines the relationship of sibling's positivity and negativity to each other and to marital conflict.

Relationship Quality, Family Subsystem, and Developmental Time

Figure 12.4 permits us to search for genetic specificity along at least two separate parameters. First, we can ask whether different qualities of relationships have common or distinct genetic influences. For example, are there separate genetic influences for mom's positivity and mom's successful control? Second, we can ask whether there are distinct genetic influences for the same relationship quality for different family subsystems. For example, do the same genetic influences that shape sibling-to-sibling negativity also shape father-child negativity?

There is also considerable theoretical relevance in examining the association of one relationship quality in one subsystem with another relationship quality in a second subsystem. For example, previous family research suggests that parental positivity might damp down sibling fighting. This would be reflected in an inverse relationship between parental warmth and sibling conflict. The two could, however, be linked entirely by common genetic factors. This would weaken the "damping-down hypothesis" because, in this instance, an alternative explanation would be more plausible: genetic factors in the child that encourage fighting with a sibling also "turn off" mother's affection for that child. In this instance the most pressing question for our hypothesis about a relationship code would be whether genetic factors also account for the disjunction between low parental warmth and sibling negativity. That is, if genes account for the association *(r)* between the two, do they also account for the lack of association $(1 - r)$?

As noted, a third crucial parameter of family process may encode genetic information: *developmental time.* That is, relationship systems develop just as individuals do. It is possible that early in its history a relationship is insensitive to a heritable characteristic of one of its members, but that later on that system is activated by this same characteristic. Data presented in Chapter 11 provide a plausible if unproven example. There is evidence suggesting that father's surveillance of his children may not be activated by the child's heritable aggressive or difficult behavior until this behavior becomes fully antisocial. At that point, we have posited, father's monitoring may be reduced and switched to the less troublesome sibling.

We have already examined data relevant to this issue in Chapter 9 without labeling it as a search for the specificity of genetic influence. These were data concerning changes in genetic influence on relationship subsystems from earlier to later adolescence. We summarize these findings at the conclusion of this chapter, where we can view them in another light.

Family Processes That May Encode Genetic Information Early in Adolescence

Figure 12.5 displays the first step in our analysis of genetic specificity and family encoding: it represents the seventy-eight correlations among the thirteen measures of family process earlier in adolescence. Let us review, briefly, the findings in each of the six sectors of this display.

In sector 1, four of the six variables show modest or substantial correlations. As expected, for moms and dads, negativity is inversely correlated with positivity. These correlations are modest (for dads, $r = -.24$, and for moms, $r = -.29$), however, suggesting that positivity is not just the inverse of negativity. Indeed, the correlations are small enough to suggest that in many families parents are high on both positivity and negativity or low on both. Figure 7.5 suggested that 122 of the families in our sample meet criteria for the latter. In this sector, much higher correlations are apparent across family subsystems than within them: mom's and dad's nega-

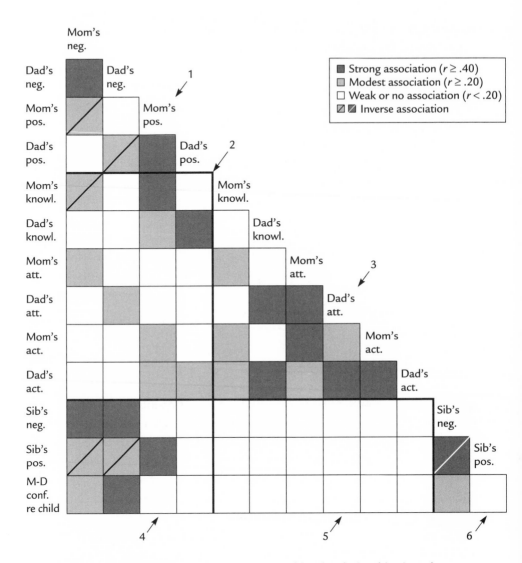

Figure 12.5   Correlations among measures of family relationships in early adolescence

tivity is highly correlated, as already noted ($r = .61$), as is mom's and dad's positivity ($r = .42$).

In sector 2 there are twenty-four comparisons made, and nine of them show significant associations. All of them fall into a simple pattern. Recall that in Chapter 7 we showed that high levels of mom's and dad's knowledge and actual control were inversely correlated with psychopathology, particularly antisocial behavior. Further, these same two variables were positively associated with adaptive outcomes, particularly cognitive agency. The reverse was true of attempted control, which was positively associated with psychopathology, particularly antisocial behavior at time 1, and inversely associated with cognitive agency. We reasoned that attempted control is actually part of a pattern of negative or ineffective parenting, and all the correlations in Figure 12.2 substantiate that hypothesis. Mom's and dad's negativity is positively correlated only with attempted control, whereas their positivity is associated with both knowledge and actual control.

Sector 3 shows a dense set of interrelations: only five of the fifteen comparisons are not significant. All of the correlations with relationship subsystems are positive; for example, mom's knowledge, attempted control, and actual control are all intercorrelated. The high positive correlations of attempted control with knowledge and actual control are surprising given that attempted control has such different consequences for adjustment. In sector 3 there are also many cross-relationship correlations. Most notable are the high correlations between parents in their attempts at and successes in controlling their children (for attempts, $r = .47$, and for success, $r = .48$). Knowledge is notable: it is the only parenting variable that is not correlated in families. This changes by later adolescence, however, when the correlation between the two is strong ($r = .40$).

Sector 4 is also dense with correlations. Mom's and dad's negativity are associated with both sibling positivity and negativity as well as parental conflict about the child, all in expected directions. Mom's and dad's positivity are also associated with sibling positivity.

Sector 5 is notable because among eighteen comparisons, none is equal to .20 or more. There are nine separate measures being compared here, and all of them show at least some genetic influence. Since the phenotypic associations among all nine of these variables are so low, we assume that most of the genetic influence is unique in each of these eighteen comparisons (see below).

Sector 6 is small, with two correlations .20 or above. As we might anticipate, sibling positivity is inversely correlated with sibling negativity. Unlike the analogous correlation for maternal positivity and maternal negativity, this correlation is more substantial ($r = -.48$). Thus, there will be fewer sibling pairs than parent-child pairs that will show both high positivity and high negativity or low positivity and low negativity. The other correlation is between the two same-generation subsystems: the marital and the sibling. A very modest correlation suggests that mom-dad conflict about a child is associated with sibling negativity ($r = .22$).

In sum, Figure 12.5 already suggests that there is a good deal of specificity in the relationship between genetic factors and family process; indeed, the number of white squares is a crude index of this specificity, since almost all of them reflect variables that have a significant genetic influence but whose correlation is quite low or zero. The pattern of white squares suggests that there is specificity for relationship quality (for example, the low correlations between dad's negativity, knowledge, and actual control) and, in one instance, for relationship subsystem (the low correlation between mom's and dad's knowledge). But we need to examine further those variables that are correlated, asking if their correlations are attributable to genetic influences. In this case specificity is unlikely. If the correlations are due primarily to environmental factors, we would expect to find that the genetic influences are specific, to some substantial degree, to each variable.

Figures 12.6 and 12.7 illustrate more fine-grained analyses of two of the correlations shown in Figure 12.5: the correlations between mom's and dad's negativity and between mom's and dad's positivity. Here we are searching for specificity of genetic influences in two different family subsystems (the father-child versus the mother-child). These two figures resemble closely those we have used to represent the analysis of stability and change across time among our adjustment variables (Chapter 8) and among our relationship variables (Chapter 9); the covariation of family variables with measures of adjustment (Chapter 10); and the analysis of specific genetic influences on domains of adjustment (earlier in this chapter). Indeed, the principles governing all three of these analyses are the

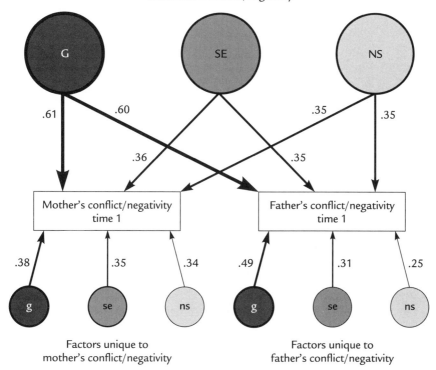

Figure 12.6 Genetic and environmental components of the covariance between mother's and father's negativity at time 1

same. The covariance we are examining here is between two different family variables measured at the same time. We can ask, once again, what is the weight of genetic and environmental factors accounting for the correlation of two variables. We can also ask again how much these genetic and environmental factors contribute to the lack of correlation between two variables, the coefficient of disjunction (1 minus the correlation coefficient).

Figure 12.6 shows analyses of the phenotypic correlation of .61 between mom's and dad's conflict/negativity. The figure clearly shows that the most significant proportion of covariance is accounted for by genetic factors (61 percent of the covariance). By comparison, the specific effects of genetic factors are quite small. The unique heritability of mother's conflict/negativity is 14 percent ($.38^2$) and for dad's it is 23 percent ($.49^2$). By comparison the common genetic influence for both variables is about 36 percent. One way to summarize these findings is to say that the heritability of association of the two variables is 36 percent and that the heritability of disjunction is 19 percent (($.38^2 + .49^2$)/2).

Figure 12.7 shows a dramatically different set of findings for mom's and dad's positivity. Here genetic factors account for none of the phenotypic correlation between the two variables ($r = .42$). Genetic factors, however, are the preponderant influence on the lack of correlation between the two. The heritability of disjunction is 34 percent (($.59^2 + .56^2$)/2). Genetic factors account for 58 percent of the lack of association ($.34/(1 - .42)$) between the two variables. These data suggest that the lack of association between positivity in two family subsystems—mother-child and father-child—may encode a good deal of genetic information. That is, father's positivity might respond to a heritable characteristic in the child that is distinct from the heritable characteristic to which mother responds. We have noted a partial example of this from Chapter 11: mother's apparent response to a heritable precursor of sociability.

Figure 12.8 summarizes these genetic analyses of covariance for all thirty-four correlations among the thirteen measures of family relationships that were .20 or above (see Appendix B). A white circle represents all those associations for which the heritability of association exceeds the heritability of disjunction. One of these six, mom's and dad's negativity, we have already noted. All the remaining pairings with common rather than specific genetic effects are for associations within each parent-child

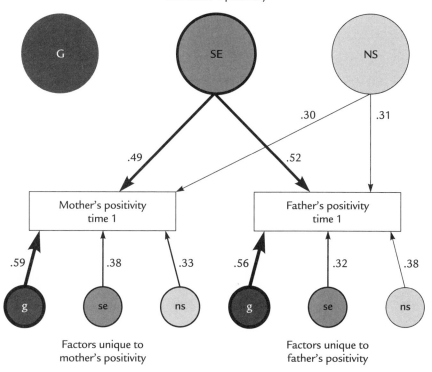

Figure 12.7   Genetic and environmental components of the covariance between mother's and father's positivity at time 1

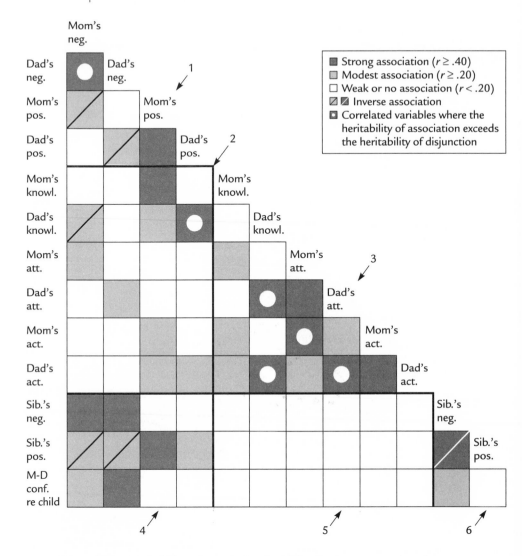

Figure 12.8 Heritability of association for all family measures at time 1

subsystem. There are no notable overlapping genetic influences across family subsystems. Of these five, four are among dimensions of monitoring/control (see sector 3). These findings suggest a common genetic origin for all three measures of parental control within each parent-child subsystem but little common genetic influence across parent-child subsystems in this domain. There is only one other instance where the heritability of overlap exceeds the heritability of uniqueness: dad's positivity with dad's knowledge. As will be shown, this instance of overlapping genetic influence provides additional insights into family relationships.

We repeated the same analyses reported here for time 2, with the exception of those involving mom's and dad's conflict about the child, which we did not measure at time 2. There were almost no changes in the overall pattern of findings (see Appendix B for details on this time 2 analysis).

Does Relationship Development Encode Genetic Information?

Figure 12.9 represents the stabilities of all twelve measures of family subsystems that we examined in earlier and later adolescence. As noted, the stabilities of parental positivity and negativity and sibling positivity and negativity are quite high. As Figure 9.9 conveyed, the stability of mother's and father's negativity is due mostly to genetic factors. The stability of parental positivity and sibling positivity and negativity is due mostly to shared environmental factors, although genetic factors make a substantial contribution. For all these variables, the genetic factors that are common across earlier and later adolescence far exceed those that are unique for each time period. Thus we can say that relatively little genetic information is coded in the development of positivity and negativity across adolescence within these relationship systems.

A very different picture emerges for the monitoring/control group of measures of parenting. For all but one of these six variables, the unique genetic effects at earlier and/or later adolescence outweigh the genetic effects that are common across time. For the sixth, mother's attempted control, the genetic effects unique to earlier or later adolescence almost equal the genetic effects that are constant across time. As Figure 9.12 showed, in some instances genetic influences that are effective in earlier adolescence turn off, whereas in other cases new genetic influences are turned on in later adolescence that were dormant earlier. For example, genetic effects that are unique to early adolescence, in mother's attempted control and actual control, disappear and are not replaced with other genetic ef-

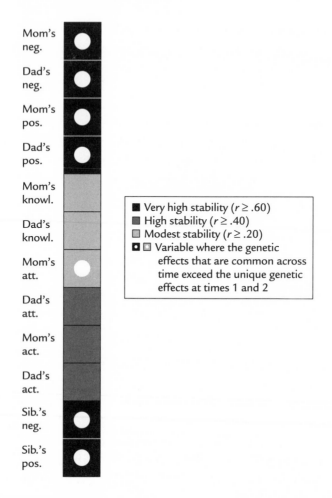

Figure 12.9 Heritability of overlap for changes in relationship subsystems from early to late adolescence

fects at time 2. This leads to a drop in the heritability of both these variables across time. In contrast, for dad's knowledge or surveillance, there is an influence of new genetic factors in late adolescence but little decline in the modest genetic influences that were effective in early adolescence. A similar, though slightly less pronounced, pattern is seen for father's attempted control.

A comparison across time of mother's and father's monitoring/control variables is instructive. Recall that genetic influence on mother's attempted and actual control declines over time and that genetic influence on father's knowledge and attempted control increases. Are mother's attempts and successes at time 1 encoding the same genetic influences as dad's knowledge and attempts at time 2?

We are calling complex relationships of this kind cross-time, cross-family subsystem and cross-relationship qualities. Analyses of these relationships are important for understanding the role of family processes and genetic influence as adolescence unfolds. What would it mean for our understanding of adolescent development if we found a substantial overlap in genetic influence between mom's attempted and actual control at time 1 and dad's knowledge and attempted control at time 2? According to our present framework, this would reflect a nonspecific genetic influence: the same genetic influence is common to four different measures of family relationships. One plausible explanation for a nonspecific finding like this, but not the only one, is that mom responds to a heritable characteristic of her young adolescent with both attempts and success in control (presumably she is responding to a genetically influenced obstreperousness in her child).

As her child gets older, however, mom "gives up": she no longer responds to this heritable and challenging quality in her child. In contrast, according to this scenario, dad at mid-adolescence responds to the same genetically influenced challenging characteristic of the child but does so three years later than mom. By later adolescence dad is less successful because what is evoked in him is knowledge of his adolescent's actions and attempts to control them, but not success in controlling them. According to this view, the adolescent is a moving target: mother shoots at the target early on and often hits a bull's eye; father takes over the shooting by later adolescence and nearly always misses. Nonspecific genetic influences of the kind we have been describing are consistent with an important evocative role for heritable characteristics in the child, but, in this instance of parental behavior control, they also imply a coordinated "dance" by the parents: moms take their adolescents on early and dads try later on, perhaps by mutual consent.

Let us suppose that the data present a contrary finding: that mother's attempts and successes at time 1 encode a very different set of genetic influences than do father's knowledge and attempts at time 2. This is a finding of genetic specificity; it suggests that two different genetic influences are required to explain this cross-time, cross-subsystem, cross-quality observation. If evocative mechanisms are important, this specific effect would not only suggest a greater range of genetic influences on parental control strategies but also require a different family theory of development: the child is not a moving and invariant provocateur for parental response. The child becomes, in effect, a *new* provocateur by later adolescence, or at least new genetic influences come into play to which father, but not mother, responds. New genetic influences do not necessarily imply that a qualitatively different provocation has emerged, for different genetic influences might lead to the same behavioral outcomes. It seems plausible, however, that with new genetic factors in play, the quality of behavior might be different at mid-adolescence from what it was at early adolescence. Perhaps the provocation is more ominous or more openly sexual than it was in early adolescence. Here the coordinated "dance between the parents," though still remotely possible, is a less plausible and unnecessary postulate. Early adolescents are setting off moms with one provocative characteristic, whereas later in adolescence they are setting off dads with, presumably, quite different characteristics. It is perhaps more accurate to think of the parents as plucked strings than as coordinated dancers.

The distinction between the first and the second scenario of parental control is important from both a family-process and a genetic point of view. The first, nonspecific scenario suggests that mothers and fathers differ only in the *timing* of their response to genetically influenced dispositions in their children. The second scenario suggests that mothers and fathers respond at different times, in part because they are responding to *different heritable characteristics* of their children. Since these characteristics emerge at different times, the issue of developmental timing in parenting processes is secondary. That is, the different periods of time in which mothers and fathers respond are secondary to the fact that the evocative characteristics of children appear at different times. According to this second scenario, it is not that mothers are handing off their obstreperous children to fathers as the children grow older, but that the characteristics to which they were responding disappear as the children grow older. If this scenario is true, it raises its own question: Why do mothers respond to one set of cues and fathers to another set?

In fact, the data from our study strongly support the second scenario. First, the correlations between mom's attempted control in earlier adolescence and dad's knowledge in later adolescence and mom's success earlier and dad's knowledge later are trivial (.04 and .15, respectively). This suggests that mom's earlier and dad's later control strategies have few or no common influences, genetic or environmental. The correlations between mother's earlier attempts and dad's later attempts and mother's earlier successes and dad's later attempts are a bit more substantial (.22 and .24, respectively). The genetic overlap, however, is .01 and .00 for these two associations. Shared environment is the nearly exclusive component of the association: 90 percent for the association of mother's and dad's attempts and 88 percent for the association of mother's actual and dad's attempted control.

These findings illumine the contrast between mothers and fathers highlighted in the last chapter. In the area of monitoring/control and antisocial behavior, mothers were seen to be reacting proactively and protectively to early signs of difficulty in their children. Fathers, by contrast, appeared to wait for full-blown antisocial behavior to develop in one of their children. Fathers then backed off and focused positive parenting on the less troublesome child. The analysis of specific genetic effects here suggests that the difference between mothers and fathers is due not to an intrinsic difference in response style, but rather to the fact that mother's control strategies respond to one set of genetic influences earlier in adolescence and father's respond to an entirely different set of genetic influences later in adolescence.

This is a good point to remind ourselves that our sample size, large as it is, is not large enough to examine the influence of the adolescent's gender on processes of control and attempted control within families. An analysis of gender influences here would almost certainly yield a more nuanced picture of mothers' and fathers' responses to their adolescents.

The Relationship Code in Developmental Theory

Our synthesis, thus far, has consisted of two basic ideas. First, there is evidence that the family relationships–effects model fits our findings. The reverse is not, however, true. Our data do not, in any sense, prove this hypothesis, which must await much more rigorous testing in an entirely different genetically informed research design. For the time being, we can say that it is plausible that the social environment may play a crucial mediating role in the ontogeny of genetic influences on adjustment. A corollary of this idea is that the nature-versus-nurture debate may no longer be applicable. If the relationships-effects model turns out to be more rigorously supported, then genetic expression of behavior will be shown to require environmental processes; nature and nurture would be inseparably intertwined.

The second idea of our synthesis is that there are highly specific links between genetic factors and family subsystems. This idea is more concrete than the first idea and will perhaps have a more galvanizing influence on efforts to understand the interplay between genes and the social environment. Our hypothesis may be more pungent than previous ideas because it proposes that specific family subsystems respond to different genetic influences with different consequences for adolescent development. Although much more evidence is needed to support it, the code hypothesis is an effort to underscore an idea that ties together all our major findings: strong associations between environmental measures and adjustment; substantial genetic influences on measures of both the environment and adjustment; the preponderant role of genetic factors in many of the observed associations between measures of the social environment and adjustment; and the high levels of specificity of genetic influences on both adjustment and environmental measures. For rhetorical emphasis we liken family process to RNA. This gives emphasis to our hypothesis that genetic factors are dependent on specific social mechanisms in order to be expressed. As social scientists, we may have very powerful tools for understanding exactly how genes express themselves in behavior, for altering adverse outcomes of genetic influence, or for enhancing positive outcomes of these influences.

But is it plausible to argue that social processes really operate to mediate genetic expression, as RNA does? Clearly there are limits to this scientific metaphor. In the case of parents, we are arguing that they respond to a profile of heritable characteristics of their children, that these responses are relatively invariant across families, and that the responses are critical to determining the ontogeny of behavior in children that we label "adjustment." Unlike RNA, however, the family is not a simple aggregate of molecules. Surely both its sensitivity to characteristics of its children and its responses to what it recognizes must be highly variable from time to time and from family to family. For example, parental psychopathology, the availability of social supports, and the presence of family stress have been documented as influences on mother's response to a temperamental, difficult child (Hetherington, 1991). Similarly, the quality of parent-child relationships can soften the negative effect of a child's difficult temperament on sibling relationships (Brody and Stoneman, 1996).

The literature linking children's temperaments with the quality of family relationships is voluminous (Blackston et al., 1994; Brody et al., 1987; Gauvain and Fagot, 1995; Lee and Bates, 1995; Stocker et al., 1989; van den Boom, 1994). For example, a meticulous study in Holland observed mothers and infants during the first six months after birth. Compared with children with easier temperaments, irritable children appeared to turn off their mothers: they got less visual and physical contact, and their mothers showed less emotional involvement with them, even though they tried to soothe them more often. Indeed, even when these fussy babies showed some positive behavior, their inured mothers seemed not to notice (van den Boom, 1994). Despite such intriguing studies, however, the area of research on temperament and family process is also a very uncertain reference point for our theory of relationship codes.

First, with very few exceptions, the literature linking temperament with family relationships has not been genetically informed. Thus, the associations between temperament and family process, reported by other investigators, could be attributed to either shared environmental or genetic factors; indeed, the data from our study suggest that these associations are much less likely to be due to nonshared factors than to shared environmental or genetic factors. The relationship code focuses only on the association between genetic factors and family process. Because temperament is only partially heritable, we can receive only partial guidance from this literature.

Second, we do not know if the temperaments that have been examined by previous research are the key intermediaries in the relationship code. To be sure, in our own formulations we have referred to the individual mediators of genetic influences on relationships as "temperaments" or "patterns of social responsiveness" in the child. But even these labels may be deceptive because they refer to stable characteristics of children that are apparent in a broad range of settings and may have relatively weak relationships with the qualities of parent-child and sibling relationships. Moreover, the factors that are most important in linking heritable characteristics and family relationships may be subtle blends of these characteristics.

Third, genetic factors may directly influence family relationships without the mediation of measurable and stable individual characteristics (Plomin, 1994). By adolescence, children have long been full participants in their family relationships. Families are not simply "reacting to" their adolescent members; the children are actively shaping their relationships using a broad range of tactics, reactions, obstructions, innuendoes, and distractions, along with a range of perceptions, sensitivities, values, and knowledge about their family. Each of these facets of their interpersonal skills and styles may have a heritable component. It is entirely plausible that the heritable behaviors relevant to the relationship code are more likely to emerge in the context of an ongoing relationship than in a formal assessment of temperament, personality, or interests in a research setting devoid of significant others.

We envision the relationship code as part of a fluid, context-dependent transduction process. In other words, we posit that the operation of the relationship code is highly dependent on the social context of the child and the family, as well as on critical features in the family itself. For example, using quantitative, behavioral genetics techniques similar to the ones we have employed in this study, other researchers have observed social conditions that favor or block the expression of genes in behavior. We have reviewed this process of gene-environment interaction in Chapter 4. Included among these conditions are stress in the family occasioned by parental psychopathology and difficulties in family interaction (Cadoret et al., 1983; Cadoret, Winokur, Langbehn, Troughton, et al., 1996; Cloninger et al., 1982; Tienari et al., 1987; Tienari et al., 1994). Indeed, characteristics of the family that moderate genetic expression—such as marital problems or parental psychopathology—may have such a pronounced effect because the ordinary and essential mechanisms of genetic expression heavily involve the family.

These studies of gene-environment interaction, however, only scratch the surface. We want to know more about how the relationship code operates under very common and salient variations in social context and in the human genome. For example, we have shown how genetic information may be encoded in differences between mother's and father's relationship behavior. But what happens to genetic transduction mechanisms in single-parent households? Is the concept of a relationship code, let alone the details of its operation, relevant for extreme environments—extreme poverty, famines, and enduring warfare—where many subtle individual differences among children may have little or no effect on their families or other significant members of their social world? Similarly, might not some genetic risk factors have profound effects on behavior without "passing through" the relationship code? This seems plausible for some of the major single-gene defects that produce profound states of mental retardation. But might it also be true for some of the most serious psychotic disorders?

Despite these potential limits of the idea, we are urging that concepts of genetic expression be broadened to include both molecular- and relational-coding processes for practical reasons: the more we know about how social and biological contexts alter gene expression, the better we can design interventions that prevent adverse consequences of certain genetic factors and promote more adaptive consequences. For the relationship code this is not a pipe dream. Some very practical strategies are available for clinical trials right now.

Evidence Needed to Support the Relationship Code Hypothesis

Two basic ideas of the relationship code need confirmation from data that we could not collect in this study. First, we need to verify that the gene-environment correlations we observed in our study (and reported in Chapter 10) reflect active or evocative influences on the family by heritable characteristics of the child. The data presented in Chapter 11 were a very small step in this direction but were severely hampered by the unanticipated stability of almost all our measures during a period notorious, but unjustly so in our experience, for its wild swings and major transitions. Second, we need to show that the family response evoked by the child has an influential role in the further development or ontogeny of the heritable behavior. For example, let us say that we can show that heritable irritability and aggressiveness in a young child evoke parental conflict and negativity. We still need to show that evoked parental conflict and negativity are a crucial ingredient of the further development of that infant reactivity into a more fully blown conduct problem somewhat later in development. More precisely, we need to show that the evoked parental component is responsible for the *genetic component* of the more fully developed conduct problem. We have to deploy the strongest feature of our research design: the true experiment. In effect, we need to "knock out" a particular developmental sequence and test its effects on the ontogeny of a particular behavior.

As indicated in Chapters 2 and 4, a prevention research design is a particularly appealing experiment for accomplishing this central aim. Although it has never been accomplished, the use of an adoption study within the context of a prevention strategy is a particularly appealing opportunity. Recall that adoption studies are ideal for teasing apart the various types of gene-environment correlations. An active/evocative effect is most clearly revealed when we note a correlation between a characteristic of a birth parent and a quality of parent-child interaction. A particularly dramatic example for adolescents has recently been reported in a sample of adopted-away adolescents and their adoptive parents. For example, the adoptive mother's disciplinary practices toward her adopted adolescent could be predicted from psychiatric assessments of the birth mother. Adopted-away children of birth parents with psychiatric disorders tended to be both hostile and antisocial, and this appeared to elicit distinctive disciplinary responses from their adoptive mothers (Ge et al., 1996). Similar findings were reported for the adoptive fathers in this study.

Conducting an adoption study without intervention would provide strong evidence for heritable characteristics in the child evoking parental responses. If such a study were made longitudinal, it might be possible to spot very early manifestations of the heritable characteristics in the child before they reliably evoke parental or sibling responses. As we have learned from our study, it is best to pick a developmental period that shows major changes and transitions. The toddler period would seem to qualify: not only is it early in development (fifteen to twenty-six months), but it is a period of enormous change in both the toddlers and the quality of their relationships with their parents. A study of this kind, so early in development, could now make use of recent research on very early manifestations of heritable precursors of problem behaviors. For example, as early as the neonatal period, marked individual differences in infant reactivity to external stimuli can be observed. Irritability and excitement can be reliably coded by trained observers, and physiological reactivity to stress stimuli can also be objectively measured. Irritable neonates and those with *low* physiological responses to stress become challenging and difficult infants (Gunnar et al., 1995; van den Boom and Hoeksma, 1994). Indeed, by the time they are ten months old, coercive patterns in infants become manifest: these infants make frequent demands of caretakers that are very hard to "turn off." These behaviors may be a prelude to serious conduct problems later in development. Also at this time infants may begin to respond to novel stimuli by retreating; such behavior may be a precursor to anxiety problems. Moreover, these patterns persist through early childhood and may become entrenched for many years after that (Biederman et al., 1990; Biederman et al., 1993; Hirshfeld et al., 1992). Recent data suggest that some of these early emotional and activity patterns may be heritable (Emde et al., 1992; Plomin et al., 1993).

The fact that mothers, who have been studied more frequently than fathers, have specific responses to some of these patterns of behavior is important for an instructive test of the relationship code hypothesis. Most mothers of difficult, challenging infants try to anesthetize themselves by becoming less attentive to their infants while at the same time expressing more irritation of their own (van den Boom, 1994; Wachs, 1992). Parents with an anxious, inhibited child may respond quite differently, becoming over-engaged with a child who retreats from novel situations (Kagan et al., 1993). Thus, on a phenotypic level there is evidence for specific responses by many parents to specific behavioral and emotional "signals" from their infants. Adoption designs have the power to weight the importance of genetic factors in these specific responses to coercive or inhibited children. If genetic factors play a role in this stimulus-response pattern, this would be supportive evidence for the idea of a relationship code. Still stronger proof would be obtained if genetic factors explaining the association of child coerciveness with parental withdrawal or irritation were different from those that explained the association between infant inhibition and parental over-involvement. In a longitudinal adoption design, it would be particularly important to show that the heritable characteristics emerged first and the parental responses followed. Parent-child relationships may be very highly reciprocal in early childhood, however, and so this sequential timing may be difficult to observe.

The strongest proof for the idea of a relationship code would come from an intervention analogous in some ways to the gene "knock-out" experiments described briefly in Chapter 4. An objective of such an intervention would be to interrupt an adverse cycle of child behavior leading to adverse parental response. Is it conceivable that we can alter parents' response to a heritable characteristic of their adopted child? Very effective interventions of this type have been developed and demonstrated for children as young as six months for birth parents rearing their own children (van den Boom, 1994). For example, mothers can be taught to become attentive to signals from their infants and to enhance their capacity to soothe them when they are accurately perceived as fussy and irritable. Indeed, interventions of this kind have shown substantial improvements in mothers' responses, parent-child relationships, and infant behavior. Mothers in the intervention group, for instance, were more attentive to their infants, the infants became more securely attached to them, and the infants themselves became better able to soothe themselves.

These results all reflect differences in means or averages between the intervention and control group, however. An adoption design would allow us to determine not only whether an intervention of this kind improved mother's response to a difficult child, but whether the intervention could reduce or eliminate the heritability of the child's behavior as well. The basic strategy from drawing inferences of this kind from an adoption design is schematically summarized in Figure 12.10.

The first panel summarizes data we might obtain from a more conventional adoption design in which no preventive intervention is introduced. Let us assume that irritability in an infant is an expression of the same genetic influences that affect antisocial behavior in a parent (legal problems, frequent fights, substance abuse, and so on). Then we could expect a simple relationship between the magnitude of antisocial problems in the birth parent and the level of irritability in the infant.

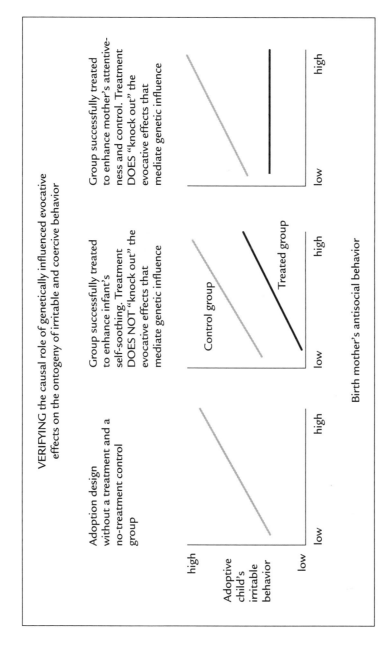

Figure 12.10 Preventive intervention as a test of the relationship-code hypothesis

The second panel shows the introduction of treatment aimed at reducing infant irritability. This may involve the parents, but no effort is made to target the active/evocative effects that are the specific mediators of genetic influence according to the relationship code idea. In this case we would expect the treated group to look better than the untreated controls: all children in the group might achieve some benefit from the treatment. But since we have not targeted the specific causal link in the chain—the evocative effects—there is no alteration in the index of heritability. In both the treated group and the control group there is still a notable correlation between the birth mother's antisocial behavior and the child's irritability.

The third panel shows something quite different. In this instance we have targeted the intervention to disrupt or "knock out" the specific evocative process that the relationship code idea posits is at the core of the ontogeny of genetically influenced patterns of adjustment. The treatment does produce a sizable mean or average improvement in the children's irritability. But the heritability is also eliminated: the association between birth parent's antisocial behavior and child's temperament is now zero. Thus, by combining an adoption and a preventive intervention study we could provide strong experimental proof, or refutation, of the relationship code idea.

Summary

In sum, the concept of a relationship code can embrace circumstances in which the family may vary in response to heritable characteristics of its members and in which the individual mediating factors are not detectable by standard psychometric procedures. The power of anonymous genetic analysis, at this point in the technology of developmental studies, is that it can explore the operation of the code even though we are ignorant about how genetic information is encoded into relationships. Even at this early stage of theory formulation, however, a number of caveats are important. To name just one, there may be extremes in both the social context of the developing child and the child's genome that constrain the importance of the relationship code in the expression of heritable behavior.

Chapter 13

SYNTHESIS III: GENETICALLY INFORMED PORTRAYALS OF ADOLESCENTS AND THEIR FAMILIES

Thus far we have made substantial progress in developing a synthetic view of adolescent adjustment. This theory has enabled us to piece together genetic and family influences on development in novel ways. Chapter 11 gave us intriguing glimpses, for example, of self-repair processes in families: the role of genetic influences on positive family processes that had a protective influence against the development of antisocial behavior during adolescence. Chapter 12 gave an even more panoramic view of the specific links between genetic influences and family processes. We discovered that genetic effects are specific for three different aspects of these subsystems. First, there is specificity of influences for specific subsystems; for example, genetic influences on mother's warmth are unrelated to genetic influences on father's warmth. Second, there is a specificity for quality of interaction within subsystems. Genetic influences on conflict and negativity are quite different from those on warmth and support for both fathers and mothers. Third, genetic influences operate at different times in the development of subsystems; we noted this phenomenon for control processes in the adolescent sample we studied. This specificity added some weight to our speculations about the ways family processes mediate genetic influence and led to the central hypothesis of this book: the relationship code.

In our effort to understand the complex interplay of genetic factors and the social environment of the family, we do not want to overlook another treasure trove: the many important environmental influences linked to adolescent adjustment by nongenetic mechanisms. In this chapter we focus on findings related to both the shared and the nonshared environment. In the former, our focus is autonomy and sociability, which are linked to family subsystems substantially by shared environmental processes. Several clues as to why this is the case reside in the role of siblings and the factors that keep family subsystems integrated. When considering the nonshared environment, we attempt to convert disappointment into contentment. There are, in fact, many clues about the nonshared environment that permit us to formulate an educated guess about what these factors might be. To illustrate one hypothesis we review, of all things, a dream of Sigmund Freud's that he reported in *Interpretation of Dreams*.

We begin our third step toward a synthesis. We have already examined the modest changes in both family relationships and adolescent adjustment that have unfolded from earlier to later adolescence. These have suggested, but not proven, that families play a crucial role in the unfolding of genetic influence on adolescent adjustment. Heritable characteristics of the child elicit responses from the family. The family response may take one of four forms: it can enhance desirable attributes of the child; exacerbate troublesome aspects; protect the child from adverse outcomes of difficult behavior; or cause parents to back off from a troubling child in an attempt to protect a sibling who appears to have better prospects. Within the context of these four patterns of reaction, we have examined all the measures of family process in our study. They have suggested that, to the extent that these four family reactions explain the genetic influences on behavior, there is an extraordinary repertoire of specific family responses to a great variety of heritable characteristics of the children. We have noted the rich encoding possibilities inherent in this repertoire of family responses: across different qualities of relationships within a subsystem, across different relationship subsystems, and across time.

We have argued that the encoding of genetic information into family processes might rival in importance, and serve in tandem with, the much better known process of RNA encoding—the critical intracellular transduction of genetic information on the road to protein synthesis. But our attempted synthesis between an environmentalist and a genetic perspective cannot end at this point. We must return to a fundamental starting point of this journey: our genetic analyses of seven measures of adolescent adjustment at earlier and later adolescence as shown in Figure 8.1. In those analyses we replicated repeated findings of other genetic studies: a genetically informed study can delineate important nongenetic, environmental influences on adjustment. Indeed, our results do confirm the importance of these influences on adolescents. These results are so impor-

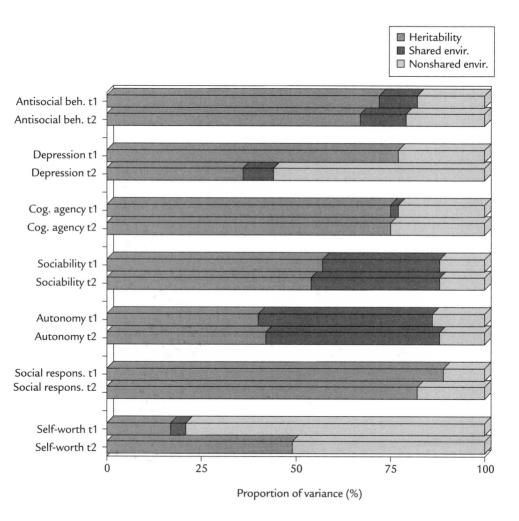

Figure 13.1 Environmental influences on adolescent adjustment

tant for our synthesis that we reproduce Figure 8.1 here (see Figure 13.1). We have changed the shading slightly to give emphasis to the topic of this chapter: environmental influences on adolescent adjustment.

Three important features deserve emphasis. First, for seven of the measures, environmental factors account for 40 percent or more of the total variance, and in five of these seven domains, they account for greater than half. Only the measures of social responsibility show a small, almost negligible, environmental influence during this stage of adolescent development. Second, as has previously been reported in many other studies, the nonshared environment accounts for almost all the environmental effects for five of our seven measures in both earlier and later adolescence: antisocial behavior, depression, cognitive agency, sociability, social responsibility, and global self-worth. Third, somewhat surprisingly, two measures show a heavy preponderance of shared environmental influences: autonomy and what we have called "sociability." In our usage the term "sociability" does not refer to a simple temperament that is detectable very early in development. Rather, in keeping with all our other measures of adjustment in adolescence, it assesses the adolescents' mastery of several aspects of their immediate social world. In our study, sociability reflects the successful involvement of the adolescents in both peer relationships and extended family relationships.

As noted, our search for specific, nongenetic, nonshared environmental influences has been frustrating. In adolescence this aspect of the environment does not appear to reside in differential parental treatment or asymmetrical sibling relationships (nor, according to our preliminary measures, in different qualities of peer-group relations). A more complete synthesis, nonetheless, needs to offer some account of the substantial influence of this aspect of the environment. We return to such an effort in the second part of this chapter. Our attention is turned first to sociability and autonomy. We did indeed find evidence for the association of both parental and sibling relationships with autonomy and sociability, and, more important, a generous percentage of these associations was due to overlapping shared environmental—not genetic—factors. A fuller understanding of these overlapping shared environmental influences provides us with an opportunity to weave together three themes: the role of the shared environment in coordinating family subsystems, the transitional role of the sibling system, and the central role of social initiative in families of adolescents. We consider each in turn.

Family Systems, Sociability, and Autonomy

We begin this section by reminding the reader of an unusual finding. We have reported that, for the most part, genetic factors account for the predominant component of the association between parenting and our measures of adjustment in adolescence. For two measures of adjustment, however, there was a distinctively different pattern: for autonomy and sociability, *shared environment* accounted for a substantial proportion of the association. Figure 13.1 shows that these are the only two adjustment measures where the most significant source of environmental influence is the shared environment. Figure 13.2 summarizes the findings on both parenting and sibling relationships and their anomalous associations with sociability and autonomy.

Figure 13.2 shows that shared environment is an important component in the association of mothering, fathering, and sibling relationships, on the one hand, and autonomy and sociability, on the other. It is the predominant component in all these associations except those involving mothering. Further, it is invariably positivity in these relationships that shows phenotypic associations with autonomy and sociability of .20 or greater. Mother's and father's positivity are only modestly correlated (.42), and each has only a modest correlation with sibling positivity (.42 and .34, respectively). Thus, it is likely that positivity within each of these three family subsystems contributes uniquely to the association with autonomy and sociability. The cumulative or total anonymous, shared environmental factors account for 31 percent and 34 percent of the variance in sociability at times 1 and 2, respectively, and 46 percent of the variance for autonomy at times 1 and 2. Therefore, it is likely that we have identified, in this study, most of the major environmental correlates of these two adjustment variables.

Why do these two measures show such strong association with the shared environment in contrast with the other five measures? And why is shared environment such an important component of their relationship with parental and sibling interaction patterns, again in contrast with the

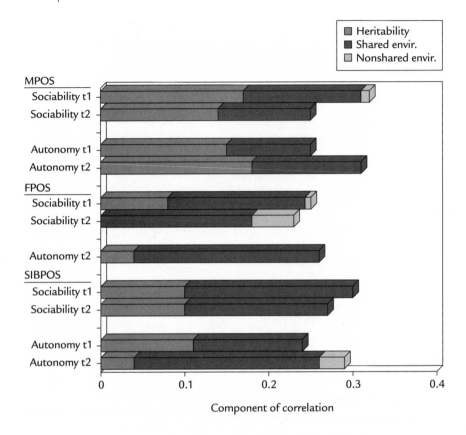

Figure 13.2   Correlates of sociability and autonomy

other five measures of adjustment? And why is positivity so much more important than negativity? Indeed, negativity has no reliable association with these adjustment measures. To explore this issue we return briefly to findings from the previous chapter.

Shared Environment and the Coordination of Family Subsystems

In Chapter 12 we focused on the relatively small role that genetic factors play in the association among measures of the family subsystems (and their correspondingly greater role as unique influences on each of the measures). Figure 13.3 returns to the same data from a somewhat different vantage point. Here we focus on the important role that shared environment plays in the seventeen associations across family subsystems that show phenotypic correlations of .20 or greater in earlier adolescence. There are no important differences in this picture in later adolescence; thus, for simplicity, we do not consider those data here.

Figure 13.3 reveals a striking pattern of findings. For every association across a family subsystem, for example, mother's positivity with father's positivity, shared environment plays a major role. For ten of these asso-

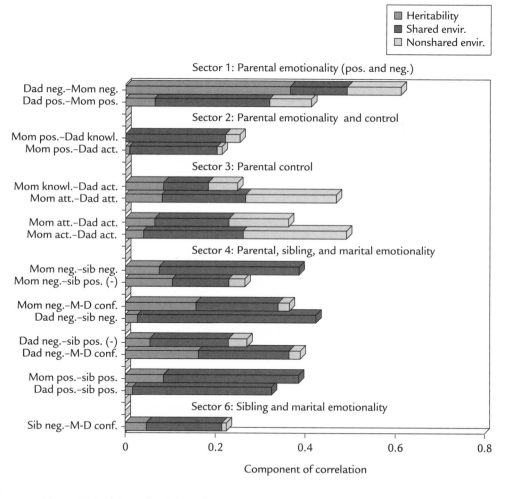

Figure 13.3 The role of shared environment in associations across family subsystems

ciations, shared environment accounts for more than half of the association between family subsystems: mother's and father's positivity; mother's positivity and dad's knowledge; mother's positivity and dad's actual control; mother's and sib's negativity; father's negativity and sib's negativity; dad's negativity and sib's positivity; dad's negativity and mother-dad conflict about a child; mother's and sib's positivity; father's and sib's positivity; and sib's negativity and mother-dad conflict about a child.

Although we could argue that the preeminence of the shared environmental role could have been confidently predicted, the findings *are* a surprise and thus they are richly informative. It would have been plausible to argue—in advance of obtaining these findings—that genetic factors might have played a major role in coordinating family subsystems. For example, sociability of adolescents in the data shows substantial heritability. Related social skills might also be influenced by genetic factors. These heritable skills might soothe the adolescent's sibling and parental relationships, and in this way genetic factors would play a preeminent role in linking family subsystems. But these heritable factors, the data tell us, do not play such a role. Similarly, nonshared factors might have played a significant role in the coordination of the parent-child subsystems; nonshared factors are major influences on both mother-child and father-child relationships. As we have observed, however, nonshared factors play a prominent role only in the area of parental control.

Thus, the preeminent role of the social environment here is both informative and important. There are three possible explanations for an important shared environmental influence on associations between any two measures. If slightly modified, these explanations can be applied to an analysis of the association between measures of two different family subsystems.

The first explanation is that some factor external to the family system, and equally influential for both siblings, jointly influences two or more family subsystems. Consider the association between mother's positive parenting toward her children and dad's positive parenting. This association is due almost entirely to shared environment. It is possible, for example, that economic factors external to the family influence this association. We may well surmise that very favorable economic circumstances would lead to higher levels of positivity in parenting relationships. Variations in circumstances across families would produce considerable variation in parental positivity; this might account for the strong association between mom's and dad's positivity. The importance of shared environment in our findings tells us that an external factor such as economic advantage is as likely to influence mom's and dad's parenting toward one child as it is to influence parenting toward the second child.

In our effort to delineate factors external to the family, we examined the effects on all our measures of family subsystems of the average parental education, of the average parental occupational status (for example, doctors receive higher ratings than manual laborers), and of social class, which weights these two factors. Their effects were trivial and hence these external factors could not explain the overall phenotypic associations across the family subsystems or the dramatic influence of shared environment on these associations. It is distinctly possible, however, that finer-grained measures of economic advantage or disadvantage (savings, inheritance, and so on) or other aspects of the shared external environment (neighborhood, extended family characteristics, and so on) might play a significant role in these findings.

A second possible explanation for the importance of the shared environment in the associations among family subsystems concerns the non-genetic influence of the adolescent siblings themselves on the relationships among the family subsystems. For example, it is conceivable that siblings in the same family are more or less equally versed in eliciting positive parenting from mothers and from fathers: if they elicit positive parenting in mothers, they are as likely to do so in fathers. Moreover, it must be some nonheritable characteristic of the siblings that is responsible for this coordination of family subsystems. If this influence of the adolescents on the family systems, the influence accounting for shared environmental effects, stemmed from heritable qualities, our analyses would show a very substantial genetic component in these associations. Of the seventeen associations examined in Figure 13.3, only one, that between dad's and mom's negativity, shows a substantial correlation attributable to genetic factors.

The data might have shown that nonshared factors were important in the association of family subsystems. Indeed, for three of the four associations in sector 3, parental control, they are important. But for the remaining fourteen associations among family subsystems, they are not. If the role of environmental factors in these associations is attributable to the children's influences on the family, and this is far from proven in these data, the importance of the *shared* environment informs us about what kind of effect this is likely to be. We know that it must be an effect that could be elicited by either child. That is, it must be due to characteristics that the two children have in common.

Many shared or similar characteristics of adolescents may have this role in the family, but here we point out only that three obvious candidates do *not* have a role: the gender of the siblings, the mean age of the siblings, and the age difference of the siblings in nontwin families. All analyses in this book have corrected for these three characteristics of children, and hence none of the results reflects their influence.

The story about the importance of the shared environment may be different in the domain of parental control. In this domain, in three of the four associations across subsystems (father-child and mother-child), non-shared factors are about as important as shared factors. This means that, in the domain of both attempted and successful control: (a) parents achieve moderate levels of similarity to each other in their level of attempted and successful control (all associations here are above .40); (b) there is differential control attempted and executed for the two different siblings; and (c) the parental agreement on this differential control is occasioned by nongenetic factors. It is tempting to speculate that a nonheritable attribute of the child stimulates this interparental agreement on control, perhaps a nonheritable quality of surliness or poor social judgment. This parental agreement on differential control may come from other sources, however. For example, parents may unfairly stereotype one child as a "bad apple" in the absence of any verifiable differences between that child and a sibling. Similarly, external to the family, one of the two adolescent sibs may suffer from differential stigmatization by neighbors, teachers, and extended family. Parental knowledge of this stigma may influence mom's and dad's differential parenting. Our analytic procedures remove simple age differences between the children as a source of these agreed-upon differential parenting practices.

Still another possibility concerns factors intrinsic to the family. For example, a family may be considered a unitary group characterized by features that are shared by the whole family. For instance, family units may differ from one another in overall levels of positive and negative transactions among all family members. A positive family ethos, born of the family's own history and development, may influence all family subsystems. This might explain much of the correlation, for example, between father's and mother's positive parenting as well as the association between positive parenting and positive sibling relationships.

What factors might contribute to differences of this kind among families? We have already ruled out differences among families in economic advantage and disadvantage and also differences in levels of education as much as our approximate measures of these family attributes would allow. In the absence of factors external to the family, we turn to intrinsic characteristics that might account for associations across subsystems. One clue to these characteristics comes from data we have already reported: we know that shared environmental influences of measures of the family are remarkably stable (see Figure 9.9). Thus, any proposed features of the family unit that can explain the data must (a) account for associations across subsystems; and (b) be stable over time.

When we turn to a careful examination of the processes that might be responsible for the important nonshared influences on adolescent adjustment, we focus our attention on the personal constructions of social reality that may be central in adolescent development. Indeed, as will be shown, many features of these highly personal and idiosyncratic constructions fit well with all our findings concerning nonshared influences. It is thus particularly intriguing that the best-documented program of research on stable differences among families considered a unitary group has focused on how the family as a social system collaborates in the development and stabilization of constructions of social reality that are implicitly shared by all members.

It is beyond the scope of this book to review in detail research on family groups. Briefly, a comprehensive and programmatic research project identified three dimensions along which families as a group differed (Reiss, 1981; Reiss and Oliveri, 1983; Reiss et al., 1983; Sigafoos et al., 1985). The first dimension concerned their belief in their ability to understand the intricacies of their social world, such as the nuances of social threat and opportunity, and to master these intricacies for the long-term benefit of the family. Some families see these intricacies as stemming from general principles of social life and as discoverable and usable, whereas other families see the social world as capricious and unpredictable. The second dimension concerned the family's belief about their own characteristics as a group: did they believe that their social environment treated them as a unitary group, or were they treated as individuals without reference to their family membership? The third dimension focused on differences among families in their belief about tradition. Families differ in the extent to which they view social convention as exemplifying long-standing tradition. Some families are highly sensitive to these innuendoes in social convention, such as class status and family traditions. As a consequence, they feel that their own predilection for maintaining tradition and stability is reinforced by social norms to which they are very sensitive. Other families are much more sensitive to novelty and changes in their social world; these perceptions reinforce adaptability and change within the family itself.

These overarching perceptions of the social world appear to determine how family rules of conduct are established, interpreted, and implemented. They are also quite stable and play a major role in shaping an emotional ethos in the family. For example, families who see their social world as capricious but feel that they are perceived as a social group (they are low on the first dimension of mastery but high on the second dimension of group solidarity) tend to have high levels of anxiety and suspiciousness about outsiders and draw firm boundaries between themselves and outside groups. In more extreme forms this suspiciousness results in an attitude of "us against the world" that regulates relationships among family subsystems. These distinctive family "world views" may be subtle reflections of cultural differences among families or may reflect how established they are in the communities in which they live, with strong contrasts, for example, between new immigrants and established families. They also may be built up over time within families and may reflect ways in which families have resolved major crises in their history together.

There are certainly other factors, beyond the family's collective construction of social reality, that distinguish one family group from another. The data presented here should provide encouragement for additional research in this area.

Sibling Influence on Adolescent Adjustment: An Echo of Parental Impact?

We have already presented data on the genetic and environmental components of the association between sibling positivity and negativity, on the one hand, and adolescent adjustment, on the other. These data have been presented in Figures 10.8, 10.15, and 10.16. We present these same data in a slightly different format in Figure 13.4. The shading in Figure 13.4 emphasizes the central role of the shared environment in almost all of these relationships. We can note, once again, the role of shared environment in the association between sibling relationships and autonomy

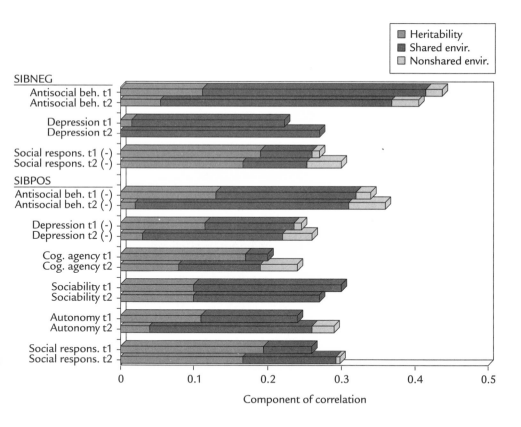

Figure 13.4 Genetic and environmental components of the association between sibling relationships and adolescent adjustment

and sociability. Particularly striking, however, is the role of shared environment in linking both sibling positivity and sibling negativity to antisocial behavior and depressive symptoms. Indeed, the role of shared environment—both in absolute terms and in relationship to other components of the associations—increases in magnitude for all eight of these associations from earlier to later adolescence. This pattern is markedly different from associations of these same adjustment measures with parenting: genetic factors are the preponderant components of the association, and, at least for mothers, the overall trend is for these genetic components to become even more important over time.

There are several possible explanations for these important differences, all of which start from the same premise: that the social influences of parents on children may decrease over time, whereas the influence of sibs and peers may increase. We see in our study only clear evidence of the increase in sibling influence, but the data are quite consistent with a declining social influence, independent of genetic effects, of parents. The explanations for sibling ascendancy can be termed "maturational," "adaptive" (Sim and Vuchinich, 1996), "contextual," and "sequestrational."

The maturational explanation argues that children acquire an increasing range of skills, interpersonal resources, and coping strategies as they get older. This makes them less dependent on their parents. For example, some of the relationships that maturing children develop with both peers and adults outside the home may serve as buffers for their adverse relationships with their parents (Jenkins and Smith, 1990). Conversely, siblings may become an increasingly important reference point for older children as they establish peer relationships outside the home. Siblings are an important source of standards for behavior with peers. They are also a source of information about social processes, events, and structures outside the home. This might be especially true of the same-gendered siblings close in age who were the focus of our study. Some phenotypic data are consistent with this finding. For example, sibling alcohol use predicts the involvement of adolescents with peer groups in which alcohol is also heavily used (Conger and Rueter, 1996).

The adaptive explanation suggests that parental influences, particularly negative ones, may have a strong impact when children first experience them, but that these influences, even if they are enduring, lose their effect over time. The declining effect may be due to various buffers the child may develop for protection from abusive situations: from spending more time outside the house to learning how to avoid abusive reactions from parents or to placating them once aroused. Recent longitudinal studies of antisocial behavior suggest that while adverse parenting may influence adjustment of both children and adolescents, the effect does wear off over the years (Sim and Vuchinich, 1996).

The contextual explanation focuses on the many different social experiences of adolescents. Although the adolescent's relationships with parents and siblings may be equally important for his or her adjustment, siblings become a much larger component of the adolescent's social world. In one study of eleven-year-old children and their siblings who were no younger than eight, the older children spent a full one-third of their after-school hours with their siblings (McHale and Crouter, 1996).

These first three explanations come from psychosocial research uninformed by genetic analyses. A fourth hypothesis, the sequestrational explanation, includes genetic factors and focuses on the response of sibling systems to genetic differences among adolescents. Indeed, the genetic influence on sibling systems is substantially less than on all other subsystems in the family, that is, the sibling system is sequestered from strong genetic influence. Since genetic influences are so slight, they cannot be responsible for the strong associations between sibling relationships and adolescent adjustment. One reason sibling relationships may be so sequestered is that the "sequestration" may merely reflect the important environmental mechanisms—maturation, adaptation, or context—that preferentially link sibling relationships to adjustment. They may simply override or drown out genetic influences on sibling relationships.

But sibling relationships may be relatively sequestered from genetic influence for a second, more specific reason intrinsic to the dynamics of sibling relationships: these relationships involve less surveillance of siblings by each other than is characteristic of parent-child or peer relationships. Parent-child relationships involve sustained care-taking across substantial periods of time. Even inept parents make an effort to adjust their parenting according to the needs, demands, and styles of their children. All of these attributes of children may be genetically influenced. The data from our study as well as from previous research on siblings suggest that the nature of sibling relationships is very different from that of parent-child relationships, especially in terms of reciprocity. Indeed, siblings tend to maintain tit-for-tat relationships, with each striving to give about as much warmth and negativity as he or she receives. Reciprocity in a sibling relationship creates an equal or shared environment for both siblings. It follows that these reciprocities must also lie at the core of the substantial shared environmental component of the associations between sibling relationships and adjustment we have been reporting.

We suggest that these reciprocities are powered by the siblings' wish for equity in their relationship with each other and with their parents. This would explain their exquisite sensitivity to perceived inequalities in their parents' treatment of them. We speculate that the tit-for-tat behavior in sibling relationships is based not on a surveillance of each other's temperaments and styles, but on a fundamental need to keep things equal. This dynamic could override strong genetic influences on sibling relationships.

This explanation is supported by additional evidence from our study. As reported in Chapter 7, there are, in fact, some important asymmetries in sibling relationships; further, these asymmetries are associated with adjustment. For example, we noted asymmetries in aggression. Adolescents who mete out punishment to their siblings are protected from antisocial behavior, whereas those who are victims are at increased risk. Similarly, siblings who are caregivers—that is, siblings who provide more of the warmth in a sibling relationship—show high levels of cognitive agency and of autonomy. The data suggest, however, that these asymmetries may be quite sensitive to genetic influence and account entirely for its association with adolescent adjustment. To put the matter another way, when siblings step out of the role of reciprocators and become either enforcers (meting out punishment) or caretakers (providing nonreciprocated warmth), their behavior is as sensitive to genetic influence as that of their parents.

One might have thought that peer relationships would show the same central characteristics of reciprocity that sibling relationships show. Indeed, more direct measures of peer relationships might reveal just such a dynamic. Our preliminary measures concerned only the quality of the entire peer group, however, and not interaction processes in dyadic friendships between the adolescent and a peer. Peer groups, considered bounded social units, also have a major surveillance function: supervising who will be admitted to the group and who extruded. Heritable characteristics of the child most likely play a significant role in this informal admission and extrusion process, which, in turn, accounts for the very high heritabilities of our measures of peer-group characteristics. Genuine reciprocity—of positive and negative behavior—within these groups may exist in the context of selection and extrusion, which may be heavily influenced by heritable characteristics of the adolescents.

In sum, we have argued that some mixture of maturation, adaptation, contextual factors, and lack of surveillance processes provides the distinctive ambiance of sibling relationships. Variations in this ambiance are associated with adolescent adjustment and account nicely for the singular importance of the shared environment in sibling relationships. One more datum presented here must be examined, however. It is not true that shared environment has no important role in the association between parent-child relationships and adolescent behavior. We have shown in Figure 13.2 that in the association between parent-child relationships and siblings' behavior toward each other, the shared environment is the preponderant influence; indeed, for positivity, it is virtually the exclusive component. That is, when we consider adolescent behavior in the frame of "adjustment," shared environment has a trivial role in its association with parenting (except for sociability and autonomy, which we consider in the next section). Most of our measures of adjustment explore the child's behavior in a broad range of environments in the home and outside of it. Thus, to be effective, parental shared environment would have to extend its influence well beyond the home: to the classroom and to peer relationships. When we consider adolescent behavior in the frame of "behavior with another family member (a sib)," shared environment plays the central role in association with parenting. Here, our measurements specify that the behavior under consideration must occur only in a narrow and well-specified context: interaction with the same-gendered sibling closest in age to the adolescent.

Given this importance of shared environment in the association of parenting and sibling relationships, it is surprising indeed that parenting does not exert at least an indirect shared environmental influence—through sibling relationships—on most measures of adjustment, particularly those measures for which shared environment is a notable contributor to individual differences. As noted, the data clearly show that it does not. We can only speculate that shared environment had a much more important role at an earlier stage of development. We speculate that parent-child and marital relationships play a profound role in the quality of sibling relationships from very early in development and that, at this earlier phase, the shared environmental impact on sibling relationships mediates a powerful shared environmental influence of parenting on child adjustment. This would explain several findings from other research. For example, for cognitive abilities we know that the effects of the anonymous shared environment are relatively strong early in childhood but decline as adolescence unfolds (Scarr and Weinberger, 1983). Our proposal about the earlier effects of parents and siblings fits with this finding. The shared environmental effects of sibling relationships may be an echo of much stronger shared environmental influences of parenting at a much earlier stage. The concept of an "environmental echo" is apt. It conveys the sense of a delayed effect but also of a modest one in comparison with the much stronger influence we are positing for the shared environmental effect of parenting earlier in development. With respect to the latter, even the largest shared environmental effect of sibling relationships is quite modest. The largest is the component of sibling negativity and antisocial behavior at time 2. The value of this component is .31 (see Figure 13.4). It accounts for only 12.7 percent of the variance in antisocial behavior (phenotypic $r = .405$; shared environmental component $= .313$; total variance of antisocial behavior accounted for by sibling negativity $= .405^2 = 16\%$; proportion of this variance accounted for by shared environment $= (.313/.405) \times 16\% = 12.7\%$).

Sociability and Autonomy

We return now to our initial question: Why is shared environment so important for individual differences in sociability and autonomy? Let us begin by taking a closer look at these two measures. Table 13.1 lists some sample items from the inventories we used to assess these two related domains of adjustment. Sociability focuses squarely on the level of social engagement that is typical for the adolescent. The social engagement includes friends, members of the opposite sex, and school groups. There is also a subtheme of productive work in these items, although in many of them there is a social component, since often the work involves others but has distinct benefits for the adolescent in terms of money or experience. Autonomy reflects initiative in self-care and family care, in managing daily activities, and in using free time for recreation. As the content of the items anticipates, we found that older children are more likely to engage in some of the activities than are younger children (see also Sigafoos et al., 1988). It is not surprising that autonomy increases with age. Indeed, no other measure of adjustment in our study shows such a distinctive increase with age. Our focus here, however, is differences among adolescents after correcting for age and the role of shared environment.

At time 1, autonomy and sociability correlate highly with each other ($r = .47$). As is characteristic of correlations among measures of *adjustment,* genetic factors account for most of this association: 60 percent. Recall that for measures of *family subsystems* the general rule is that genetic factors account more for the disjunction than for the association between measures. Shared environment accounts for 33 percent of the association between autonomy and sociability, and nonshared environment accounts for the rest, 7 percent. However, the balance between common and unique shared environmental influences on sociability and autonomy favors the unique influences. Of the 34 percent of shared environmental influences responsible for sociability, for example, 14 percent are shared with autonomy and 19 percent are unique. There is a similar balance for autonomy: 19 percent of shared environmental influences are shared with sociability, but 27 percent are unique. Figure 13.2 is not helpful in delineating the shared environmental influences that are distinctive for sociability and for autonomy. The shared environmental component of mother's, dad's, and sibling's positivity contributes in equal measures to autonomy and sociability. "Positivity" is, of course, an omnibus rubric for a set of more specific measures of warmth and support. Perhaps a more nuanced set of indexes for positivity would reveal that some forms of positivity favor sociability and others favor autonomy. Perhaps the closeness component of "positivity" is differentially associated with sociability and the assertiveness component with autonomy.

The larger question is why these two variables are so strongly associated

Table 13.1 Sample items from measures of sociability and autonomy

Scale	Sample items
Sociability Child Behavior Checklist—social subscale	• List organizations, clubs, teams, groups (up to three). Compared with other children, is yours less active, average, more active? • About how many close friends does your child have? About how many times a week does your child do things with them? • Compared with other children of his/her age, how well does your child get along with brothers and sisters, other children, parents?
Autonomous Functioning Checklist (AFC)—social and vocational subscale	• Close relationships with opposite-sex teenagers • Casual friendships with adults outside the family • Active in recreational groups of teenage friends • Has many friendships • Active in extra-curricular groups (school-based) • Close relationship with member of extended family • Worked to earn money by providing a service • Volunteer work with pay for school, political, or social agency • Explored career interests by visiting worksites

with shared environments and in particular with an atmosphere of family positivity that is shared equally by the two sibs, in sharp contrast to other measures of adjustment. We suggest that both autonomy and positivity measure the positive quality of the adolescents' participation in group life as their social worlds expand. These social worlds include the family household itself, the extended family, peer groups, and the school and work setting. Sociability reflects the levels of the adolescents' engagement in these settings. Autonomy reflects the adolescents' increasing awareness of their own role and effectiveness in these settings. The measures of family process included in the global construct of "positivity" include feelings

Table 13.1 *(continued)*

Scale	Sample items
Autonomy	
AFC—self- and family-care subscale	• Keeps personal belongings in order
	• Performs simple medical care and first aid for him/herself
	• Does designated household chores
	• Purchases own clothing
AFC—self-management subscale	• Uses telephone and telephone directories
	• Uses library services
	• Plans transportation to and from special activities
	• Manages own budget from allowance or earnings
AFC—recreation subscale	• Listens to music
	• Writes letters to friends, relatives, acquaintances
	• Takes lessons to improve artistic or academic skills
	• Pursues activities related to career interests
	• Attends club meetings

Note: Wording of items is condensed for simplification. Scores used in this study reflect cumulated values from separate scales administered to mothers, fathers, and adolescents.

of affection and involvement, but also comfort with assertiveness and with the ability to make one's own position clearly understood. These attributes of assertiveness and clarity in family life have been termed "individuation" (Grotevant and Cooper, 1985, 1986). Thus some of the seeds of sociability and autonomy in the adolescent may rest in the balance between involvement and individuation in the patterns of family life they experience. Our research design permits us to take this interpretation two steps further.

First, we know that parents do treat children differently when it comes to positivity (see Chapter 7, particularly Figure 7.1). Moreover, there are notable asymmetries in sibling relationships: within sibling pairs, one sibling is often the caregiver (giving more warmth than is received), and the other is the care-receiver (receiving more than is given). Our data tell us, however, that in the area of autonomy and sociability these parental differences and sibling asymmetries are unimportant. If they were important, we would see more distinctive associations with the nonshared component of parental and sibling positivity. Figure 13.2 tells us that these effects are negligible, meaning that an adolescent may benefit from the special blend of closeness and individuation in his or her family whether it is directed at the adolescent or at the sib. In a critical sense the level of autonomy and sociability of adolescents both in the household and in many domains of their social world is an excellent index of the balance of closeness and individuation that characterizes not any specific family relationship but the entire nexus of relationships in the family as a whole.

Second, autonomy and sociability also reflect genetic differences among adolescents. Indeed, these are the only two domains of adjustment in which genetic factors and shared environment have equal influence. If our overall theory of the relationship code is correct, the family has two distinct roles in the ontogeny of sociability and autonomy. First, genetic differences among children elicit different levels of maternal and sibling warmth, which subsequently enhance or promote sociability and autonomy. This sequence is suggested for maternal and sibling positivity by the data presented in Figures 11.13 and 11.16. Second, the overall atmosphere of family closeness and individuation contributes equally to the autonomy and sociability of both siblings.

In sum, autonomy and sociability take a special place in our synthesis. Adolescents who show large measures of both are, by all accounts, headed for good long-term outcomes as life unfolds for them (Vaillant and Vaillant, 1981; Werner and Smith, 1982). This fortunate position may be the joint product of their genes and the atmosphere of affection and individuation in their families.

Some Clues to the Nonshared Environment

At the conclusion of Chapter 10 we acknowledged the silence of the nonshared environment in the data from our study. There, we mourned the absence of evidence of *specific* nonshared effects. We can say with confidence that, on the basis of the data we collected, the following family characteristics do not reflect nongenetic, nonshared influences on the adolescent: differential marital conflict about the adolescent versus the sib, differential parenting toward siblings, and asymmetrical relationships the sibs construct with each other. We noted only three exceptions to this pattern. In addition to this surprising paucity of findings, our preliminary data on differential peer relationships produced no evidence of influences of nongenetic, nonshared environment. Given that our very large twelve-year study was designed to identify nongenetic, nonshared factors, this dearth of findings is not only disappointing but galvanizing. The data we have reported in this chapter, particularly in Figure 13.1, confirm previous findings: the nonshared environment is the predominant environmental influence on individual differences among adolescents. Thus, we are motivated to re-examine closely what the data do tell us about the nonshared environment. Indeed, a closer look at our findings suggests that they can teach us a good deal about anonymous nonshared factors. This, in turn, might produce a sturdy hypothesis about these important influences. In all, we observed three attributes of the anonymous non-shared environment in our study.

First, as noted, we confirmed the findings of many previous studies: the nonshared environment is the most important nongenetic component for most of our measures of adjustment (see Figure 13.1). Second, nonshared factors that influence our measures of adjustment are more unstable than stable. We presented data on this first in Chapter 8 and summarize it here in a somewhat different form in Figure 13.5. The figure compares the role of nonshared factors in the stability of the seven measures of adjustment with their role in change in these same measures from earlier to later adolescence. For each of these domains of adjustment, we examine the balance of stability versus instability of nonshared factors. For all seven of the domains we measured, instability exceeds stability. For four of the seven, instability exceeds stability by a ratio of well over two to one.

The third characteristic of the nonshared environment is reflected in the next two figures. Figure 13.6 is an analysis of associations among the seven measures of adjustment. It is analogous to analyses we presented in Chapter 12 searching for the balance between specific and overlapping genetic factors in associations among measures of family subsystems. Here we are interested in the balance between overlapping and unique

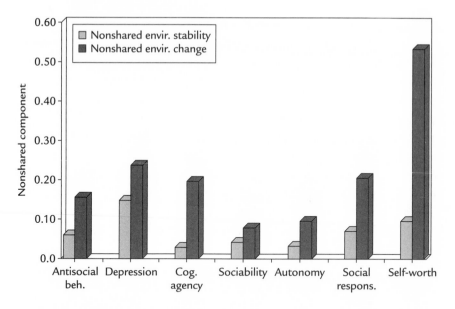

Figure 13.5 Balance of nonshared influences comparing stable and unstable components

nonshared influences on measures of adjustment. Figure 13.6 shows these balances, from data earlier in adolescence, for all measures of adjustment that had a correlation of .20 or higher. In thirteen of the seventeen associations the balance was in favor of unique nonshared influences. In ten of the seventeen associations the magnitude of the overlapping nonshared influences was trivial (less than .05).

This pattern is even clearer in later adolescence. Figure 13.7 shows an analysis of eighteen associations that were .20 or greater. In all but one of these associations, unique nonshared influences were greater than overlapping influences. In most instances the influence of overlapping nonshared factors was trivial and the influence of unique effects was very substantial. What we are seeing in these two figures is very unusual. As a rule, the associations between environmental factors, even when they are narrowly defined, and behavior tend to be broad. For example, negativity and conflict in parent-child relationships have been associated with reduced resourcefulness under stress (Sroufe, 1991), reduced self-esteem

(Isberg et al., 1989), reduced self-confidence, and increased antisocial behavior (Conger et al., 1993).

A fourth characteristic of anonymous or unmeasured nonshared environmental factors is also important. Whatever these factors might be, they are not likely to have substantial associations with the specific family and peer measures we included in our study. If these correlated but unmeasured influences were important, then the family measures we did use would serve, in part, as their surrogates and might show more notable nonshared effects in the analyses we conducted for Chapter 10.

For example, let us say that the quality of the adolescent's most impor-

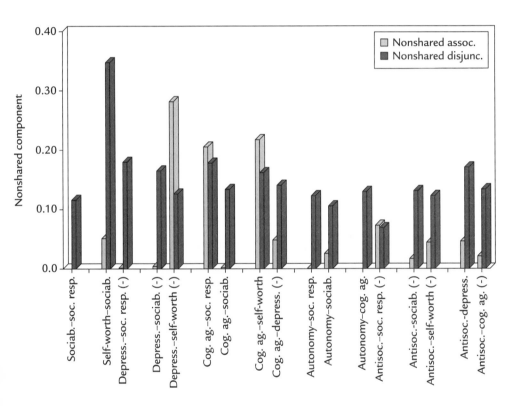

Figure 13.6   Comparison of common and unique nonshared influences in the associations among measures of adjustment in earlier adolescence where phenotypic correlations equated or exceeded .20

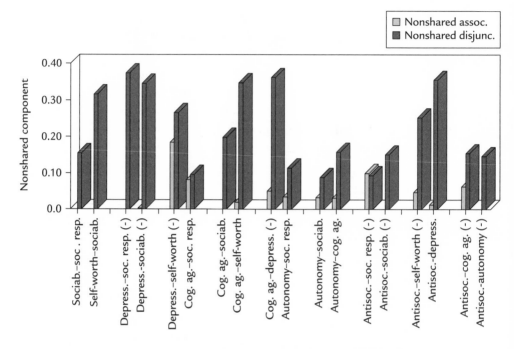

Figure 13.7   Comparison of common and unique nonshared influences in the associations among measures of adjustment in later adolescence where phenotypic correlations equated or exceeded .20

tant heterosexual peer relationship was a crucial nonshared influence on adjustment. This is one environmental variable we did not measure. Let us also suppose that this variable was moderately or highly correlated with the adolescent's relationship with mother or sibling. These are variables, of course, that we did measure. Let us also suppose that the heterosexual peer relationship is associated with antisocial behavior. We already know that maternal and sibling relationships are associated with this index of adjustment. If all these conditions held, then the measures of maternal or sibling relationships could serve as a surrogate in the analysis of non-genetic, nonshared environment. That is, their nonshared association with an adjustment index such as antisocial behavior might reflect, to some degree, the unmeasured association of girlfriend/boyfriend relation-ship and antisocial behavior.

Ordinarily, we would say that the former association—between parenting or sib relationship and antisocial behavior—is confounded by the latter, the heterosexual tie, and that it should be corrected. Here we take a very different tack. We suggest that it is likely that the associations between family subsystems and adjustment we did measure reflect (or are confounded by), to some extent, the relationship between unmeasured variables and adjustment. If this is the case, then the virtual absence of any nongenetic, nonshared associations between family measures and adjustment has implications for the environment outside the family. The paucity of findings for family variables diminishes our chances of finding such associations in unmeasured, nonfamily variables that are both correlated with the family measures and also affect adjustment. To put the matter yet another way, the deafening silence of nonshared environment within the family suggests that the nongenetic, nonshared influence of a number of nonfamily variables also may be muted.

One might argue that we cannot tell whether or not the social environment outside the family is an important source of systematic nonshared environment unless it is directly measured. Perhaps systematic measures of adolescents' intimate heterosexual relationships or peer relationships or a careful delineation of their relationships with senior mentors (camp counselors, teachers, or priests) might be revealing. Before we conduct a massive new hunt for these variables, however, we must consider the four attributes of the nonshared environment we have just enumerated.

We suggest that no commonly measured attributes of the adolescent's social environment have all four of these properties: decisive importance in accounting for individual differences in many domains of adjustment, substantial instability across time, very specific relationships with particular measures of adjustment, and a lack of correlation with any of a broad range of measures of family relationships and qualities of peer groups. Thus our hunt for these highly significant influences must take us into new areas. The most parsimonious approach would be to search for adolescent experiences that conform to this foursome of unusual characteristics as well as meet the central criterion for nonshared environment: they are distinctively different for siblings in the same family.

Any interpretation of these data on nonshared environment must be tempered by a reminder that errors arising from the imprecision of our measures make some contribution to these estimates. We cannot readily separate out the contribution of this error from the contribution of non-shared environmental factors themselves. Error will, however, contribute to all four of the attributes we have listed: variance among adolescents, instability, specificity of influence, and lack of correlations between environmental and adjustment variables. The only way to safeguard against the serious incursion of error into the delineation of nonshared effects is to use reliable measures. The substantial reliability of our measures is documented in Chapter 5.

Setting error of measurement aside, we are prompted by the data to dust off an old concept in psychology first developed by Freud and used successfully by clinical psychoanalysis up to the present time. Freud's approach to delineating principles of psychological development stands in sharp contrast to the strategies used in our study and in most contemporary studies of development conducted by either geneticists or psychosocial researchers. Freud had relatively little concern with the impact of verifiable experiences and social circumstances on the developing individual. Although he referred occasionally to hereditary influences on development, he had an imprecise understanding of genetics, and it played at most a marginal role in the evolution of his psychology of development. Instead, Freud focused on how individuals *represent* their experiences of themselves and others. Representations are, in essence, individual constructions of past and current experience built up over time. They are highly idiosyncratic or unique for each person.

In the Freudian psychology of representations, the quintessential psychological phenomenon was the dream. In contrast to dream interpreters since antiquity, Freud took a decidedly individual perspective. Dream contents could not be interpreted, for example, according to the biblical methods of Joseph and Daniel, who regarded the visual contents of dreams as universal symbols and hence deciphered them using a lexicon that connected dream content to life experience. For example, when Joseph and the Pharaoh's cup-bearer were both in prison, the cup-bearer told Joseph about a dream in which he had seen a grape vine with three branches. He had pressed the grapes to make wine for Pharaoh. Joseph, confident in a divinely authored dream lexicon, told the cup-bearer that each branch stood for a day, and that in three days he would be forgiven by Pharaoh and returned to his service as cup-bearer. The fact that the dream came true validates Joseph's special relationship with God rather than his method of dream interpretation.

Freud took a different approach. For each dreamer, the dream represents a set of inexpressible ideas that Freud called "dream thoughts." These dream thoughts are built around powerful and idiosyncratic wishes of the dreamer that are both expressed and disguised by visual images that can be understood only in the context of the particular life experience of the dreamer. For example, in the *Interpretation of Dreams,* Freud reported many of his own dreams, including one in which his close friend and family physician appears ill. The illness was graphic: Freud saw his friend Otto with brown skin and protruding eyes. The key to interpreting this dream, experienced at a time when Freud was struggling to make his early ideas about psychoanalysis acceptable to a very hostile professional audience, lay in highly idiosyncratic experiences and memories of Freud. This constellation of memories hinged on the name "Basedow." Basedow was one of the discoverers of hyperthyroidism, whose cardinal symptoms are brown skin and protruding eyes. Indeed, the illness that is often referred to as Graves' disease in the United States was known as Basedow's disease in Freud's time. But there was no simple correspondence between Freud's own spontaneous thought about the dream and any circumstance of his current life. Otto had never experienced this illness, nor was his last name Basedow.

Basedow was part of Freud's own system of representation. Indeed, when Freud allowed himself to think freely about the dream—an early form of free association—his first thought was of Basedow. This led him in two directions. First, he recalled an incident that occurred when he was in the company of a famous professor, Professor R. Freud and Professor R. had been in a serious carriage accident. Freud escaped unharmed, but Professor R. had been injured; he had been helped by a courtly gentleman with the unmistakable skin and eye changes of Basedow's disease. The second train of thought led Freud to another famous Basedow, a nineteenth-century educator. At this point Freud remembered that Otto was not just his family's physician but someone whom Freud had asked for help in educating his children, just as Professor R. had asked for help from the courtly gentleman with Basedow's disease. Once these two lines of thinking became explicit, Freud began to detect the underlying dream thoughts of his own dream. With evident embarrassment, Freud acknowledged that in his dream he had put himself in the place of Professor R. That is, in dreaming of Otto he was constructing a scenario that included himself and a man with Basedow's disease. He was asking for help from a man with thyroid problems. Freud was forced to acknowledge, first to himself and then to his readers, his own envy of Professor R. and his wish for fame and grandeur, which early in his career he could not fully admit to himself. The envy of famous professors and his own towering ambitions were Freud's inexpressible dream thoughts. These were held in check by a wish to appear to himself and others as modest and free of envy.

The dream is a paradigm of the representational process because in highly condensed form it reflects an exceptional array of both early and recent experience. (On the day of the dream Otto had visited Freud and had looked tired, but he had not shown any signs of Basedow's disease—Freud added those symptoms in his own processes of representation.) The dream illustrates how distinctively each individual can organize these experiences. What is most pertinent for our understanding of nonshared environment is how this idiosyncratic congeries can filter verifiable experience. Had we been present in Freud's home on the Bergasse in Vienna on the afternoon of Otto's visit, we could have verified that Otto had made a call and perhaps also that he had looked peaked, perhaps on the verge of a minor illness. We could not know that, through a cascade of subsequent psychological events within Freud, Otto's visit would stimulate feelings of both ambition and anxious shame in his then obscure patient.

Research is beginning to suggest that idiosyncratic experiences such as these and their residues in memory can serve as powerful filters for social experiences (Derryberry and Reed, 1996). This may explain why the same social experiences may lead to depression in one adolescent, to an outburst of antisocial behavior in another, or even to sociable behavior in a third. Let us suppose that these idiosyncratic experiences, built up over time (and reflected most dramatically in remembered dream content), are a major contributor to nonshared experience. This would certainly explain differences between siblings in the same family—even identical twins. Equally important, these idiosyncratic filters may explain why, in a population of individuals, nonshared representations cannot account for any correlation among measures and hence will appear as unique to each.

But the data from our study strongly suggest that nonshared factors are also unstable. Can Freudian representations account for these features as well? We can see in Freud's handling of his own Otto dream the seeds of an important idea that has since been more fully developed: the idea of "intrapsychic conflict." Freud's own wish for fame and prestige is in conflict with his more socially acceptable ideas about the modest and simple rewards of taking care of his patients and attempting to understand them more fully. In recent research in clinical settings this central idea of intrapsychic conflict has been developed in ways that are pertinent to an understanding of the instability of the nonshared environment. Careful methods have been developed to delineate quite distinct representations of self and other that can co-exist within one individual. Using reliable coding of detailed records of exchanges between therapists and patients, researchers have found compelling evidence that people can experience, at various times, radically different representations of themselves and others (Horowitz, 1989; Horowitz and Eells, 1993; Horowitz et al., 1995). This phenomenon is not restricted to exotic cases of multiple personality disorders but is present in a broad range of people with rather mundane and modest psychological difficulties.

Evidence from this research suggests that some of these representations of self and other are dreaded. Though they are deeply held, people tend to hide them not only from others but from themselves. These dreaded representations include feelings of worthlessness and vulnerability to abuse, as well as the thought that significant others are exploitative and untrustworthy. At the other extreme are unrealistic feelings of strength, capability, and invulnerability to disappointments; these representations include images of others as admiring and gratifying. A third category of representations consists of more ordinary images in which people represent themselves as a mixture of competence and modest defect. The most crucial finding in this line of research is that these representations shift over time, and sometimes the shifts can occur quite rapidly. If these representations do constitute an important filter through which experience influences behavior, then their changes over time can account for the instability we have identified in the nonshared environment. It may be that in adolescence, in particular, representations of this kind are subject to dramatic and frequent shifts.

As Freud's Otto dream makes clear, intrapsychic conflicts are generally not easily accessible because they are related to ideas and feelings that produce shame, guilt, or anxiety. In Freud's case his own grandiose aspirations probably produced a mixture of all three feelings. Thus if we are to explore this explanation for the nonshared environment, we will need better methods for accessing these shifting self-representations in nonclinical settings.

Chapter 14

EPILOGUE:

THE FAMILY

This study was shaped by a simple idea: that it was possible to sort influences on psychological development into three categories: genetic factors, environmental factors shared by siblings in the same family, and environmental factors unique to each sibling. This idea could be made even simpler: data from studies with genetically informed research designs strongly suggested that shared environmental factors played a trivial role in psychological development and in psychological adjustment. Hence, the important factors in development could be divided quite simply into genetic and nonshared. To be sure, this idea had its roots in a thoughtful and well-documented literature from cross-sectional studies on individual differences in adjustment in children, adolescents, and adults and from genetically informed longitudinal studies of development. It had already been tested by impressive research designs and by thoughtful examination of data. But like all simple ideas about complex subjects, it has its limits. Two of them were apparent even before we undertook our study, and we have tried to stress their importance from the very beginning.

First, at its core, the idea of three categories of developmental influences is a statistical one. It follows from additive analytic models developed by behavioral genetics researchers to analyze data from twin and adoption designs. By itself, it does not inform us at all about the mechanisms by which psychological development unfolds. It gives us no conceptual tools for understanding the specific sequences of development, for preventing developmental misfortunes, for remedying them once they occur, or for enhancing the development of adaptive skills.

Second, as an additive model, the idea does not take account of complex blends of its genetic and environmental components. We discussed at length in Chapter 4 the phenomenon of gene-environment interaction. Genetic factors may express themselves in behavior only under certain environmental circumstances. For example, data suggest that this may be the case for schizophrenia. As noted, adoptees at genetic risk for schizophrenia are more likely to develop the full-blown syndrome if they are reared in troubled families (Tienari et al., 1985; Tienari et al., 1987; Tienari et al., 1994). A similar case may hold for antisocial behavior, substance abuse, and depression (Cadoret and Cain, 1981; Cadoret, Winokur, Langbehn, Troughton, et al., 1996; Cadoret et al., 1983; Cadoret et al., 1987; Cadoret et al., 1995). In Chapter 4 we argued that these relatively simple interactions between genetic risk and environmental circumstances were highly informative but might be atypical. Although these gene-environment interactions are unlikely in the samples we studied and using the measures we selected, the continuing search for them is important. If they are more plentiful than we now suspect, they may require us to revise our analyses and conclusions.

Moreover, other interactions—such as those between shared and non-shared factors—are possible. For example, the presence of a child with a chronic disability may alter a family's "shared environment"; that is, the presence of chronic disability may alter the family situation for all members (Gonzalez et al., 1989; Reiss et al., 1993). It is not just the disability itself that has this effect, but all the consequences of dealing with it. For example, these families must focus time, energy, and expense on the child with the disability. Moreover, the social networks of these families often include other families facing the same difficulty. Many of these families report feeling that they live in a world that is very different from that experienced by families whose children are healthy. It would not be surprising to find that all family members' behavior toward one another is shaped, to some degree, by this shared construct or self-identification as a family organized to serve the needs of a disabled child.

It is intriguing that in families of this kind differential parental treatment has very different associations with adjustment than it does in families in which all children and adults are healthy. In effect, there is an interaction between the quality of the shared environment and the apparent impact of the nonshared environment. For example, in families in which the younger sib is disabled and the older sib is healthy, more parental positivity directed toward the younger sib than toward the older sib is associated with *less* depression and anxiety in the older, healthy sib. By contrast, in families in which there is no disability, more positivity toward the younger sibling is associated with *more* depression and anxiety in the older sib (McHale and Pawletko, 1992). This second finding is comparable to those obtained in our study (see Figure 7.13). Apparently, the presence of a child with a disability provides the family with a distinctive frame (see Chapter 6) for interpreting its own behavior: when more positive feelings are directed at a younger child in a family without a disability frame, this treatment is regarded as "favoritism." In a family with a disability frame, it is regarded as "legitimate support."

In sum, we were well aware from the outset of the conspicuous statistical armature of our idea about development and of the model's insensitivity to interaction among its components. Moreover, our sibling research design was ill-equipped to detect interactions of this kind. We did detect a few major interactions, such as the much higher heritability of antisocial behavior in girls versus boys, but because they failed to reach statistical significance, we have not reported them. Despite its limits, our simple idea about three classes of influences is a useful model for reviewing the data from our study. Many of our findings were surprising and challenged our initial ideas about the influence of genetic factors as well as the shared and nonshared environment.

Most important, we have revised our thinking about the genetic component of our model almost completely. For example, we were struck by the magnitude of change in genetic influence on measures of adjustment across the small span of three years between earlier and later adolescence. The most conspicuous example was genetic effects on sociability. Its overall heritability was substantial and stayed about the same from earlier to later adolescence. But fully half the genetic factors that influence sociability in early adolescence disappear, only to be replaced with new genetic influences later in adolescence.

We found similar changes, though smaller in magnitude, for genetic influences on family relationships. These changes were most notable for the limit-setting efforts of parents: surveillance, attempted control of adolescent behavior, and actual control. The data suggested that genetic *change* is a major engine in the many transformations affecting both adolescents and their families during this critical period of development. In retrospect, this finding is hardly surprising. Biological changes, from the endocrine changes of puberty (Brooks-Gunn and Reiter, 1990) to brain maturation (Benes et al., 1994; Buchsbaum et al., 1992), are a hallmark of the adolescent period. Moreover, there is a very intimate connection between some of these biological changes and family functioning (Steinberg, 1987; Steinberg, 1989). But the relationship between these phenotypic biological changes and the genetic changes we have observed in our own work, if any, remain to be deciphered.

In addition to suggesting the central role of genetic change, the data indicated a substantial blurring of the boundary of what we called "genetic" and "environmental" influences. The expression of genetic influences in many domains of adjustment may depend, as we have speculated, on specific links between heritable characteristics of children and particular family relationships. If this is the case, we can no longer think of these as genetic influences but as gene-initiated sequences, heavily involving the family and, perhaps, requiring specific family responses for their full expression in adjustment. This idea conveys a great deal more than the common notion that genes need the environment to express themselves in behavior. According to this more common and unassailable view, if children are not clothed, fed, taught to speak their native language, and taught the basic behavioral requirements of their culture, few genetic influences can express themselves in behavior, development, and adjustment. What is new in our formulation is the possibility of far more specific links among genetic factors, family processes, and adjustment. Our proposal is not simply that the environment has a general and nonspecific facilitative or preparatory role in the behavioral expression of genetic influences, but rather that specific family processes may have *distinctive and necessary roles* in the actual mechanisms of genetic expression.

We have also revised our conception of the shared environment. Our findings suggest that it is much more important than our original formulation held. First, shared environmental factors appear to regulate associations among family subsystems. For example, positive attributes of a marital relationship are highly correlated with warmth in parenting, which in turn is correlated with positive sibling relationships. The same set of findings applies to conflict in family relationships: if there is conflict in one, there is likely to be conflict in all. Shared environmental factors, perhaps intrinsic to the family, are the most significant component of all these associations.

Second, shared environmental factors are an important link between sibling relationships and adolescent adjustment. We have speculated that these effects may echo earlier shared environmental effects of parents: the sibling relationship may be a conservator of earlier parent-child relationships and their impact on children's psychological adjustment. We have termed this effect the "environmental echo."

Third, shared environment is a major component of the links between mother's, father's, and sibling's warmth and adolescent autonomy and sociability. We have suggested that shared environmental factors may be responsible for shaping the family's perception of itself and reflect its comfort, cohesion, confidence, and initiative. The autonomy and sociability of the child are perhaps, among all the domains of adjustment, the clearest expression of the family's views of its social position.

Finally, much to our surprise, the nonshared environment remains a mystery, although many clues have emerged in the data. We had expected, of course, something more obvious. Surely, we thought, the nonshared environment that influences adjustment would be reflected in differential treatment of the children by their parents, or differential sibling experiences. We did indeed find these differential parenting and sibling experiences, and as expected, they had powerful associations with adolescent adjustment. Much to our surprise, however, virtually none of these associations was independent of genetic differences between the children in the same family. These associations of differential experience and adjustment clearly represent gene-environment correlations, not the nonshared, nongenetic influences our study was designed to identify. Even our look outside the family was not rewarding. For example, we found that differential qualities of peer groups did not contribute to any domain of adjustment independent of genetic differences between siblings.

Nonetheless, we did discover important clues about the nonshared environment. First, nonshared factors are the most important source of environmental influence on most domains of adolescent adjustment. This finding was not surprising and indeed confirms the findings of almost all previous studies. Second, the nonshared environment is distinctive for each of the adjustment domains; the nonshared factors that influence one domain of adjustment do not overlap very much with those influencing other domains. Third, it is unstable from earlier to later adolescence. Fourth, by implication, whatever these nonshared factors are, they are not very highly correlated with our measures of family relationships, peer-group quality, or stressful life events.

The nonshared environment might simply turn out to be the residue of random but influential events that pile up in the course of most people's lives. According to this view, the individual is influenced by genetic and shared environmental forces that can be estimated and, in some specific cases, directly measured. Individual development is here conceived as a Brownian particle shaped by random hits of unpredictable events. Although this explanation "fits" all our findings on the nonshared environment, it is highly unsatisfactory. It is the precise logical equivalent of the statement, "We don't know what the nonshared environment is and have no good ideas about it, either."

The psychoanalytic perspective provided a more positive frame and the sinews of a more genuine explanation for experiences that are unique to each sibling in the family, that are unique for each domain of adjustment, and that change rapidly across time. The central idea drawn from psychoanalysis is that children develop very distinctive systems of representations of themselves and their social world over time; some of these cause them shame and guilt. For example, a healthy child with a severely disabled younger sibling can see him- or herself in a number of different ways: as fortunate, unfairly advantaged, neglected, expert in caretaking, subtly deformed, beatifically radiant, unusually autonomous, vulnerable to serious disability like the afflicted sibling, exceptionally fit, unfit to be a spouse, particularly suited to being a spouse, a member of a close and valiant family, a member of a stigmatized family, a prisoner in a "chronic disease ward," a carrier of a latent illness that will be passed on to children, or an experienced caretaker unusually qualified to raise children. As noted, representations of this kind could well explain differences between siblings in the same family. In this particular example they could explain the different responses of two healthy siblings who share the experience of having a sibling with a disability. Because these unique representations have different meanings for different children, this perspective could also explain why nonshared experiences do not account for correlations across domains of adjustment.

The psychoanalytic perspective provides a good way of thinking about how these distinctive representations develop as a reflection of accumulated but not necessarily random life experiences. Equally important, this perspective offers a way of thinking about how individuals may acquire several of these self-representations at the same time and how they may engender feelings of inner conflict. For example, many healthy children of siblings with a severe disability might feel guilty about their self-perceived success as survivors of a suffocating, illness-oriented family system. Such a representation might be in conflict with a second self-representation, that of a loyal caretaker of a stricken sibling. This concept of inner conflict—central to the psychoanalytic perspective—is a good explanation of the unstable influences of these representations. Clinically based research suggests that conflicting self-images of this kind may ascend at different times in response to different stimuli (Horowitz, 1989; Horowitz and Eells, 1993; Horowitz et al., 1995).

The psychoanalytic perspective also offers clues to powerful methods for detecting these individual and conflicting representations of self and of the social world. Developmental researchers are hobbled by their brief encounters with the children, adolescents, and adults they study. Even researchers who do extended longitudinal studies typically encounter their research subjects for at most a few hours a year. Moreover, these encounters are constrained by strict requirements that they be as uniform as possible across subjects. As noted, highly successful and objective methods for studying conflicting representations have been developed based on a much more intense, prolonged, and individually tailored contact between the researcher and the subject. Typically, these relationships are focused by the researcher's intense curiosity about the individual meanings the subjects ascribe to the nuances of their relationships with others and to their own behavior and feelings. Currently these methods are most appropriate in clinical settings, where the relationship is framed by therapeutic objectives. But the social science literature is replete with examples of interviews of comparable sensitivity and flexibility conducted in nonclinical settings. Our findings on both the importance and the elusiveness of the nonshared environment should provide broad encouragement to research theories and methods that help elucidate the unique perceptions and experiences of individuals as they unfold across time (Mishler, 1986).

Even now, as this book draws to a close, the tripartite scheme shapes our thinking. Indeed, the modifications of the simple idea with which we started this effort are still phrased as changes to the "genetic," the "shared," and the "nonshared" components of this idea. Perhaps a more important transformation of this simple idea is yet to come: a set of empirically based concepts that spans these three domains; a set of ideas that permits us to understand how genetic, shared, and nonshared influences are organized as a coherent context of development. The idea of the relationship code, the principal synthesis offered in this book, does not address this issue. It best explains only a portion of the genetic component of our simple developmental idea. Even in the data we found evidence that a substantial component of genetic influence on adolescent adjustment may not be involved in the specific gene-environment correlations we could analyze. That is, even if the mediating role of family process accounted for all the gene-environment correlations we observed, there might be plenty of genetic influence on adjustment left over. Nor does this idea have any relevance to the important findings on the shared and nonshared environment we have reported. Finally, the idea is still unproven.

Thus, if we continue to work within the tripartite structure, we will soon need an idea of how to bind these parts together. Even now the data suggest, in vague outline, a very distinctive view of psychological development. At its core this view centers on the family, not on the genes. For it seems that the family and our understanding of its remarkable virtuosity provide the key to grasping the three major influences on adolescent adjustment evident in the data. Indeed, we have found that the family's response to heritable characteristics in its members allows the expression of a range of genetic influences. Evolutionary process may somehow have seen to it that families are more uniform than not in this regard. It is far from clear that a family's response to heritable characteristics of its children depends on the responder's actually being in a family. For example, recall that mother's positivity and father's positivity have completely different genetic influences. Perhaps this difference has little to do with parents' roles in the family and simply reflects gender differences between adults in their responses to children. Similarly, the distinctive genetic influences on siblings, in contrast with parents, might simply reflect the substantial age differences between the two.

Perhaps the response of adults to infant cries is a simple paradigm of the parental and sibling responses we are proposing in the relationship code. Infant cries elicit strong and reliable emotional and physiological responses from adults whether or not they are parents and whether or not they are responding to their own children or to someone else's. Cries are a distinctive characteristic of individual infants, and adults who do not know the children can reliably distinguish the cries of one infant from those of another (Gustafson and Green, 1994). Moreover, there are important differences across infants in adults' responses to their cries. For example, children at higher risk for developmental disorders emit higher-pitched cries, which convey a greater sense of urgency and evoke more changes in heart rate in listening adults (Zeskind, 1987). Likewise, cries resulting from hunger can be distinguished from those prompted by pain even if they do not always lead to differences in behavioral responses (Gustafson and Harris, 1990). Although there are important cultural differences in responses to crying, there is also considerable consistency across cultures in differential responses to different types of cries (Zeskind, 1983). Thus, adults are somehow equipped to respond with high levels of discrimination to emotional signals from infants. As children mature, this responsiveness of adults may be elaborated into a reliable and complex relationship code.

Thus, as a collection of attuned individuals, the versatile family we are conceiving may come equipped from the start with sensitivities to important signals from its children. But the versatile family can go far beyond that. As the relationships within the family unfold across time, the group's experiences of its own children may become an integral part of its collective life. They become part of its collective memory of times passed and a shared basis for viewing its place in the world (Reiss, 1967; Reiss, 1981; Reiss, 1982; Reiss and Klein, 1987).

In addition to this shared group perspective, individuals within families have considerable latitude to embroider their own constructions of themselves and others. The strong influence of the nonshared environment may reflect this fundamental feature of family life. The central importance of nonshared environment in psychological development, an importance firmly endorsed by the findings of this study, suggests that this individual latitude is a property of most families. Perhaps there are some dictatorial families in which a common view is enforced on all members, but if this were the rule, there might be little room for the nonshared environment in promoting both developmental success and developmental failure.

This view of the family is consistent with the data from our study but requires firmer evidence to support it. Nonetheless, as a conception it represents a serious threat to simple ideas of either biological or environmental determinism. From the environmental perspective, we are conceiving of capacities in the family that may be present from the very outset: capacities to respond to distinctive, heritable characteristics of its children—and perhaps to signals from adult members as well. This feature of family life must be widespread, despite differences among families, if it is to account for the pervasive and strong gene-environment correlations observed in this study. In this book we have focused our interpretations on the constraints imposed by genetic factors in the children. But there is increasing evidence from other studies that genetic factors in parents may influence parental (Kendler, 1996; Perusse et al., 1994) and marital relationships (McGue and Lykken, 1992). Thus, these genetic influences on relationships limit the impact of distinctive features of the family itself. From the genetic perspective, if genetic factors require specific family processes for their expression, then biology is not destiny either. According to our hypotheses, many genetic factors, powerful as they may be in psychological development, exert their influence only through the good offices of the family.

Appendixes

Glossary

References

Index of Tables and Figures

General Index

EXPLANATION OF METHODS

FOR DATA PRESENTED

IN CHAPTERS 8 THROUGH 13

This appendix is intended to provide a more detailed explanation of the analyses and methods than is presented in the text. We also include many tables of results that appear in the text as bar graphs, thereby allowing the reader to examine the actual values represented in the graphs. The tables included in this appendix and in Appendix B concern only the results of analyses examining genetic and environmental influences on family relationships and/or adolescent development (Chapters 8 through 13). There are no additional explanations or tables of results for the analyses presented in Chapter 7 because the figures for that chapter contain the same data that would be included in such tables. Moreover, we have published elsewhere a full account of the statistical methods used in Chapter 7 (Reiss et al., 1995). All tables and figures follow the text of the appendix.

Data Preparation

Prior to any calculations, we corrected all scores for age, sex, and their interaction by computing standardized partial residuals from the regression of scores on these variables (McGue and Bouchard, 1984). We corrected nontwin sibling scores for age differences within sibling pairs. Because all twin pairs are the same age, any effects of the age difference of the sibling pairs could decrease the similarity of nontwin sibling pairs, thereby inflating the estimate of genetic effects.

Correlational Analyses

Sibling univariate intraclass correlations were computed by correlating the score of sibling 1 with the score of sibling 2, separately for sibling type. For example, to compute the sibling intraclass correlation for maternal conflict/negativity at time 1, we correlated mother's negativity toward sibling 1 with mother's negativity toward sibling 2. This correlation was computed separately for MZ twin pairs, DZ twin pairs, full sibling pairs in nondivorced families, full sibling pairs in step-families, half-sibling pairs in step-families, and blended sibling pairs in step-families. If genetic influences are the only source of sibling similarity, we would expect the magnitude of the correlations to match the degree of genetic relatedness for the six sibling groups: MZ twins (100%) > DZ twins (50%) = full siblings, nondivorced (50%) = full siblings, step (50%) > half-siblings (25%) > blended siblings (0%). If shared environmental factors are the only source of sibling similarity, then the correlations should be of approximately the same magnitude and be high across the sibling types. Finally, if nonshared environmental influences are important, the correlations should be uniformly low across all six sibling types.

Sibling similarity can also be represented for bivariate associations (the correlation between two measures). In this case, cross-sibling correlations are computed. The logic behind cross-sibling correlations follows from intraclass sibling correlations: if genetic influences are important, the pattern of correlations will follow the pattern of genetic relatedness; if shared environmental influences are important, the correlations should be uniformly large across sibling types; and if nonshared environmental influences are important, the correlations will be low for all sibling types. The computation of cross-sibling correlations is also similar to the computation of intraclass correlations. Sibling 1's score on measure X is correlated with sibling 2's score on measure Y, separately by sibling type. In other words, how well can sibling 1's score on measure X predict sibling 2's score on measure Y? For example, using mother's conflict/negativity and adolescent depression, mother's negativity toward sibling 1 is correlated with sibling 2's score for depression. Cross-sibling correlations represent the degree to which the *covariance between two measures* can be explained by genetic and environmental influences. For this reason, cross-sibling correlations are constrained by the magnitude of the phenotypic correlation.

Sibling cross-correlations are presented in a format corresponding to their presentation in the graphics used in the book. The longitudinal sibling correlations are presented in Table A1 for adolescent development (see Chapter 8) and Table A2 for family relationships (see Chapter 9). The associations between family relationships and adolescent development at times 1 and 2 are presented in Tables A3 and A4, respectively (see Chapter 10).

Table A5 contains the correlations for the cross-lagged associations between the environmental composites at time 1 and the adolescent adjustment composites at time 2. Table A6 contains the correlations for the cross-lagged associations between adolescent adjustment at time 1 and the environmental composites at time 2 (see Chapter 11). It is important to note that the correlations presented in Tables A5 and A6 represent the *raw* cross-lagged correlations, not the *partialed* cross-lagged correlations. In other words, the effects of stability and of the contemporaneous associations at times 1 and 2 have not been accounted for. Although these tables of correlations provide an approximation of the genetic and environmental contributions to these cross-lagged associations, they do not control for the effects of stability and for the associations between the measures at each measurement occasion (see detailed description of cross-lagged model below). Nonetheless, they are a useful starting point for anticipating the results of the genetic model-fitting. Figures 11.11, 11.12, and 11.13 provided the partialed cross-lagged correlations and are an important tool for the interpretation of the model-fitting results as explained in detail in Chapter 11.

The sibling cross-correlations representing the interrelationships among the environmental composite measures at time 1 are presented in Table A7 and in Table A8 for time 2. These are the most important data reported in Chapter 12. Finally, the interrelationships among the adolescent adjustment composites are presented in Table A9 for time 1. The model-fitting results for these correlations are also presented in Chapter 12.

Model-Fitting Analyses

Bivariate Models

Bivariate genetic analyses can be used to decompose the phenotypic co-variance between measures into genetic, shared environmental, and non-shared environmental components (Plomin et al., 1997). The bivariate genetic models employed for most of the analyses were designed to assess genetic and environmental contributions to the covariance of two measures of interest (Figure A1). Nine latent parameters are illustrated in Figure A1. The latent factors *G, SE,* and *NS* represent the genetic and environmental components of covariance among the three measures. The two paths leading to each latent factor are set to be equal. In other words, the path leading from *G* to measure 1 is equal to the path leading from *G* to measure 2. This condition is necessary to make the model identifiable. The bivariate model illustrated in Figure A1 also estimates genetic and environmental contributions that are *unique to each measure.* In other words, the genetic and environmental influences on measure 1 that do not correlate with measure 2 and the genetic and environmental influences on measure 2 that are not correlated with measure 1 are also estimated. The latent factors *g, se,* and *ns* represent genetic and environmental influences that are unique to measure 1 and to measure 2.

These bivariate models have a second use in the book. They are helpful in estimating the contribution of genetic, shared environmental, and nonshared environmental factors to variation on single measures of adjustment and of family relationships. In order to understand our general approach to computation, consider a bivariate analysis of the covariance between measure *x* and measure *y.* Suppose we are interested in the components of genetic and environmental influence on measure *x.* We would add the component of variance that was unique to *x* to the component of variance that was shared by measure *x* with measure *y.* For example, to estimate the heritability of antisocial behavior, we used the bivariate analysis of antisocial behavior and mother's negativity (the use of any bivariate model involving antisocial behavior would have produced identical results). The heritability of antisocial behavior is that which is unique to antisocial behavior, in this bivariate analysis, plus that which is shared with maternal negativity. The results of computations like these were displayed in Figures 8.1, 9.4, and 9.5.

Genetic Cross-Lagged Model

Figure A2 represents the genetic model that includes the cross-lagged associations between family processes at time 1 and adolescent adjustment at time 2; this model is described in detail below. It should be noted that the alternate model represented by Figure A3 operates in the same way (adolescent adjustment at time 1 and family processes at time 2).

The cross-lagged genetic model was designed to operate in two steps. First, genetic and environmental influences that contribute to stability in family processes and adolescent adjustment and to the contemporaneous associations between the two at each time of measurement were estimated. Second, genetic and environmental influences on the cross-lagged association were estimated. This second step resulted in an estimate of genetic and environmental contributions to the cross-lagged association that is independent of stability and the contemporaneous associations. In other words, using Figure A2 as an example, to what degree are the same genetic and environmental factors that influence family process the same genetic and environmental factors that influence the adolescent's adjustment three years later?

The models illustrated in Figures A2 and A3 represent only one member of the sibling pair. If both siblings were illustrated in the model, all the genetic factors would be connected by double-headed arrows with values corresponding to the degree of genetic similarity (1.0 for MZ; .5 for DZ, full sib, nondivorced, and full sib, step; .25 for half-sib, step; and 0 for blended sib, step). The shared environmental factors are set to be the same for both members of the twin and sibling pairs because they reside in the same household and the nonshared environmental factors are, by definition, uncorrelated for the two siblings.

Figures A2 and A3 contain twelve latent factors. The latent factors $G1$, E_s1, and E_n1 represent genetic, shared, and nonshared environmental contributions to stability in family process and in the contemporaneous association between family process and adolescent adjustment at time 1. Genetic, shared, and nonshared environmental influences on stability in adolescent adjustment and the contemporaneous associations between family process and adolescent adjustment at time 2 are represented by the latent factors $G2$, E_s2, and E_n2. Our focus here is the genetic, shared environmental, and nonshared environmental contributions to the cross-lagged associations. In Figure A2, the cross-lagged association between family process at time 1 and adolescent adjustment at time 2 is represented by latent factors $G3$, E_s3, and E_n3. Finally, in Figure A2, any genetic and environmental influences on adolescent adjustment at time 2 that are independent of stability and family process are represented by the latent factors $g4$, e_s4, and e_n4.

Figure A3 represents the *alternate* model, the cross-lagged association of adolescent adjustment at time 1 and family process at time 2. The first two sets of latent factors are the same; however, the latent factors $G3$, E_s3, and E_n3 now represent genetic and environmental influences on the association between adolescent adjustment at time 1 and family process at time 2, and the residual genetic and environmental influences ($g4$, e_s4, and e_n4) represent unique genetic and environmental influences on family process at time 2. It is important to note that the genetic cross-lagged models represented by Figures A2 and A3 are not nested. Therefore, it is impossible to establish which model best represents the data, as is typically done in phenotypic cross-lagged analyses. It is possible, however, to compare the two models to evaluate which model provides a more parsimonious fit (the fewest number of paths and the smallest c^2 and RMSEA [Root Mean Square Error of Approximation]; see explanation of fit indexes below).

This genetic model is a derivation of Cholesky factoring or triangular decomposition (Neale and Cardon, 1992). This approach is most often used for longitudinal analysis of a single measure or for a multivariate analysis of simultaneously measured variables that have been ordered according to a hypothesis (Loehlin, 1996). The cross-lagged genetic model is a combination of these two functions. The variables are ordered such that common variance due to stability and contemporaneous associations is decomposed into genetic and environmental components before genetic and environmental contributions to the cross-lagged association are estimated. Because there was no hypothesis for weighting stability differently from the contemporaneous associations, these associations were considered simultaneously in this model.

Fit Indexes

The overall fit of the model was tested by χ^2 and the RMSEA (Steiger, 1990). A nonsignificant χ^2 and a RMSEA that is low (below .10) indicate that the model accurately represents the data. Previous research has found that χ^2 is likely to reject a model that fits the data well but imperfectly, is very sensitive to sample size, and improves when more parameters are added to the model (Mulaik et al., 1989; Neale and Cardon, 1992; Tanaka, 1993). Because RMSEA also incorporates parsimony into its assessment of fit, it is often used as an alternative fit index (Browne and Cudeck, 1993). For a general discussion of quantitative genetic model-fitting, see Loehlin (1992b), Plomin et al. (1990), or Neale and Cardon (1992). Variance/covariance matrixes, computed separately for each of the sibling types, were used for the model-fitting analyses (Cudeck, 1989; Neale and Cardon, 1992).

Table A1 Longitudinal phenotypic and sibling correlations for stability in measures of adolescent development (Chapter 8: Antithesis I)

	r_p	Sibling cross-correlations					
		MZ	DZ	FN	FS	HS	BS
Negative adjustment:							
Antisocial behavior	.61	.66	.40	.19	.34	.37	.08
Depressive symptoms	.60	.33	.31	.16	.11	.04	.18
Positive adjustment:							
Cognitive agency	.71	.71	.33	.22	.26	.17	−.01
Sociability	.65	.65	.64	.33	.38	.48	.29
Autonomy	.60	.58	.45	.41	.43	.16	.50
Social responsibility	.71	.77	.20	.07	.20	.17	.09
Global self-worth	.34	.40	.12	−.07	.04	.03	.10

Note: r_p = phenotypic correlation for the whole family; MZ = monozygotic twins; DZ = dyzygotic twins; FN = full siblings who have not experienced divorce; FS = full siblings in step-families; HS = half-siblings in step-families; BS = blended (genetically unrelated) siblings in step-families.

Table A2 Phenotypic and sibling cross-correlations for stability in family subsystems (Chapter 9: Antithesis II)

	r_p	Sibling cross-correlations					
		MZ	DZ	FN	FS	HS	BS
Mother's parenting:							
Negativity	.66	.60	.51	.33	.32	.37	.10
Positivity	.60	.59	.56	.45	.41	.60	.23
Monitoring/knowledge	.32	.11	.20	.13	.20	.21	.33
Attempted	.38	.30	.46	.25	.32	.39	.13
Actual	.43	.46	.43	.35	.34	.33	.25
Father's parenting:							
Negativity	.62	.60	.57	.22	.32	.24	.30
Positivity	.64	.67	.52	.51	.42	.57	.47
Monitoring/knowledge	.34	.22	.45	.21	.13	.42	.22
Attempted	.41	.46	.46	.39	.25	.27	.41
Actual	.46	.44	.44	.57	.37	.34	.37
Sibling relationships:							
Negativity	.63	.57	.61	.42	.44	.57	.53
Positivity	.65	.69	.51	.28	.57	.60	.52

Note: r_p = phenotypic correlation for the whole family; MZ = monozygotic twins; DZ = dyzygotic twins; FN = full siblings who have not experienced divorce; FS = full siblings in step-families; HS = half-siblings in step-families; BS = blended (genetically unrelated) siblings in step-families.

Table A3 Phenotypic and sibling cross-correlations for associations between family relationships and adolescent development at time 1 (Chapter 10: Antithesis III)

	r_p	Sibling cross-correlations					
		MZ	DZ	FN	FS	HS	BS
Mother's negativity with:							
Antisocial behavior	.59	.54	.25	.31	.24	.30	.12
Depressive symptoms	.35	.26	.02	.07	.10	.20	.11
Cognitive agency	−.31	−.35	−.22	−.10	.00	−.17	−.14
Social responsibility	−.48	−.46	−.20	−.20	−.16	−.16	−.02
Global self-worth	−.20	−.22	.03	.05	−.05	−.17	−.13
Mother's positivity with:							
Antisocial behavior	−.32	−.40	−.15	−.20	−.07	−.24	−.16
Depressive symptoms	−.20	−.11	−.11	−.08	−.04	−.12	−.03
Cognitive agency	.24	.32	.17	.23	−.01	.25	.09
Sociability	.32	.31	.35	.21	.13	.21	.12
Autonomy	.25	.26	.26	.13	.09	.19	.11
Social responsibility	.35	.40	.19	.25	.06	.28	.05
Mother's monitoring/knowledge with:							
Antisocial behavior	−.27	−.44	−.09	−.21	−.13	−.22	−.10
Cognitive agency	.23	.34	.19	.25	.10	.22	.04
Sociability	.23	.27	.17	.15	.15	.21	.03
Social responsibility	.26	.34	.13	.28	.11	.19	.03
Father's negativity with:							
Antisocial behavior	.56	.44	.31	.27	.17	.29	.15
Depressive symptoms	.38	.24	.00	.06	.21	.15	.26
Cognitive agency	−.30	−.34	−.14	−.07	−.05	−.08	−.09
Social responsibility	−.47	−.45	−.21	−.20	−.16	−.18	−.07
Global self-worth	−.21	−.23	.04	−.05	−.01	−.09	−.07

Table A3 *(continued)*

	r_p	\multicolumn{6}{c}{Sibling cross-correlations}					
	r_p	MZ	DZ	FN	FS	HS	BS
Father's positivity with:							
Antisocial behavior	−.26	−.38	−.04	−.15	−.03	−.16	−.06
Sociability	.24	.34	.16	.11	.09	.11	.37
Social responsibility	.29	.36	.17	.33	.00	.08	.15
Father's monitoring/knowledge with:							
Antisocial behavior	−.21	−.29	−.07	−.20	−.05	−.21	−.07
Social responsibility	.22	.33	.18	.21	.03	.08	.10
Sibling negativity with:							
Antisocial behavior	.45	.40	.40	.48	.34	.36	.37
Depressive symptoms	.23	.28	.09	.21	.29	.29	.31
Social responsibility	−.33	−.26	−.27	−.37	−.33	−.17	−.22
Sibling positivity with:							
Antisocial behavior	−.37	−.41	−.15	−.24	−.20	−.29	−.31
Depressive symptoms	−.28	−.23	−.01	−.16	−.17	−.19	−.26
Cognitive agency	.25	.32	.03	.27	.10	.12	.12
Sociability	.30	.44	.11	.28	.17	.14	.34
Autonomy	.23	.32	.14	.16	.08	.10	.24
Social responsibility	.37	.41	.16	.35	.19	.28	.23

Note: r_p = phenotypic correlation for the whole family; MZ = monozygotic twins; DZ = dyzygotic twins; FN = full siblings who have not experienced divorce; FS = full siblings in step-families; HS = half-siblings in step-families; BS = blended (genetically unrelated) siblings in step-families.

Table A4 Phenotypic and sibling cross-correlations for associations between family relationships and adolescent development at time 2 (Chapter 10: Antithesis III)

	r_p	Sibling cross-correlations					
		MZ	DZ	FN	FS	HS	BS
Mother's negativity with:							
Antisocial behavior	.63	.50	.41	.33	.27	.26	.06
Depressive symptoms	.35	.28	.11	.32	.13	.11	.14
Cognitive agency	−.33	−.15	−.19	−.08	−.09	−.06	−.05
Social responsibility	−.55	−.37	−.30	−.05	−.16	−.03	−.03
Global self-worth	−.23	−.20	−.03	−.22	.00	−.22	−.06
Mother's positivity with:							
Antisocial behavior	−.36	−.40	−.23	−.14	−.07	−.25	−.29
Cognitive agency	.25	.30	.17	−.08	.01	.27	.15
Sociability	.24	.40	.15	.11	.09	.17	.14
Autonomy	.32	.38	.24	.14	.23	.16	.15
Social responsibility	.36	.45	.26	.07	.03	.32	.11
Mother's monitoring/knowledge with:							
Antisocial behavior	−.39	−.40	−.42	−.08	−.26	−.15	−.11
Cognitive agency	.25	.22	.11	.19	.13	.12	.06
Social responsibility	.37	.40	.20	.02	.27	.19	.03
Father's negativity with:							
Antisocial behavior	.54	.46	.51	.19	.19	.23	.10
Depressive symptoms	.35	.15	.08	.24	.16	.07	.22
Cognitive agency	−.33	−.23	−.30	−.01	−.15	−.15	−.11
Social responsibility	−.50	−.35	−.29	.08	−.10	−.12	−.15

Table A4 *(continued)*

	r_p	MZ	DZ	FN	FS	HS	BS
				Sibling cross-correlations			
Father's positivity with:							
Antisocial behavior	−.21	−.36	.01	−.16	−.04	−.16	−.11
Sociability	.21	.28	.01	.05	.19	.33	.25
Autonomy	.24	.27	.12	.01	.28	.37	.35
Social responsibility	.26	.34	−.08	.10	.05	.28	.23
Father's monitoring/knowledge with:							
Antisocial behavior	−.30	−.37	−.10	−.27	−.23	−.12	−.08
Cognitive agency	.16	.22	.04	.13	.05	.19	.04
Social responsibility	.29	.32	.08	.14	.15	.16	.20
Sibling negativity with:							
Antisocial behavior	.42	.56	.35	.24	.29	.33	.53
Depressive symptoms	.27	.26	.35	.32	.30	.34	.27
Social responsibility	−.40	−.42	−.32	−.30	−.26	−.22	−.42
Sibling positivity with:							
Antisocial behavior	−.38	−.38	−.33	−.32	−.24	−.39	−.39
Depressive symptoms	−.27	−.20	−.20	−.24	−.22	−.11	−.32
Cognitive agency	.27	.21	.15	.08	.17	.37	.11
Sociability	.24	.49	.17	.10	.15	.23	.22
Autonomy	.28	.29	.19	.23	.24	.22	.25
Social responsibility	.40	.50	.29	.25	.21	.43	.27

Note: r_p = phenotypic correlation for the whole family; MZ = monozygotic twins; DZ = dyzygotic twins; FN = full siblings who have not experienced divorce; FS = full siblings in step-families; HS = half-siblings in step-families; BS = blended (genetically unrelated) siblings in step-families.

Table A5 Phenotypic and sibling cross-correlations for cross-lagged associations between family relationships at time 1 and adolescent development at time 2 (Chapter 11: Synthesis I)

		Sibling cross-correlations					
	r_p	MZ	DZ	FN	FS	HS	BS
Mother's negativity with:							
Antisocial behavior	.49	.62	.27	.29	.25	.27	.06
Depressive symptoms	.27	.31	.10	.12	.14	.12	.21
Cognitive agency	−.30	−.24	−.16	−.18	−.07	−.14	−.19
Social responsibility	−.47	−.54	−.19	−.07	−.12	−.09	−.15
Mother's positivity with:							
Antisocial behavior	−.28	−.40	−.25	−.24	−.16	−.25	−.17
Sociability	.25	.35	.28	.24	.11	.12	.20
Autonomy	.25	.34	.26	.20	.18	.11	.20
Social responsibility	.28	.38	.19	.20	.17	.35	.15
Mother's monitoring/knowledge with:							
Antisocial behavior	−.20	−.34	−.19	−.13	−.07	−.14	−.03
Cognitive agency	.20	.06	.16	.23	.04	−.04	.18
Father's negativity with:							
Antisocial behavior	.42	.53	.31	.09	.21	.26	.00
Depressive symptoms	.25	.32	−.02	.05	.18	.09	.28
Cognitive agency	−.29	−.38	−.15	−.02	−.14	−.11	−.17
Social responsibility	−.42	−.40	−.21	−.04	−.14	−.09	−.13
Father's positivity with:							
Antisocial behavior	−.19	−.40	−.08	−.18	−.15	−.17	−.06
Sociability	.22	.36	.09	.10	.17	.24	.31
Autonomy	.20	.24	.14	.14	.19	.24	.36
Social responsibility	.21	.29	.08	.18	.17	.20	.27
Father's monitoring/knowledge with:							
Antisocial behavior	−.14	−.18	−.18	−.11	−.12	−.04	−.07
Sibling negativity with:							
Antisocial behavior	.33	.48	.34	.17	.26	.37	.20
Social responsibility	−.28	−.40	−.25	−.10	−.11	−.18	−.33
Sibling positivity with:							
Antisocial behavior	−.31	−.44	−.24	−.14	−.19	−.31	−.23
Depressive symptoms	−.22	−.20	−.15	−.19	−.12	−.08	−.28
Cognitive agency	.20	.26	.00	−.02	.06	.17	.10
Sociability	.24	.41	.09	.02	.17	.09	.41
Autonomy	.23	.37	.01	.04	.17	.07	.35
Social responsibility	.37	.58	.22	.17	.08	.27	.15

Note: r_p = phenotypic correlation for the whole family; MZ = monozygotic twins; DZ = dyzygotic twins; FN = full siblings who have not experienced divorce; FS = full siblings in step-families; HS = half-siblings in step-families; BS = blended (genetically unrelated) siblings in step-families.

Table A6 Phenotypic and sibling cross-correlations for cross-lagged associations between adolescent development at time 1 and family relationships at time 2 (Chapter 11: Synthesis I)

	r_p	Sibling cross-correlations					
		MZ	DZ	FN	FS	HS	BS
Mother's negativity with:							
Antisocial behavior	.48	.42	.33	.22	.20	.28	.11
Depressive symptoms	.24	.20	.02	.17	.15	.15	−.02
Cognitive agency	−.23	−.19	−.20	−.02	−.09	−.02	−.07
Social responsibility	−.40	−.33	−.27	−.19	−.17	−.17	.01
Mother's positivity with:							
Antisocial behavior	−.25	−.44	−.15	−.07	.02	−.26	−.30
Sociability	.26	.36	.31	.24	.12	.13	.08
Autonomy	.23	.18	.23	.01	.20	.16	.15
Social responsibility	.25	.38	.24	.16	−.06	.18	.12
Mother's monitoring/knowledge with:							
Antisocial behavior	−.27	−.38	−.27	−.19	−.13	−.17	−.28
Cognitive agency	.23	.33	.24	.26	−.02	.16	.13
Social responsibility	.26	.39	.29	.18	.13	.11	.18
Father's negativity with:							
Antisocial behavior	.46	.40	.43	.13	.18	.31	.00
Depressive symptoms	.22	.05	.12	.06	.15	.16	.15
Cognitive agency	−.26	−.21	−.35	−.07	−.13	−.10	−.06
Social responsibility	−.39	−.31	−.30	−.13	−.15	−.22	.07
Father's positivity with:							
Social responsibility	.23	.23	.07	.14	−.11	.23	.20
Father's monitoring/knowledge with:							
Antisocial behavior	−.24	−.38	−.03	−.18	−.10	−.31	−.12
Sibling negativity with:							
Antisocial behavior	.35	.33	.39	.18	.23	.37	.33
Social responsibility	−.30	−.30	−.27	−.23	−.31	−.22	−.16
Sibling positivity with:							
Antisocial behavior	−.24	−.30	−.16	.03	−.13	−.29	−.29
Depressive symptoms	−.21	−.20	.02	−.13	−.03	−.25	−.21
Cognitive agency	.22	.27	.14	−.03	.11	.21	.03
Sociability	.26	.49	.22	.20	.08	.21	.24
Social responsibility	.27	.31	.27	.03	.23	.30	.09

Note: r_p = phenotypic correlation for the whole family; MZ = monozygotic twins; DZ = dyzygotic twins; FN = full siblings who have not experienced divorce; FS = full siblings in step-families; HS = half-siblings in step-families; BS = blended (genetically unrelated) siblings in step-families.

Table A7 Phenotypic and sibling cross-correlations for interrelationships among measures of adolescent development at time 1 (Chapter 12: Synthesis II; Chapter 13: Synthesis III)

	r_p	Sibling cross-correlations					
		MZ	DZ	FN	FS	HS	BS
Antisocial beh.–cognitive agency	−.44	.48	.26	.18	.02	.14	.16
Antisocial beh.–depressive symp.	.45	.37	.07	.14	.17	.18	.18
Antisocial beh.–global self-worth	−.21	.26	.01	.05	.01	.09	.05
Antisocial beh.–sociability	−.22	.31	.06	.00	.05	.13	.09
Antisocial beh.–soc. responsibility	−.69	.63	.28	.26	.19	.12	.07
Autonomy–cognitive agency	.24	.15	−.02	.02	.06	.06	.04
Autonomy–sociability	.47	.41	.29	.19	.33	.26	.00
Autonomy–soc. responsibility	.25	.28	.01	.13	.02	.19	.06
Cognitive agency–depressive symp.	−.43	.31	.10	.14	.00	.03	.01
Cognitive agency–global self-worth	.33	.28	.02	.01	.11	.05	.01
Cognitive agency–sociability	.39	.43	.23	.26	.09	.18	.04
Cognitive agency–soc. responsibility	.45	.54	.25	.21	.03	.24	.07
Depressive symp.–global self-worth	−.39	.38	.00	.03	.01	.02	.04
Depressive symp.–sociability	−.26	.23	.04	.03	.02	.12	.02
Depressive symp.–soc. responsibility	−.39	.34	.01	.21	.12	.12	.12
Global self-worth–sociability	.20	.24	.08	.04	.03	.09	.60
Sociability–soc. responsibility	.25	.37	.17	.18	−.01	.16	.07

Note: r_p = phenotypic correlation for the whole family; MZ = monozygotic twins; DZ = dyzygotic twins; FN = full siblings who have not experienced divorce; FS = full siblings in step-families; HS = half-siblings in step-families; BS = blended (genetically unrelated) siblings in step-families.

Table A8 Phenotypic and sibling cross-correlations for interrelationships among family subsystems at time 1 (Chapter 12: Synthesis II)

		Sibling cross-correlations					
	r_p	MZ	DZ	FN	FS	HS	BS
Mother's parenting with mother's parenting:							
Negativity-positivity	.29	.38	.11	.25	.09	.31	.20
Negativity-monitoring/knowledge	.21	.44	.08	.19	.14	.20	.10
Negativity-attempts	.32	.10	.31	.17	.18	.06	.09
Positivity-monitoring/knowledge	.41	.32	.21	.39	.26	.32	.21
Positivity-actual	.27	.14	.11	.18	.24	.26	.10
Monitoring/knowledge-attempts	.25	.19	.10	.24	.04	.19	.12
Monitoring/knowledge-actual	.43	.29	.28	.33	.14	.28	.20
Attempts-actual	.65	.46	.46	.52	.23	.38	.20
Father's parenting with father's parenting:							
Negativity-positivity	.24	.25	.03	.25	.06	.10	.09
Negativity-attempts	.30	.08	.28	.08	.18	−.07	.22
Positivity-monitoring/knowledge	.46	.37	.19	.34	.28	.34	.18
Positivity-actual	.39	.26	.17	.24	.34	.37	.19
Monitoring/knowledge-attempts	.41	.14	.27	.27	.34	.34	.27
Monitoring/knowledge-actual	.54	.25	.26	.37	.42	.40	.34
Attempts-actual	.67	.37	.26	.36	.45	.32	.33
Father's parenting with mother's parenting:							
Dad positivity–mom positivity	.42	.43	.33	.35	.20	.34	.28
Dad attempts–mom attempts	.47	.21	.04	.32	.21	.13	.20
Dad attempts–mom actual	.38	.23	.20	.29	.20	.18	.05
Dad actual–mom mon./knowledge	.24	.19	.00	.11	.11	.21	.13
Dad actual–mom attempts	.36	.17	.34	.31	.13	.14	.15
Dad actual–mom actual	.48	.20	.28	.37	.23	.29	.11
Dad negativity–mom negativity	.61	.53	.33	.21	.23	.31	.19
Mom positivity–dad mon./knowledge	.29	.19	.16	.21	.17	.29	.26
Mom positivity–dad actual	.21	.21	−.01	.11	.21	.40	.19
Mother's and father's parenting with sibling relationship:							
Dad negativity–sibling negativity	.40	.30	.37	.49	.33	.43	.42
Dad negativity–sibling positivity	.26	.27	−.04	.26	.09	.22	.33
Dad positivity–sibling positivity	.34	.47	.28	.24	.25	.31	.44
Mom negativity–sibling negativity	.43	.43	.43	.46	.39	.42	.32
Mom negativity–sibling positivity	.27	.32	−.04	.17	.11	.16	.26
Mom positivity–sibling positivity	.42	.47	.31	.30	.40	.40	.34
Other associations:							
Mom negativity–parent conflict about child	.39	.37	.39	.23	.26	.14	.13
Dad negativity–parent conflict about child	.40	.40	.39	.20	.21	.15	.24
Sibling negativity–parent conflict about child	.22	.23	.27	.25	.22	.10	.10
Sibling negativity–sibling positivity	.48	.52	.46	.48	.53	.48	.37

Note: r_p = phenotypic correlation for the whole family; MZ = monozygotic twins; DZ = dyzygotic twins; FN = full siblings who have not experienced divorce; FS = full siblings in step-families; HS = half-siblings in step-families; BS = blended (genetically unrelated) siblings in step-families.

Table A9 Phenotypic and sibling cross-correlations for interrelationships among family subsystems at time 2 (Chapter 12: Synthesis II)

		Sibling cross-correlations					
	r_p	MZ	DZ	FN	FS	HS	BS
Mother's parenting with mother's parenting:							
Negativity-positivity	.29	.29	.07	.20	.14	.16	.28
Negativity-monitoring/knowledge	.24	.31	.16	.09	.11	−.05	−.02
Negativity-attempts	.24	.05	.24	.14	−.01	.16	.11
Positivity-monitoring/knowledge	.44	.32	.26	.26	.25	.27	.19
Positivity-actual	.29	.09	.14	.23	.21	.31	.20
Monitoring/knowledge-attempts	.27	−.04	.10	.21	.16	.22	.15
Monitoring/knowledge-actual	.42	.02	.19	.40	.35	.32	.41
Monitoring/attempts-actual	.60	.42	.46	.40	.32	.49	.21
Father's parenting with father's parenting:							
Negativity-attempts	.21	.17	.23	−.01	.09	−.11	−.05
Positivity-monitoring/knowledge	.54	.36	.25	.38	.36	.33	.27
Positivity-actual	.39	.16	.16	.42	.28	.40	.18
Monitoring/knowledge-attempts	.37	.18	.30	.23	.22	.40	.14
Monitoring/knowledge-actual	.52	.32	.25	.43	.26	.50	.29
Attempts-actual	.71	.64	.52	.44	.43	.45	.42
Mother's parenting with father's parenting:							
Dad negativity–mom negativity	.58	.39	.43	.15	.14	.16	.12
Dad negativity–mom positivity	.21	.43	.25	.08	.07	.14	.28
Dad negativity–mom monitoring/knowledge	.23	.27	.34	.05	.18	.09	−.02
Dad positivity–mom positivity	.42	.41	.28	.28	.17	.53	.46
Dad positivity–sibling positivity	.32	.51	.29	.24	.16	.42	.43
Dad positivity–mom monitoring/knowledge	.21	.23	.04	.07	.19	.04	.15
Mom negativity–dad monitoring/knowledge	.19	.23	.12	.05	.09	.12	−.05
Mom positivity–dad monitoring/knowledge	.25	.30	.26	.17	.10	.30	.16

Table A9 *(continued)*

	r_p	Sibling cross-correlations					
		MZ	DZ	FN	FS	HS	BS
Mother's parenting with father's parenting (continued):							
Dad mon./knowledge–mom mon./knowledge	.40	.37	.28	.19	.21	.02	.22
Dad mon./knowledge–mom actual	.26	.03	.12	.30	.14	.22	.20
Dad attempts–mom mon./knowledge	.20	.05	.06	.13	.27	.15	.15
Dad attempts–mom attempts	.37	.29	.28	.15	.16	.17	.34
Dad attempts–mom actual	.38	.29	.30	.13	.29	.26	.22
Dad actual–mom mon./knowledge	.24	.18	.05	.13	.21	.26	.01
Dad actual–mom attempts	.35	.30	.24	.14	.13	.31	.03
Dad actual–mom actual	.46	.44	.34	.30	.30	.35	.12
Mother's and father's parenting with sibling relationship:							
Dad negativity–sibling negativity	.37	.39	.42	.29	.29	.33	.30
Dad negativity–sibling positivity	.24	.12	.18	.18	.10	.20	.31
Mom negativity–sibling negativity	.47	.48	.51	.38	.36	.47	.52
Mom negativity–sibling positivity	.27	.04	.26	.29	.22	.18	.20
Mom positivity–sibling positivity	.42	.43	.52	.33	.32	.54	.33
Sibling negativity–sibling positivity	.44	.42	.58	.45	.46	.33	.32
Sibling positivity–dad monitoring/knowledge	.22	.27	.24	.11	.01	.28	.24
Sibling positivity–mom monitoring/knowledge	.25	.29	.22	.06	.29	.19	.25

Note: r_p = phenotypic correlation for the whole family; MZ = monozygotic twins; DZ = dyzygotic twins; FN = full siblings who have not experienced divorce; FS = full siblings in step-families; HS = half-siblings in step-families; BS = blended (genetically unrelated) siblings in step-families.

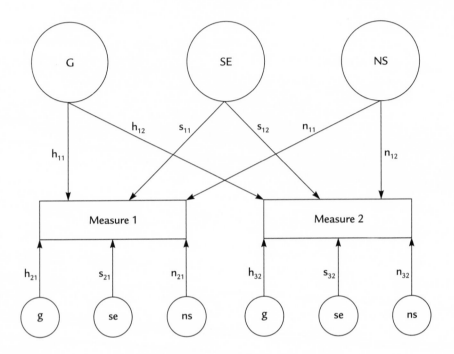

Figure A1   Diagram of a bivariate model

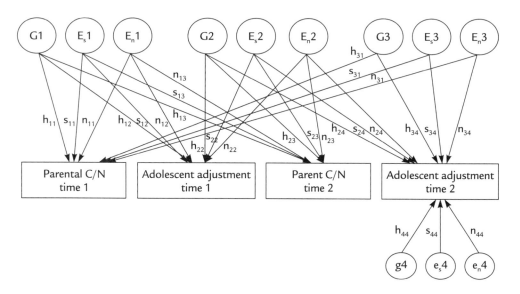

Figure A2 Diagram of a genetic cross-lagged model: family process, time 1 → adolescent adjustment, time 2

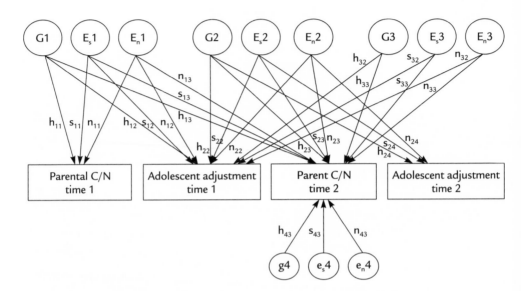

Figure A3 Diagram of a genetic cross-lagged model: adolescent adjustment, time 1 → family process, time 2

EXPLANATION OF RESULTS

Tables B1 through B9 contain the results of bivariate genetic model-fitting analyses. The first two columns contain the fit indexes (RMSEA and χ^2), and the remaining columns contain the standardized parameter estimates. Significance is indicated for those paths that were significant as estimated by standard errors and t-values. The results presented in Tables B1 through B8 can be interpreted in terms of Figure A1. The labeled paths in Figure A1 correspond to the columns in Tables B1 through B8 for each bivariate association. For example, in the first row of Table B3, mother's negativity and antisocial behavior, the parameter estimate for path h_{11}, leading from the *G1* factor to measure X (in this case measure 1 = mother's positivity) can be found in the column labeled h_{11} (.71). Similarly, the next column contains the parameter estimate for h_{12}, the path leading from *G1* to measure Y (antisocial behavior, in this example). As described in the text, the parameter estimates for the common factors can be multiplied and added to compute an estimate of the phenotypic correlation. Using the same example, mother's positivity and antisocial behavior, the common genetic paths are multiplied (.71 × .58), the common shared environmental paths are multiplied (.35 × .29), and the common nonshared environmental parameter estimates are multiplied (.28 × .23). These multiplied estimates can then be summed (.41 + .10 + .06) to produce the estimated phenotypic correlation of .57, which is within rounding error of the actual phenotypic correlation reported in column 1 of Table A3.

Univariate Results

The univariate heritabilities and environmentalities presented in Chapters 8 and 9 were drawn from the results of the family relationships and adolescent development bivariate analyses reported in Tables B3 and B4 for times 1 and 2, respectively. The heritability of each measure in any bivariate analysis can be computed in the following way. All the paths that connect the measure to a genetic factor are squared and then summed. For example, the heritability of mother's negativity is estimated by squaring the path from $G1$ to mother's negativity (measure 1 in Figure A1) and adding this to the square of the path from $g2$ to mother's negativity ($.71^2 + .30^2 = .59$). The estimates of shared environmental and nonshared environmental contributions to the total variance in mother's positivity are computed in the same way. Shared environmental influences account for 22 percent and nonshared environmental influences account for 17 percent of the total variance in mother's positivity. The genetic, shared environmental, and nonshared environmental univariate estimates should all add up to 100 percent of the total variance (within rounding error).

Bivariate Results

The bivariate model (see Figure A1) was applied only when the phenotypic correlation between two measures was .20 or greater. Tables B1 and B2 contain the standardized parameter estimates for stability and change from time 1 to time 2 for adolescent development (Chapter 8) and family relationships (Chapter 9), respectively. Model-fitting results for the associations between family relationships and adolescent adjustment at time 1 and time 2 are presented in Tables B3 and B4, respectively (Chapter 10). Table B7 contains the results for models examining the associations among the adolescent adjustment variables at time 1 (Chapter 12). Finally, Tables B8 and B9 contain the results of models testing the relationship code, interrelationships among the family relationship measures, at time 1 and time 2, respectively (Chapter 12).

Cross-Lagged Results

The cross-lagged genetic model (see Chapter 11) was applied when the phenotypic cross-lagged correlations were .20 or greater and the partialed cross-lag was .15 or greater. As explained in Chapter 11, there were eighteen cross-lagged associations that met these criteria, therefore only those associations were examined using the cross-lagged genetic model. Standardized parameter estimates and fit indexes are presented in Tables B5 and B6. The parameter estimates are labeled according to the labeled paths in Figures A2 and A3. The standardized parameter estimates for the contemporaneous associations and stability are located in the top two sections of Tables B5 and B6. The parameter estimates for the cross-lagged associations are located in the third section, and the residuals are located in the fourth section of Tables B5 and B6.

Time 1 Environment and Time 2 Adjustment

Table B5 contains the results of the models that examined the cross-lagged associations between environment measured at time 1 and adjustment measured at time 2. The standardized parameter estimates represent the labeled paths in Figure A2. For example, $h31$ represents the path from the cross-lagged genetic factor $(G3)$ to the environmental measure at time 1, and $h34$ represents the path from the $G3$ to the adjustment measure at time 2. As described above, the standardized parameter estimates for each path are contained in Table B5.

Time 1 Adjustment and Time 2 Environment

Table B6 contains the results of the cross-lagged models that examined the cross-lagged associations between adjustment measured at time 1 and environment measured at time 2. The standardized parameter estimates represent the labeled paths in Figure A3. Using an example equivalent to that presented above, $h32$ represents the path from the cross-lagged genetic factor $(G3)$ to the adjustment measure at time 1, and $h33$ represents the path from $G3$ to the environmental measure at time 2. The standardized parameter estimates for the cross-lagged genetic models examining the association from time 1 adjustment at time 2 environment are presented in Table B6.

Fit Indexes

For all the models, the χ^2 values were significant; however, all the RMSEA values were .05 or below (see the first two columns in Tables B1–B9). Taken together, these two fit indexes indicate that the fit of the models was acceptable.

Table B1 Stability in adolescent development from time 1 to time 2: Standardized parameter estimates from bivariate genetic model-fitting (Chapter 8: Antithesis I)

	RMSEA	χ^2	Common factors							Unique factors				
			b_{11}**	b_{12}	S_{11}	S_{12}	n_{11}	n_{12}	b_{21}	S_{21}	n_{21}	b_{32}	S_{32}	n_{32}
Negative adolescent adjustment:														
Antisocial behavior	.05	107.05*	.66*	.64*	.35*	.34*	.25*	.24*	.49*	—	.38*	.51*	—	.41*
Depressive symptoms	.07	144.04*	.67*	.66*	—	—	.39*	.38*	.55*	—	.30*	.00	.18	.62*
Positive adolescent adjustment:														
Cognitive agency	.03	74.62*	.81*	.82*	—	—	.17	.17	.44*	—	.35*	.08	.14	.52*
Sociability	.04	83.97*	.53*	.51*	.57*	.54*	.21*	.20*	.51*	.00	.28*	.53*	.23	.28*
Autonomy	.04	84.90*	.52*	.47*	.59*	.53*	.19*	.17*	.38*	.30*	.32*	.46*	.41*	.30*
Social responsibility	.07	160.86*	.80*	.76*	—	—	.27*	.26*	.40*	-.02	.36*	.28*	—	.53*
Global self-worth	.02	62.09*	.49*	.49*	—	—	.31*	.31*	—	.09	.81*	.50*	—	.64*

*Path co-efficient is significant at $p < .05$ by t-test.
**See Figure A1 for an explanation of these path symbols.

Table B2 Stability in family environment from time 1 to time 2: Standardized parameter estimates from bivariate genetic model-fitting (Chapter 9: Antithesis II)

			Common factors						Unique factors					
	RMSEA	χ^2	h_{11}	h_{12}	s_{11}	s_{12}	n_{11}	n_{12}	h_{21}	s_{21}	n_{21}	h_{32}	s_{32}	n_{32}
Mother's parenting:														
Negativity	.02	62.22*	.69*	.67*	.40*	.39*	.13	.12	.33	.27	.41*	.30	.19	.50*
Positivity	.06	124.87*	.46*	.44*	.61*	.58*	.12	.12	.33	.19	.50*	.27	.21	.58*
Monitoring/knowledge	.00	49.82	—	—	.44	.43*	.31*	.30*	.48*	.42*	.55*	.44*	.36*	.63*
Attempts	.00	50.19	.32*	.32*	.53*	.53*	—	—	.44*	.29	.58*	—	.36*	.70*
Actual	.00	32.20	.18	.18	.57*	.58*	.20	.20	.40	.22	.63*	—	.40*	.65*
Father's parenting:														
Negativity	.04	86.08*	.69*	.69*	.38*	.38*	.00	.00	.19	.29*	.50*	.24	.21	.53*
Positivity	.06	118.74*	.43*	.44*	.64*	.66*	.07	.07	.45*	—	.45*	—	.26*	.55*
Monitoring/knowledge	.00	69.61*	.28	.29	.45*	.46*	.19	.20	.24	.56*	.56*	.60*	.17	.53*
Attempts	.03	51.64	.30*	.30*	.57*	.57*	—	—	.34	.33	.60*	.54*	—	.54*
Actual	.03	69.37*	.26*	.26*	.61*	.62*	—	—	.41*	.30	.55*	.44*	.26	.53*
Sibling relationship:														
Negativity	.06	115.98*	.37*	.36*	.66*	.65*	.14*	.13*	.00	.56*	.30*	—	.59*	.28*
Positivity	.03	73.48*	.37*	.37*	.67*	.67*	.08	.08	.30*	.49*	.30*	.10	.52*	.36*

*Path co-efficient is significant at $p < .05$ by t-test.

Table B3 Associations between family relationships and adolescent development at time 1: Standardized parameter estimates from bivariate genetic model-fitting (Chapter 10: Antithesis III)

	RMSEA	χ^2	Common factors						Unique factors					
			b_{11}	h_{12}	s_{11}	s_{12}	n_{11}	n_{12}	b_{21}	s_{21}	n_{21}	h_{32}	s_{32}	n_{32}
Mother's parenting with adolescent adjustment:														
Negativity–antisocial beh.	.03	75.27*	.71*	.58*	.35*	.29*	.28*	.23*	.30*	.32*	.31*	.62*	.12	.36*
Negativity–depressive symp.	.04	109.51*	.60*	.46*	.21	.16	.25*	.19*	.50*	.42*	.33*	.72*	—	.45*
Negativity–cognitive agency	.02	60.05	.54*	.46*	.18	.15	.15*	.13*	.57*	.41*	.39*	.79*	—	.35*
Negativity–soc. responsibility	.02	68.04*	.60*	.75*	—	—	.17*	.22*	.51*	.45*	.39*	.58*	—	.25*
Negativity–global self-worth	.02	62.49*	.15	.34	.12	.27	.21*	.46*	.77*	.44	.36*	—	—	.77*
Positivity–antisocial beh.	.05	123.72*	.48*	.34*	.41*	.29*	.24*	.17*	.42*	.44*	.42*	.78*	.10	.39*
Positivity–depressive symp.	.06	156.75*	.44*	.30*	.18	.12	.21	.14	.46*	.57*	.44*	.82*	—	.45*
Positivity–cognitive agency	.04	119.83*	.44*	.33*	.29*	.22*	—	—	.47*	.52*	.49*	.83*	—	.38*
Positivity–sociability	.04	113.36*	.45*	.38*	.40*	.34*	.10	.08	.44*	.46*	.48*	.64*	.45*	.34*
Positivity–autonomy	.04	115.71*	.39*	.39*	.31*	.31*	.07	.07	.50*	.52*	.48*	.50*	.60*	.36*
Positivity–soc. responsibility	.05	142.60*	.47*	.53*	.17	.19	.15	.16	.46*	.55*	.47*	.76*	—	.29*
Mon./knowledge–antisocial beh.	.03	90.19*	.47*	.32*	.40*	.27*	—	—	.07	.46*	.64*	.78*	.17	.43*
Mon./knowledge–cognitive agency	.02	69.40*	.43*	.31*	.32*	.23*	—	—	.21	.50*	.64*	.84*	—	.38*
Mon./knowledge–sociability	.02	61.20*	.36*	.29*	.36*	.29*	.04	.03	.29	.49*	.64*	.69*	.49*	.35*
Mon./knowledge–soc. responsibility	.03	79.80*	.36*	.38*	.22	.23	.13	.13	.33	.55*	.63*	.83*	—	.30*
Father's parenting with adolescent adjustment:														
Negativity–antisocial beh.	.04	100.09*	.70*	.56*	.36*	.28*	.28*	.22*	—	.40*	.38*	.64*	.15	.36*
Negativity–depressive symp.	.05	140.57*	.55*	.42*	.37*	.28*	.26*	.20*	.45*	.36*	.39*	.69*	—	.47*
Negativity–cognitive agency	.02	63.02*	.61*	.51*	—	—	—	—	.38*	.50*	.48*	.78*	—	.37*
Negativity–soc. responsibility	.04	94.40*	.62*	.76*	.08	—	.14*	.18*	.34*	.50*	.48*	.56*	—	.28*
Negativity–global self-worth	.03	75.79*	.20*	.44*	.08	.17	.21*	.45*	.68*	.50*	.44*	—	—	.76*

458 | APPENDIX B

Table B3 (continued)

	RMSEA	χ^2	b_{11}	b_{12}	s_{11}	s_{12}	n_{11}	n_{12}	b_{21}	s_{21}	n_{21}	b_{32}	s_{32}	n_{32}
					Common factors						Unique factors			
Father's parenting with adolescent adjustment (continued):														
Positivity–antisocial beh.	.04	113.16*	.52*	.40*	.21	.16	.09	.07	.40*	.57*	.42*	.75*	.27	.42*
Positivity–sociability	.04	110.98*	.30*	.27*	.43*	.38*	.09	.08	.59*	.43*	.42*	.71*	.41*	.33*
Positivity–soc. responsibility	.04	117.94*	.44*	.52*	.13	.15	.07	.09	.51*	.58*	.43*	.77*	—	.32*
Mon./knowledge–antisocial beh.	.04	91.41*	.41*	.30*	.32*	.23*	.00	.00	.00	.63*	.58*	.80*	.21	.42*
Mon./knowledge–soc. responsibility	.03	70.91*	.35*	.39*	.15	.17	.03	.03	.20	.68*	.59*	.85*	—	.33*
Sibling behavior with adolescent adjustment:														
Negativity–antisocial beh.	.05	135.86*	.34*	.33*	.56*	.54*	.15*	.15*	—	.67*	.31*	.60*	—	.46*
Negativity–depressive symp.	.05	156.60*	.13	.12	.47*	.44*	—	—	.27*	.75*	.36*	.65*	—	.60*
Negativity–soc. responsibility	.05	144.82*	.36*	.54*	.21*	.32*	.10*	.14*	—	.84*	.33*	.70*	—	.30*
Positivity–antisocial beh.	.04	104.14*	.37*	.36*	.44*	.43*	.15*	.15*	.29*	.70*	.27*	.70*	—	.42*
Positivity–depressive symp.	.05	136.44*	.35*	.33*	.35*	.34*	.11	.11	.31*	.75*	.29*	.72*	—	.50*
Positivity–cognitive agency	.04	99.51*	.40*	.42*	.18	.19	—	—	.23*	.81*	.32*	.80*	—	.38*
Positivity–sociability	.04	111.52*	.29*	.34*	.41*	.48*	—	—	.35*	.73*	.32*	.67*	.29*	.35*
Positivity–autonomy	.04	98.06*	.28*	.38*	.30*	.42*	—	—	.36*	.78*	.31*	.51*	.53*	.37*
Positivity–soc. responsibility	.05	141.64*	.36*	.54*	.21*	.31*	.00	.00	.33*	.78*	.32*	.70*	—	.34*
Parental conflict about child:														
Conflict–antisocial beh.	.05	187.86*	.58*	.22*	.65*	.25*	-.23*	.09*	—	.43*	—	.80*	.30*	.39*
Conflict–depressive symp.	.05	159.22*	.58*	.21*	.33*	.19*	.12	.04	—	.57*	.19	.83*	—	.48*
Conflict–cognitive agency	.04	93.77*	.57*	.36*	.00	.00	.15*	.09*	.11	.78*	.17*	.86*	.00	.36*
Conflict–soc. responsibility	.05	135.69*	.53*	.49*	.15	.14	.20*	.18*	.25*	.76*	.11*	.80*	.00	.27*

*Path co-efficient is significant at $p<.05$ by t-test.

Table B4 Associations between family relationships and adolescent development at time 2: Standardized parameter estimates from bivariate genetic model-fitting (Chapter 10: Antithesis III)

	RMSEA	χ^2	Common factors							Unique factors				
			b_{11}	b_{12}	S_{11}	S_{12}	n_{11}	n_{12}	b_{21}	S_{21}	n_{21}	b_{32}	S_{32}	n_{32}
Mother's parenting with adolescent adjustment:														
Negativity–antisocial beh.	.06	60.39*	.72*	.58*	.35*	.28*	.37*	.30*	.24	.19	.35*	.58*	.19	.35*
Negativity–depressive symp.	.06	71.16*	.58*	.46*	.29*	.23*	.22	.18	.50*	.27	.45*	.37	.17	.73*
Negativity–cognitive agency	.00	51.48*	.52*	.46*	—	—	.33*	.29*	.55*	.42*	.39*	.74*	—	.41*
Negativity–soc. responsibility	.03	65.80*	.63*	.77*	—	—	.24*	.30*	.40*	.45*	.43*	.47*	—	.31*
Negativity–global self-worth	.04	59.45	.28	.62	-.03	-.06	.15	.33	.72	.38	.48	.30	—	.64*
Positivity–antisocial beh.	.05	95.90*	.42*	.31*	.48*	.35*	.30*	.22*	.34	.38*	.50*	.76*	—	.40*
Positivity–cognitive agency	.05	91.08*	.50*	.41*	.23	.18	.00	.00	.23	.55*	.58*	.74*	.00	.50*
Positivity–sociability	.05	89.89*	.41*	.34*	.36*	.30*	.05	.05	.34	.50*	.59*	.65*	.51*	.34*
Positivity–autonomy	.04	84.78*	.44*	.42*	.37*	.35*	—	—	.29	.48*	.59*	.50*	.58*	.35*
Positivity–soc. responsibility	.06	96.71*	.54*	.61*	—	—	—	—	.24	.56*	.58*	.66*	—	.45*
Mon./knowledge–antisocial beh.	.03	56.62	.58	.40	.45	.31	.15	.10	—	.20	.63	.69	.21	.6
Mon./knowledge–cognitive agency	.02	58.20	.45*	.34*	.28	.21	.19	.14	—	.48*	.67*	.76*	—	.49*
Mon./knowledge–soc. responsibility	.05	82.06*	.58*	.60*	.03	.03	—	—	—	.47*	.67*	.64*	—	.47*

Table B4 *(continued)*

	RMSEA	χ^2	Common factors						Unique factors					
			b_{11}	b_{12}	S_{11}	S_{12}	n_{11}	n_{12}	b_{21}	S_{21}	n_{21}	b_{32}	S_{32}	n_{32}
Father's parenting with adolescent adjustment:														
Negativity–antisocial beh.	.03	60.57*	.70*	.54*	.35*	.27*	.30*	.23*	.00	.32*	.44*	.61*	.23	.40*
Negativity–depressive symp.	.04	66.24*	.33	.25	.39*	.30*	.44*	.33*	.60*	.29	.32*	.53*	.00	.68*
Negativity–cognitive agency	.03	69.88*	.53*	.45*	.22	.19	.25*	.22*	.56*	.31	.44*	.71*	—	.46*
Negativity–soc. responsibility	.04	74.66*	.63*	.75*	—	—	.24*	.28*	.41*	.44*	.43*	.52*	.28	.32*
Positivity–antisocial beh.	.02	45.51	.45*	.32*	.30	.21	.08	.06	—	.64*	.54*	.75*	.28	.46*
Positivity–sociability	.03	67.76*	—	—	.47*	.38*	.25*	.20*	.42*	.54*	.50*	.74*	.43*	.27*
Positivity–autonomy	.01	53.80	.20	.18	.49*	.45*	.03	.03	.37	.53*	.55*	.63*	.50*	.34*
Positivity–soc. responsibility	.04	76.77*	.46*	.50*	—	—	—	—	.37	.70*	.55*	.75*	—	.44*
Mon./knowledge–antisocial beh.	.03	60.71	.55*	.38*	.36*	.24*	—	—	.34	.35	.58*	.72*	.25	.46*
Mon./knowledge–cognitive agency	.00	52.88	.40*	.30*	.23	.17	—	—	.52*	.43*	.57*	.79*	—	.51*
Mon./knowledge–soc. responsibility	.03	69.90*	.51*	.54*	—	—	—	—	.46*	.44*	.57*	.72*	—	.44*
Sibling behavior with adolescent adjustment:														
Negativity–antisocial beh.	.07	117.64*	.24	.23	.57*	.55*	.19*	.19*	.31*	.66*	.25*	.63*	—	.46*
Negativity–depressive symp.	.06	105.44*	—	—	.52*	.52*	.00	.00	.36*	.70*	-.32*	.00	—	.86*
Negativity–soc. responsibility	.08	135.57*	.34*	.49*	.24*	.36*	.18*	.26*	.31*	.80*	.24*	.61*	—	.44*
Positivity–antisocial beh.	.01	53.07	.13	.12	.56*	.52*	.23*	.21*	.37*	.63*	.29*	.69*	—	.44*
Positivity–depressive symp.	.00	50.93	.19	.18	.45*	.42*	.20	.19	.33	.72*	.31*	.38	—	.78*
Positivity–cognitive agency	.03	63.27*	.28	.28	.33*	.34*	.21*	.22*	.28	.78*	.30*	.72*	—	.48*
Positivity–sociability	.05	87.24*	.30*	.33*	.39*	.43*	—	—	.21	.76*	.37*	.66*	.40*	.34*
Positivity–autonomy	.02	58.99	.18	.22	.42*	.52*	.16*	.19*	.33*	.74*	.33*	.62*	.43*	.29*
Positivity–soc. responsibility	.05	84.06*	.34*	.49*	.30*	.42*	.06	.09	.26	.76*	.37*	.42*	—	.63*

*Path co-efficient is significant at $p<.05$ by t-test.

Table B5 Results of genetic cross-lagged analyses examining environment at time 1 and adolescent adjustment at time 2: Fit indexes and standardized maximum-likelihood model-fitting estimates (Chapter 11: Synthesis I)

	Mother's negativity with:			
	Antisocial beh.	Depressive symp.	Cognitive agency	Soc. responsibility
$\chi^2_{(189)}$	390.61*	314.08*	253.79*	580.11*
RMSEA	.05	.04	.03	.07
Contemporaneous association at time 1 and stability in environment				
Ga11**	−.76*	−.76*	.73*	−.72*
Ga12	−.50*	−.29*	−.20	.35*
Ga13	−.55*	−.69*	.70*	−.53*
Es11	.32*	.41*	.40	.44*
Es12	.00	.00	−.09	−.09
Es13	.33*	.27	.33	.43*
En11	.20*	−.09	.08	−.09
En12	.44*	−.48*	−.39*	.36*
En13	−.11	.40*	−.21	.22
Contemporaneous association at time 2 and stability in adolescent adjustment				
Ga22	−.72*	.75*	−.87*	−.84*
Ga23	−.02	.07	.07	.17
Ga24	−.50*	.60*	.75*	.76*
Es22	.18	.00	−.17	−.19
Es23	.38*	.37*	−.31	−.30
Es24	.44*	.28*	−.07	−.07
En22	−.12	.35*	−.16	−.07
En23	−.66*	.40*	.51*	.60*
En24	−.45*	.33*	.31*	.38*
Cross-lagged association between environment at time 1 and adolescent adjustment at time 2				
Ga31	−.39*	−.27	.29	.31
Ga34	−.45*	−.27	.32	.33
Es31	.00	.00	−.19	−.10
Es34	.00	.00	−.15	−.08
En31	.36*	.42*	.43*	.42*
En34	−.02	.06	.07	.15*
Unique to adolescent adjustment at time 2				
Ga44	.00	.00	—	—
Es44	.00	.00	—	—
En44	.39*	.62*	.45*	−.37*

Table B5 *(continued)*

	Mother's positivity with:			
	Antisocial beh.	Sociability	Autonomy	Soc. responsibility
$\chi^2_{(189)}$	413.26*	292.34*	336.03*	476.07*
RMSEA	.06	.04	.05	.06
Contemporaneous association at time 1 and stability in environment				
Ga11	−.53*	.30*	−.18	−.42
Ga12	.21	.59*	−.53*	−.01
Ga13	−.48*	.06	.08	−.43*
Es11	.58	−.61*	−.67*	.62*
Es12	−.08	−.09	−.09	.14
Es13	.52	−.67*	−.64*	.52*
En11	−.13	.18*	−.21*	−.23*
En12	.44*	.01	−.11	−.43*
En13	.14	.69*	−.64*	−.14
Contemporaneous association at time 2 and stability in adolescent adjustment				
Ga22	.86*	.43*	.35*	−.82*
Ga23	.00	.25	.38*	−.26*
Ga24	−.61*	.67*	.65*	−.85*
Es22	.13	.58*	−.68*	−.30*
Es23	−.33	.05	−.04	.31*
Es24	−.35	.52*	−.46*	.02
En22	.09	.35*	−.34*	−.20
En23	−.62*	−.07	.19	.60*
En24	−.24*	.14*	−.12	.01
Cross-lagged association between environment at time 1 and adolescent adjustment at time 2				
Ga31	−.25	−.32	.32	.17
Ga34	−.43*	−.31	.18	.15
Es31	.20	−.26	.19	−.24
Es34	.16	−.26*	.46*	−.15
En31	.51*	.57*	.58*	.55*
En34	−.02	−.01	.00	.02
Unique to adolescent adjustment at time 2				
Ga44	—	—	—	—
Es44	—	—	—	—
En44	.48*	.32*	.33*	.48

Table B5 *(continued)*

	Mother's mon./knowledge with:		Father's mon./knowledge with:
	Antisocial beh.	Cognitive agency	Antisocial beh.
$\chi^2_{(189)}$	299.16*	258.42*	315.02*
RMSEA	.04	.03	.045
Contemporaneous association at time 1 and stability in environment			
Ga11	.28	−.32	−.13
Ga12	−.59*	−.31	−.45*
Ga13	−.20	.28	.39*
Es11	−.44	−.49	.54*
Es12	.10	−.17	.11
Es13	−.47*	−.46	.50*
En11	.19*	.21*	.18*
En12	−.04	−.04	.13
En13	.70*	.71*	.59*
Contemporaneous association at time 2 and stability in adolescent adjustment			
Ga22	.58*	.84*	−.68*
Ga23	−.40*	.21	−.43*
Ga24	−.74*	.77*	−.59*
Es22	.34*	−.04	−.36*
Es23	−.29	−.35	−.17*
Es24	−.34	−.12	−.37*
En22	.44*	−.40*	.42*
En23	.00	−.18	−.19
En24	−.15	−.14	.17*
Cross-lagged association between environment at time 1 and adolescent adjustment at time 2			
Ga31	−.30	−.28	.08
Ga34	−.29	−.29	.51*
Es31	.40	.32	−.49*
Es34	.13	.14	−.12*
En31	.67*	.66*	.65*
En34	.03	.06	.00
Unique to adolescent adjustment at time 2			
Ga44	—	—	.00
Es44	—	—	.00
En44	.45*	−.52*	.46*

Table B5 *(continued)*

	Father's negativity with:			
	Cognitive agency	Depressive symp.	Soc. responsibility	Antisocial beh.
$\chi^2_{(189)}$	294.88*	338.93*	571.84*	364.33*
RMSEA	.04	.05	.07	.05
Contemporaneous association at time 1 and stability in environment				
Ga11	−.71*	.70*	−.70*	−.65*
Ga12	.23*	.19	.40*	−.41*
Ga13	−.68*	.67*	−.50*	−.53*
Es11	−.41	−.50*	.44	−.48*
Es12	.07	−.12	−.06	−.15
Es13	−.31	−.27	.43	−.33*
En11	−.04	−.14*	.08	−.20*
En12	.36*	−.56*	−.33*	−.52*
En13	.28	.22	−.16	.04
Contemporaneous association at time 2 and stability in adolescent adjustment				
Ga22	−.88*	.74*	−.83*	.68*
Ga23	.04	.00	.14	.37*
Ga24	.72*	.62*	.76*	.77*
Es22	−.01	−.12	.14	.29*
Es23	−.34	.37*	.31	−.34*
Es24	−.22	.00	.10	.00
En22	.21	.27	.10	.02
En23	−.50*	.54*	−.65*	.60*
En24	−.28*	.41*	−.38*	−.22*
Cross-lagged association between environment at time 1 and adolescent adjustment at time 2				
Ga31	−.28	−.20	.25	−.39
Ga34	−.31	−.23*	.26	−.43
Es31	.21	.15	.20	.00
Es34	.12	.13	.12	.00
En31	.44*	.43*	.45*	.38*
En34	.12	.06	.19*	.02
Unique to adolescent adjustment at time 2				
Ga44	—	.00	—	.00
Es44	—	.00	—	.00
En44	.48*	.61*	−.39*	.42

Table B5 *(continued)*

	Father's positivity with:		
	Antisocial behav.	Sociability	Soc. responsibility
$\chi^2_{(189)}$	401.98*	325.88*	807.28*
RMSEA	.05	.04	.09

Contemporaneous association at time 1 and stability in environment

Ga11	.28*	−.59*	−.57*
Ga12	−.60*	−.23*	−.25
Ga13	.03	−.35*	−.36*
Es11	−.67*	−.58*	.63*
Es12	.03	−.05	.05
Es13	−.75*	−.69*	.64*
En11	.15*	.02	−.01
En12	−.06	.20*	−.12
En13	.60*	−.41*	.50*

Contemporaneous association at time 2 and stability in adolescent adjustment

Ga22	−.55*	−.73*	.89*
Ga23	.25*	.15	.06
Ga24	.75*	−.38*	.69*
Es22	.36*	.54*	−.14
Es23	.05	.21	.35
Es24	−.32	.57*	.08
En22	.45*	.28*	.33*
En23	.10	.41*	.30*
En24	−.16*	.19*	.31*

Cross-lagged association between environment at time 1 and adolescent adjustment at time 2

Ga31	−.46*	.10	.19
Ga34	−.35	.60*	.41*
Es31	.18	.32*	.23
Es34	.06	.19	.12
En31	.46*	.45*	.44*
En34	−.13	.00	−.07

Unique to adolescent adjustment at time 2

Ga44	—	—	—
Es44	—	—	—
En44	.41*	.31*	.49*

Table B5 *(continued)*

	Sibling's negativity with:		Sibling's positivity with:	
	Antisocial beh.	Soc. responsibility	Antisocial beh.	Depressive symp.
$\chi^2_{(189)}$	439.59*	576.54*	357.49*	319.73*
RMSEA	.06	.07	.05	.04

Contemporaneous association at time 1 and stability in environment				
Ga11	−.45*	−.42*	−.49*	−.48*
Ga12	−.33*	.38*	.41*	.34*
Ga13	−.39*	−.34*	−.34*	−.28*
Es11	−.77*	−.78*	.73	−.69
Es12	−.29*	.17	−.07	.05
Es13	−.47*	−.51	.57	−.66
En11	.02	−.03	.04	.04
En12	.42*	−.26*	−.35*	.06
En13	−.08	−.20*	−.18	.31*
Contemporaneous association at time 2 and stability in adolescent adjustment				
Ga22	−.76*	−.82*	.80*	.76*
Ga23	.09	−.04	.10	.23*
Ga24	−.60*	.66*	−.60*	−.40*
Es22	.08	.07	−.09	−.16
Es23	.72*	−.72*	.62	.54
Es24	.38*	−.32*	.38	.36
En22	−.21	−.27*	.25	−.52*
En23	−.30*	.28*	−.36*	.22*
En24	−.27*	.41*	−.25*	.43*
Cross-lagged association between environment at time 1 and adolescent adjustment at time 2				
Ga31	.12	.11	−.31*	−.28
Ga34	.38*	.23	−.47*	−.24*
Es31	.31	−.32	−.19	.36
Es34	.19	−.12	−.09	.16
En31	.31*	.32*	.29*	.30*
En34	.03	.10	−.06	.00
Unique to adolescent adjustment at time 2				
Ga44	—	—	—	—
Es44	—	—	—	—
En44	.49*	−.46*	.45*	.67*

Table B5 *(continued)*

	Sibling's positivity with:			
	Cognitive agency	Sociability	Autonomy	Soc. responsibility
$\chi^2_{(189)}$	30.523*	289.16*	289.43*	539.92*
RMSEA	.04	.04	.04	.07

Contemporaneous association at time 1 and stability in environment				
Ga11	−.43*	.47	.51*	−.46*
Ga12	−.31*	.46	.31*	−.28*
Ga13	−.34*	.20	.25*	−.35*
Es11	−.78	−.56*	.54*	−.74
Es12	−.05	.00	−.17*	−.09
Es13	−.59	−.85*	.84*	−.56
En11	.00	−.17*	−.07	.00
En12	.29*	.39*	.27*	.20*
En13	−.19*	−.20	−.28*	−.27*

Contemporaneous association at time 2 and stability in adolescent adjustment				
Ga22	.86*	.52	−.57*	.88*
Ga23	−.03	.21	.01	−.07
Ga24	.72*	.68	−.41*	.65*
Es22	−.05	.60*	−.65*	−.04
Es23	−.62	.12	−.22*	−.63
Es24	−.25	.48*	−.51*	−.29
En22	−.28*	.05	.23*	−.30*
En23	−.34*	.36*	.33*	−.30*
En24	−.26*	−.05	.17*	−.35*

Cross-lagged association between environment at time 1 and adolescent adjustment at time 2				
Ga31	.33*	−.22	−.23*	−.32*
Ga34	.35*	−.29	−.49*	−.36*
Es31	.06	.55*	.54*	−.21
Es34	.02	.13	−.02	−.08
En31	.30*	.30*	.30*	.32*
En34	.00	.05	.03	.06

Unique to adolescent adjustment at time 2				
Ga44	—	—	—	—
Es44	—	−.29*	.45*	—
En44	−.49*	.35*	.32*	.48*

*Path co-efficient is significant at $p < .05$ by t-test.
**See Figures A2 and A3 for an explanation of these path symbols.

Table B6 Results of genetic cross-lagged analyses examining adolescent adjustment at time 1 and environment at time 2: Fit indexes and standardized maximum-likelihood model-fitting estimates (Chapter 11: Synthesis I)

	Antisocial beh. with mom neg.	Depressive symp. with mom neg.	Cog. agency with mom neg.	Soc. responsibility with mom neg.
$\chi^2_{(189)}$	445.18*	349.94*	289.42*	653.51*
RMSEA	.061	.049	.037	.082
Contemporaneous association at time 1 and stability in environment				
Ga11	−.75*	.73*	.74*	−.73*
Ga12	−.41*	.11	.09	−.24*
Ga13	−.49*	.66*	.63*	−.53*
Es11	.48*	.50*	−.49*	.49*
Es12	.13	.04	−.05	.11
Es13	.27*	.28*	−.33	.36*
En11	.45*	.47*	−.46*	.06*
En12	.20*	.25*	−.46*	.06
En13	.03	.01	.00	−.04
Contemporaneous association at time 2 and stability in adolescent adjustment				
Ga22	.31*	.61*	.76*	.65*
Ga23	.33*	.21	.07	.27*
Ga24	.81*	.58*	.84*	.88*
Es22	.32*	−.12	.04	.04
Es23	.16	−.09	−.18	−.37*
Es24	.37*	−.31*	.14	.05
En22	.18*	.23*	.10*	.17*
En23	.26*	.08	.22*	.27*
En24	.46*	.75*	.53*	.47*
Cross-lagged association between adolescent adjustment at time 1 and environment at time 2				
Ga31	.63*	.49*	.47*	.61*
Ga34	−.10	.00	.05	−.13
Es31	−.06	.11	.15	.00
Es34	.28*	−.32*	−.28	.00
En31	.39*	.48*	.40*	.33*
En34	.03	−.10	−.03	−.08
Unique to environment at time 2				
Ga44	.35*	.20	.28	.13
Es44	.00	.00	.00	.00
En44	.52*	.53*	.50*	.52*

Table B6 *(continued)*

	Sociability with mom pos.	Autonomy with mom pos.	Soc. responsibility with mom pos.
$\chi^2_{(189)}$	280.43*	322.33*	425.35*
RMSEA	.037	.045	.061

	Contemporaneous association at time 1 and stability in environment		
Ga11	.49*	−.50*	−.44*
Ga12	.11	−.13	−.04
Ga13	.46*	−.40*	−.41*
Es11	.67*	.66*	.67*
Es12	.06	.04	.08
Es13	.52*	.51*	.49*
En11	.57*	.56*	.59*
En12	.06	.04	.12*
En13	.00	.01	.04

	Contemporaneous association at time 2 and stability in adolescent adjustment		
Ga22	.37*	.36*	−.73*
Ga23	.15	.29*	−.26*
Ga24	.73*	.65*	−.84*
Es22	.54*	.46*	.07
Es23	−.02	.00	−.31
Es24	.60*	.68*	.05
En22	.14*	.10*	.15*
En23	.02	−.05	.00
En24	.34*	.34*	.54*

	Cross-lagged association between adolescent adjustment at time 1 and environment at time 2		
Ga31	−.62*	.50*	.41*
Ga34	.06	−.13	−.07
Es31	−.23*	.51*	−.25
Es34	−.21	.06	.17
En31	.31*	.34*	.43*
En34	.03	.02	−.09

	Unique to environment at time 2		
Ga44	.08	.00	.00
Es44	.27	.31*	.00
En44	.61*	.61*	.62*

Table B6 *(continued)*

	Antisocial beh. with mom mon./knowledge	Cog. agency with mom mon./knowledge	Soc. responsibility with mom mon./knowledge
$\chi^2_{(189)}$	280.81*	258.56*	525.00*
RMSEA	.04	.032	.073
Contemporaneous association at time 1 and stability in environment			
Ga11	.42*	.42*	.40*
Ga12	.34	.13	.07
Ga13	−.20	−.17	−.18
Es11	.59*	.59*	.60*
Es12	.02	.08	.11
Es13	.30*	.35*	.32*
En11	.69*	.69*	.69*
En12	−.04	−.06	.04
En13	.20*	.18*	.18*
Contemporaneous association at time 2 and stability in adolescent adjustment			
Ga22	.50*	.76*	.71*
Ga23	.39*	.15	.39*
Ga24	.81*	.84*	.86*
Es22	.34*	.03	.12
Es23	.22	.29	.44*
Es24	.36*	.12	.09
En22	.16*	.12*	.17*
En23	−.04	.05	−.08
En24	.46*	.52*	.51*
Cross-lagged association between adolescent adjustment at time 1 and environment at time 2			
Ga31	−.56*	−.48*	.58*
Ga34	−.01	−.02	−.10
Es31	−.11	−.11	.00
Es34	−.41*	−.31	.00
En31	.41*	.38*	.31*
En34	.06	.09	−.09
Unique to environment at time 2			
Ga44	.00	.37	.00
Es44	.00	.00	.00
En44	.68*	.68*	.68*

Table B6 *(continued)*

	Antisocial beh. with dad neg.	Depressive symp. with dad neg.	Cog. agency with dad neg.	Soc. responsibility with dad neg.
$\chi^2_{(189)}$	418.62*	382.39*	323.60*	710.98*
RMSEA	.058	.053	.042	.088

Contemporaneous association at time 1 and stability in environment

Ga11	.72*	−.69*	.71*	.69*
Ga12	.44*	−.10	.15*	.36*
Ga13	.44*	−.66*	.65*	.47*
Es11	−.51*	−.53*	.51*	.53*
Es12	−.08	−.10	.00	.03
Es13	−.34*	−.26*	.28*	.36*
En11	.47*	−.50*	.49*	.50*
En12	.13*	−.22*	.00	.00
En13	.06	.00	.00	.00

Contemporaneous association at time 2 and stability in adolescent adjustment

Ga22	.33*	.60*	.77*	.63*
Ga23	.34*	.00	.00	.24*
Ga24	.80*	.60*	.82*	.87*
Es22	.37*	.13	.00	.17
Es23	.09	.21	.30	.18
Es24	.38*	.30	.17	.05
En22	.17*	.23*	.11*	.19*
En23	.15*	.20*	.20*	.28*
En24	.46*	.74*	.55*	.49*

Cross-lagged association between adolescent adjustment at time 1 and environment at time 2

Ga31	.58*	.52*	.44*	.54*
Ga34	−.05	−.09	.00	−.30*
Es31	.00	.00	.12	.04
Es34	.23	−.25	.19	.30
En31	.42*	.48*	.40*	.33*
En34	.00	.04	.00	.07

Unique to environment at time 2

Ga44	−.44*	.28	.25	−.20
Es44	—	.00	.00	.00
En44	.55*	.51*	.53*	.51*

Table B6 *(continued)*

	Soc. respons. with dad pos.	Soc. resp. with dad mon./knowl.	Antisoc. beh. with sibling neg.	Soc. resp. with sibling neg.
$\chi^2_{(189)}$	373.89*	346.76*	389.85*	397.85*
RMSEA	.056	.05	.06	.06
	Contemporaneous association at time 1 and stability in environment			
Ga11	−.51*	−.14	.40*	.23*
Ga12	−.14	−.04	.28*	.39*
Ga13	−.29*	−.56*	.35*	.17
Es11	−.70*	−.72*	.86*	.88*
Es12	−.03	.00	.25*	.13*
Es13	−.59*	−.24*	.43*	.48*
En11	.50*	.68*	.33*	.41*
En12	.07	.07*	.03	.04
En13	.05	.06	.06	.15*
	Contemporaneous association at time 2 and stability in adolescent adjustment			
Ga22	.73*	.81*	.48*	.73*
Ga23	.19*	.20	.00	.02
Ga24	.82*	.79*	.81*	.74*
Es22	−.11	.02	.19	−.11
Es23	.41*	.43*	.66*	−.75*
Es24	−.10	.13	.36*	−.27*
En22	.17*	.11*	.17*	.17*
En23	.03	.05	.09*	.12*
En24	.57*	.60*	.47*	.62*
	Cross-lagged association between adolescent adjustment at time 1 and environment at time 2			
Ga31	.38*	.00	.59*	.00
Ga34	−.12	.00	−.17*	.00
Es31	.27*	−.33*	−.06	−.23*
Es34	−.05	−.10	.36	.05
En31	.44*	.46*	.45*	.45*
En34	.06	−.08	.09*	−.01
	Unique to environment at time 2			
Ga44	.00	.00	.00	.00
Es44	.00	.00	.00	.00
En44	.58*	.62*	.29*	.37*

Table B6 *(continued)*

	Antisocial beh. with sibling positivity	Depressive symp. with sibling positivity	Cognitive agency with sib. positivity
$\chi^2_{(189)}$	343.16*	349.85*	283.69*
RMSEA	.050	.05	.04

Contemporaneous association at time 1 and stability in environment			
Ga11	−.57*	.56*	.57*
Ga12	−.12	.16	.07
Ga13	−.31*	.33*	.33*
Es11	.69*	.69*	−.69*
Es12	.06	−.02	−.03
Es13	.61*	.61*	−.61*
En11	.45*	.45*	.45*
En12	.08	.03	.03
En13	.01	.01	.01
Contemporaneous association at time 2 and stability in adolescent adjustment			
Ga22	.47*	.72*	.78*
Ga23	.19*	.13	.08
Ga24	.80*	.58*	.85*
Es22	.37*	−.10	.08
Es23	−.13	.03	.14
Es24	.38*	−.33*	−.19
En22	.18*	.21	.10*
En23	.00	−.03	−.04
En24	.47*	.75*	.52*
Cross-lagged association between adolescent adjustment at time 1 and environment at time 2			
Ga31	−.65*	−.40	−.46*
Ga34	.02	.08	.06
Es31	.00	−.01	.04
Es34	.38*	.40*	−.37
En31	.40*	.50*	.39*
En34	.03	.01	−.02
Unique to environment at time 2			
Ga44	.00	−.05	.18
Es44	.00	.00	.00
En44	.58*	.57*	.56*

Table B6 *(continued)*

	Sociability with sibling positivity	Soc. responsibility with sibling positivity
$\chi^2_{(189)}$	396.76*	380.61*
RMSEA	.04	.06
Contemporaneous association at time 1 and stability in environment		
Ga11	−.57*	−.51*
Ga12	−.09	−.14
Ga13	−.35*	−.29*
Es11	.69*	.70*
Es12	.05	.04
Es13	.59*	.59*
En11	.45*	.49*
En12	.04	.07
En13	−.01	.05
Contemporaneous association at time 2 and stability in adolescent adjustment		
Ga22	.37*	.72*
Ga23	−.05	.19*
Ga24	.73*	.82*
Es22	.54*	−.11
Es23	.10	.41*
Es24	.59*	−.09
En22	.14*	.17*
En23	.18*	.03
En24	.34*	.56*
Cross-lagged association between adolescent adjustment at time 1 and environment at time 2		
Ga31	.62*	−.38*
Ga34	−.02	.12
Es31	.23	−.27*
Es34	−.01	.05
En31	.32*	.44*
En34	.00	.06
Unique to environment at time 2		
Ga44	−.12	.00
Es44	−.41*	.00
En44	.55*	.58*

*Path co-efficient is significant at $p<.05$ by t-test.

Table B7 Specificity in outcome at time 1: Standardized parameter estimates from bivariate genetic model-fitting (Chapter 12: Synthesis II; Chapter 13: Synthesis III)

			Common factors						Unique factors					
	RMSEA	χ^2	h_{11}	h_{12}	s_{11}	s_{12}	n_{11}	n_{12}	b_{21}	s_{21}	n_{21}	b_{32}	s_{32}	n_{32}
Antisocial behavior–cognitive agency	.04	95.83*	.63*	.66*	—	—	.14	.15	.61*	.25	.39*	.65*	—	.34*
Antisocial behavior–depressive symptoms	.05	144.73*	.60*	.58*	.26	.25	.22*	.21*	.63*	.06	.36*	.59*	—	.46*
Antisocial behavior–global self-worth	.04	408.64*	.17	.26	.11	.17	.17*	.26*	.84*	.29	.37*	.85*	—	.33*
Antisocial behavior–sociability	.04	93.64*	.38*	.45*	.16	.19	.12	.14	.76*	.26	.40*	.60*	.53*	.32*
Antisocial behavior–social responsibility	.03	87.31*	.62*	.95*	—	—	.22*	.33*	.57*	.33*	.37*	—	—	—
Autonomy–cognitive agency	.03	84.77*	.60*	.46*	—	—	—	—	.16	.70*	.36*	.81*	—	.36*
Autonomy–sociability	.03	88.01*	.58*	.49*	.43*	.37*	.17*	.15*	.24	.52*	.33*	.56*	.43*	.32*
Autonomy–social responsibility	.03	83.86*	.46*	.52*	—	—	.02	.02	.44*	.68*	.37*	.79*	—	.33*

Table B7 *(continued)*

	RMSEA	χ^2	Common factors						Unique factors					
			b_{11}	b_{12}	s_{11}	s_{12}	n_{11}	n_{12}	b_{21}	s_{21}	n_{21}	b_{32}	s_{32}	n_{32}
Cognitive agency–depressive symptoms	.05	128.99*	.66*	.60*	—	—	.23*	.21*	.65*	—	.28*	.58*	.25	.45*
Cognitive agency–global self-worth	.02	93.01*	.15*	.38*	.11*	.27*	.29*	.75*	.86*	.19	.32*	—	-.05	.47*
Cognitive agency–sociability	.03	90.78*	.57*	.63*	.06	.07	.04	.05	.73*	—	.38*	.42*	.55*	.35*
Cognitive agency–social responsibility	.04	260.18*	.38*	.54*	—	—	.38*	.54*	.64*	—	.55*	.61*	—	.23*
Depressive symptoms–global self-worth	.06	152.78*	.15	.42	.06	.18	.32*	.88*	.75*	.25	.49*	—	—	.11
Depressive symptoms–sociability	.05	137.78*	.48*	.58*	—	—	.06	.07	.72*	.17	.47*	.44*	.60*	.33*
Depressive symptoms–social responsibility	.05	135.95*	.49*	.79*	—	—	.03	.05	.72*	—	.50*	.51*	—	.33*
Global self-worth–sociability	.05	67.43*	.53*	.23*	—	—	.34*	.15*	—	—	.77*	.70*	.57*	.32*
Sociability–social responsibility	.02	86.85*	.41*	.54*	—	—	—	—	.64*	.55*	.35*	.77*	—	.33*

*Path co-efficient is significant at $p < .05$ by t-test.

Table B8 Specificity in environment at time 1: Standardized parameter estimates from bivariate genetic model-fitting (Chapter 12: Synthesis II)

	RMSEA	χ^2	Common factors						Unique factors					
			h_{11}	h_{12}	S_{11}	S_{12}	n_{11}	n_{12}	h_{21}	S_{21}	n_{21}	h_{32}	S_{32}	n_{32}
Mother's parenting with mother's parenting:														
Negativity-positivity	.04	113.01*	.31*	.35*	.38*	.42*	.19*	.21*	.71*	.29*	.37*	.53*	.43*	.44*
Negativity-mon./knowledge	.02	69.26*	.26*	.31*	.32*	.38*	—	—	.73*	.35*	.41*	.38*	.46*	.63*
Negativity-attempted	.00	48.95	.43*	.51*	.24*	.29*	.14	.17	.65*	.40*	.40*	.24	.52*	.55*
Positivity-mon./knowledge	.04	86.29*	.35*	.38*	.46*	.49*	.19	.21	.53*	.39*	.45*	.31	.36*	.60*
Positivity-actual	.00	93.35*	.33*	.34*	.36*	.37*	.15	.15	.54*	.48*	.46*	.40*	.42*	.62*
Mon./knowledge-actual	.02	62.59*	.37*	.36*	.44*	.44*	.34*	.33*	.34*	.40*	.53*	.42*	.32*	.53*
Attempted-actual	.04	62.17*	.46*	.46*	.52*	.51*	.43*	.42*	.32*	.28*	.39*	.26	.24	.48*
Mon./knowledge-attempted	.02	46.16	.37*	.37*	.29*	.29*	.20	.20	.32	.53*	.60*	.44*	.51*	.54*
Father's parenting with father's parenting:														
Negativity-positivity	.03	98.07*	.44*	.46*	.20	.21	.07	.08	.57*	.46*	.48*	.48*	.58*	.43*
Negativity-attempted	.06	73.89*	.27	.31	.31*	.37*	.29*	.34*	.66*	.39*	.39*	.35*	.55*	.48*
Positivity-mon./knowledge	.04	66.41*	.42*	.45*	.44*	.46*	.23*	.24*	.49*	.44*	.38*	—	.52*	.50*
Positivity-actual	.04	88.54*	.34*	.38*	.45*	.50*	.19*	.21*	.56*	.41*	.39*	.38*	.43*	.48*
Mon./knowledge-attempted	.03	68.26*	.39*	.41*	.45*	.48*	.19	.20	.08	.55*	.55*	.24	.44*	.55*

Tale B8 *(continued)*

			Common factors						Unique factors					
	RMSEA	χ^2	b_{11}	b_{12}	S_{11}	S_{12}	n_{11}	n_{12}	b_{21}	S_{21}	n_{21}	b_{32}	S_{32}	n_{32}
Mon./knowledge-actual	.03	73.63*	.40*	.41*	.53*	.55*	.30*	.31*	—	.47*	.50*	.36*	.35*	.42*
Attempted-actual	.03	83.77*	.46*	.45*	.56*	.55*	.41*	.40*	0.00	.35*	.42*	.30*	.35*	.33*
Mother's parenting with father's parenting:														
Mom negativity–dad negativity	.04	82.01*	.61*	.60*	.36*	.35*	.35*	.34*	.38*	.35*	.34*	.49*	.31*	.25*
Dad positivity–mom positivity	.04	153.40*	.25	.26	.49*	.52*	.30*	.31*	.59*	.38*	.33*	.56*	.32*	.38*
Dad attempted–mom actual	.02	65.32*	.33*	.33*	.35*	.35*	.38*	.38*	.33*	.56*	.45*	.44*	.42*	.51*
Dad actual–mom mon./knowledge	.03	84.57*	.29	.29	.31*	.32*	.25*	.26*	.45*	.58*	.46*	.39*	.51*	.58*
Dad actual–mom attempted	.02	63.55*	.25	.26	.40*	.41*	.36*	.37*	.47*	.52*	.38*	.51*	.42*	.44*
Dad actual–mom actual	.03	71.35*	.20	.20	.47*	.47*	.48*	.48*	.50*	.47*	.22*	.49*	.32*	.42*
Mom positivity–dad mon./knowledge	.05	92.20*	—	—	.47*	.47*	.18*	.18*	.63*	.38*	.46*	.38*	.54*	.56*
Mom positivity–dad actual	.03	114.86*	.09	.10	.43*	.45*	.10	.11	.63*	.42*	.48*	.53*	.48*	.51*
Dad attempted–mom attempted	.01	51.29	.29	.28	.43*	.43*	.45*	.45*	.36*	.50*	.38*	.48*	.41*	.37*

Mother's and father's parenting with sibling relationships:

Mom negativity–sibling negativity	.04	100.16*	.30*	.25*	.61*	.51*	—	—	.57*	—	.46*	.18	.72*	.37*
Mom negativity–sibling positivity	.03	89.28*	.36*	.29*	.39*	.32*	.21*	.17*	.69*	.26	.36*	.35*	.77*	.26*
Dad negativity–sibling positivity	.05	104.37*	.26*	.21*	.47*	.37*	.23*	.18*	.67*	.16	.42*	.40*	.75*	.26*
Dad positivity–sibling positivity	.02	106.76*	.15	.11	.63*	.49*	—	—	.62*	—	.44*	.44*	.67*	.31*
Mom positivity–sibling positivity	.04	129.97*	.34*	.25*	.64*	.47*	—	—	.47*	—	.50*	.39*	.68*	.32*
Dad negativity–sibling negativity	.04	95.44*	.18	.15	.64*	.52*	—	—	.47*	—	.58*	.27*	.71*	.36*

Other environmental associations:

Sib. negativity–sib. positivity	.04	114.96*	.19*	.19*	.66*	.67*	.16*	.17*	.21	.60*	.31*	.40*	.51*	.27*
Sib. negativity–parent's conflict about child	.04	111.08*	.13*	.36*	.25*	.67*	.06	.17	.27*	.85*	.35*	.45*	.41*	.16
Mom negativity–parent's conflict about child	.04	121.42*	.27*	.58*	.29*	.63*	.11*	.23*	.71*	.40*	.41*	—	.46*	—
Dad negativity–parent's conflict about child	.03	146.57*	.28*	.58*	.31*	.65*	.11*	.23*	.60*	.45*	.50*	—	.43*	—

*Path co-efficient is significant at $p < .05$ by t-test.

Table B9 Specificity in environment at time 2: Standardized parameter estimates from bivariate genetic model-fitting (Chapter 12: Synthesis II)

	RMSEA	χ^2	Common factors									Unique factors		
			h_{11}	b_{12}	S_{11}	S_{12}	n_{11}	n_{12}	b_{21}	S_{21}	n_{21}	b_{32}	S_{32}	n_{32}
Father's parenting with father's parenting:														
Negativity-attempted	.05	85.28*	.44*	.52*	.00	.00	.12	.14	.31	.58*	.53*	.59*	.44*	.50*
Positivity-mon./knowledge	.03	64.10	.45*	.48*	.48*	.50*	.27*	.29*	.44*	.00	.50*	.00	.52*	.47*
Positivity-actual	.03	68.99*	.27*	.30*	.47*	.51*	.23*	.25*	.45*	.40*	.46*	.30	.55*	.51*
Mother's parenting with father's parenting:														
Dad negativity–mom mon./knowledge	.04	71.83*	.37*	.42*	.25	.29	.00	.00	.00	.48*	.71*	.65*	.34	.52*
Dad positivity–mom mon./knowledge	.01	52.24	.12	.13	.33*	.35*	.28*	.30*	.42*	.44*	.64*	.39*	.64*	.48*
Dad attempted–mom attempted	.01	54.05	.00	.00	.47*	.46*	.40*	.39*	.07	.50*	.62*	.63*	.29	.36*

*Path co-efficient is significant at $p < .05$ by t-test.

Appendix C

ADDITIONAL GENETIC ANALYSES

We performed one set of additional analyses in order to examine the appropriateness of including step-families with half-siblings and step-siblings in a genetic design. As described in some detail in Chapter 5, we took into consideration the length of time that the siblings had been living together in the same home. In this way, we could explore the potential impact on sibling similarity of simply living together for long periods of time.

Chapter 5 describes in some detail the issues that are unique to samples that include step-families. These unique circumstances of step-families include the following: (1) the length of the remarriage is variable, although our sampling constraints required that it be at least five years in duration; (2) at least one of the parents is rearing at least one child to whom he or she is genetically unrelated; and (3) there is the possibility that the children have some impact on the selection of the step-parent and on the success of the remarriage. We will deal only with the first issue here.

In order to examine the potential impact of the length of time that the siblings had been living together, we conducted a series of analyses. The length of the time of co-residence for the sibling pairs reported in Chapter 5 was computed in the following way. First, the number of years that the two parents had been living together in the same household, including the length of the marriage and the length of cohabitation prior to marriage, was computed. The second step was to establish whether the children had been residing with this parent since the divorce, and if not, to calculate when they had begun living at least half-time with this parent. (By "this parent" we mean the parent who participated in the NEAD study.) It was also necessary to determine whether both children had been living together in the same household; we based this determination on whether one or both children had been living at least half-time with the parent who participated in the NEAD study. As a third and final step, we computed the length of co-residence for the sibling pairs. Because full siblings were the result of mother's bringing two children from her previous marriage, in most cases the full siblings had been residing together all their lives, which we set to be equal to the younger sibling's age. There were a few cases in which one of the siblings had been living with the father at least part time. In such cases, however, the length of co-residence was adjusted for the length of time the child had lived in a different household from the one in which the sibling had lived. An additional factor for full siblings in step-families is the length of time in the step-family, which is independent from the length of co-residence. If the length of the marriage and cohabitation was less than the age of the younger child, as was the case most often, this value was used as the length of time in the step-family household. Half-siblings were someone less difficult. In all cases, half-siblings were the result of mother's or father's having brought a child from a previous marriage into a new family and then having had a child once remarried. The length of half-sibling co-residence and the length of time in the step-family were identical and were equal to the age of the younger child. We recognize that the older child had been residing in the step-family longer than the younger child, but for these analyses, we were interested more in the shared sibling environments than in individual-specific residence. Finally, genetically unrelated step-siblings were the result of mother's having brought a child from a previous marriage and father's having brought a child from a previous marriage. In this case, as with half-siblings, the length of co-residence was equal to the length of time in the step-family, and in all cases was equal to the length of the marriage and cohabitation of the parents. In no case did the length of co-residence of time in the step-family have a systematic impact on sibling similarity. These findings are presented in detail in Chapter 5.

Active gene-environment correlation. *See Gene-environment (GE) correlation*

Anonymous influences. Unmeasured but indirectly estimated influences of genetic or environmental factors on measures of behavior. Techniques of quantitative behavioral genetics are used to estimate the balance or weight of such influences. Specific genes and their influence are not measured; nor are specific aspects of the environment. Thus these influences remain anonymous.

Blended sibs. Sibling pairs who are not similar genetically. Each has a different father and mother.

Child-effects model. A hypothetical sequence beginning with genetic factors, which then influence child or adolescent development, which in turn influences family relationships. This model is distinguished from the *parent-effects model.*

Cluster analysis. A statistical procedure for identifying subgroups of individuals or families within a sample. Members of a subgroup, defined by this procedure, are more similar to one another on a set of measured attributes than they are to members of other groups.

Correlation coefficient. A statistic that quantifies the association between one *variable* and another.

Covariance. A statistic reflecting the degree of association between two measures.

Covariance model. A statistical procedure designed for this study that estimates both the direct effect of an adolescent's family experience on the adolescent's adjustment and the indirect influence of the sibling's experience of the family on the adolescent's adjustment. The statistical term for the former is *specific correlation,* and for the latter, *cross-correlation.*

Cross-correlation. A term with two distinct meanings. First, the cross-correlation is used in the *covariance model* to estimate the indirect effects of the sibling's family experience on the adolescent's adjustment. In this instance, the cross-correlation, as used in this book, is a *path coefficient* estimating the association between two *latent variables.* Second, cross-correlations are used as the main statistics in computing the weight of genetic and environmental influences on the *covariance* between any two measures, say, measure *o* and measure *p*. Computationally, the cross-correlation is the association between measure *o* in one twin (or sib) and measure *p* in the co-twin (or sib). In this instance the cross-correlation is an *intraclass correlation* between two directly measured variables.

Cross-lagged correlation. The association between a measure at time 1, say, measure *m,* and another measure at time 2, say, measure *n.* This simple statistic is the basis of a more complex procedure for estimating, in a *longitudinal study,* whether changes in *m* precede changes in *n,* or vice versa. Cross-lagged correlations can be computed only in longitudinal studies.

Cross-sectional study. A study design that measures the association between two measures, say, measure *x* and measure *y,* at a single point in time. This design cannot distinguish causal ordering or processes: changes in *x* may precede or follow changes in *y.* This study design is distinguished from a *longitudinal study* and an *experimental study.*

Dependent variable. A variable that, in a theory, is thought to be responsive to or influenced by another variable or process known as the *independent variable.*

Dizygotic (DZ) twins. Twins that result from the simultaneous fertilization of two separate eggs, or zygotes, by two different sperm; also known as fraternal twins. Genetically speaking, DZ twins are the same as *full sibs.*

Environmentality. A quantitative estimate of the impact of nongenetic factors on any measure of behavior. More specifically, an indication of the percentage of variation among individuals in a sample, with a possible range of 0 to 100, that can be attributed to environmental factors.

Evocative gene-environment correlation. *See Gene-environment (GE) correlation*

Experimental study. A study design that requires a planned intervention or treatment by the researcher. Typically, families or individuals are assigned, at random, either to a group that receives the treatment or to a control group that does not. The distinction between treatment and no treatment can be called measure *x,* and the outcome of the treatment can be called measure *y.* If variation in *x* is associated with variation in *y,* then the treatment is said to be effective. Experimental studies are often used to explore causal hypotheses in psychological development. If treatment *x* always leads to changes in variable *z,* and changes in variable *z* always precede changes in variable *y,* then a hypothesis that *z* causes *y* is strengthened. Experimental studies, like other kinds of studies, can be corrupted by *third variable* problems.

Full sibs. Sibling pairs who have a common mother and common father and share approximately 50 percent of the genes responsible for individual differences.

Gene-environment (GE) correlation. The degree of association between genetic and environmental influences on a particular measure of behavior. There are three types of GE correlation. *Passive gene-environment correlation* is attributed exclusively to overlap in genes between members of the same family. For example, parental negativity toward a child and the child's own genetic characteristics may enhance the probability that the child will become antisocial. The child's genes may be associated with parental negativity because the same genes that in children put them at risk for antisocial behavior act, in the parent, to increase negative behavior toward children. *Evocative gene-environment correlation* arises when variation in heritable characteristics among individuals elicits different responses from the environment. *Active gene-environment correlation* occurs when heritable characteristics of individuals lead them to change or shape their environment.

Gene-environment interaction (GxE).   The influence of variation in the environment on the expression of genetic influence on a measure of behavior. For example, the genetic influence on the evolution of schizophrenia may be reduced by a well-functioning family environment and enhanced by a disturbed family environment.

Genetic cascade.   The progressive decrease in *intraclass correlations* as the genetic relatedness of siblings declines. The slope or steepness of the genetic cascade, a term developed for this study, reflects the *heritability* of a measure of behavior. For example, for a highly heritable measure of behavior, one would expect the following genetic cascade, or decline in *intraclass correlations,* across sibling types: monozygotic twins → dizygotic twins = full sibs, non-divorced = full sibs, steps → half-sibs, steps → unrelated sibs, steps.

Genetic overlap.   An informal term used to describe the association of genetic influences on two different traits. Wherever there is a complete overlap, the same genetic influences are responsible for the heritable portion of both traits.

Half-sibs.   Siblings who have only one parent in common. In this study, half-sib pairs had a common mother who had been remarried and had one sib in the pair by her former husband and one by her new husband.

Heritability.   A quantitative estimate, varying from 0 to 100 percent, of the influence of genetic factors on a measure of behavior. Heritability of association is an estimate of the importance of genetic factors in accounting for the *covariance* of two measures of behavior. Heritability of disjunction is an estimate of the importance of genetic factors in the lack of association of two measures of behavior. Heritability of stability is an estimate of the importance of genetic factors in the association of measures of behavior at two different points in time. Heritability of change is an estimate of the importance of genetic factors in the lack of association between a measure of behavior at two different points in time.

Independent variable.   In a theory, the variable or process that is thought to initiate or cause a subsequent variable or process known as the *dependent variable.*

Indicated preventive intervention.   A preventive intervention that is provided to a group of people who show some indications that they are at risk for psychological difficulties or for psychiatric disorders. For example, young children who have irritable temperaments or become regularly distressed at any limits set by their parents are at future risk for developing antisocial behavior. An indicated preventive intervention would focus on just these children.

Intraclass correlation.   A statistic reflecting the association between a measure of behavior in one twin (or sibling) and that same measure in the co-twin (or co-sibling). The measure differs from the more typical correlation coefficient because it does not require one sibling to be arbitrarily assigned to one class of individuals (say, "older siblings") and the other to a second class (say, "younger siblings"). The computation of this statistic treats each twin or sibling as part of the same class of individuals. The same statistic is used to judge the reliability of a coding procedure. Here the association between coding values assigned to the same set of families by two different coders is computed.

Latent variables.   Variables that are estimated but not directly measured. In contrast to *anonymous influences,* which can never be measured, latent variables can be estimated both by a set of specific measures and by their association with other variables. In this book, this is true for the *covariance model,* which uses latent variables to estimate the association among the following: the family experience of the adolescent, the family experience of the sib, and the adolescent's adjustment. But such variables cannot always be delineated. In the statistical models used to estimate *heritability,* "G," "SE," and "NSE" are latent variables that cannot be further delineated and hence are anonymous.

Longitudinal study.   A study design in which at least two different variables, x and y, are measured on at least two different occasions. This design permits a tentative estimate as to whether variation in x precedes or follows variation in y.

Mediating variable.   A variable or process thought to explain the association between an *independent variable* and a *dependent variable.*

Model-testing heritability.   The estimate of *heritability* obtained by using a statistical model that integrates many computations. In this study model-testing heritabilities were most often used for making comparisons among all six groups of sibling types: *MZ and DZ twins, full sibs* in nondivorced families, *full sibs* in step-families, *half-sibs,* and *blended sibs.* The computational procedures of these models must make certain assumptions in order to produce their results, and the estimates yielded by these models are only as valid as those assumptions. For example, the models used to compare the six groups of sibling types assume that *shared environment* is equal for all groups and that *GxE interaction* is minimal or nonexistent.

Monozygotic (MZ) twins.   Twins who develop from the fertilization of a single egg, or zygote; also known as identical twins because they are genetically identical.

Multiple-systems hypothesis.   A theory positing that children and adolescents are capable of monitoring or responding to social processes in multiple subsystems in their families, such as the marital relationship and the father-sib relationship. According to this hypothesis, processes in these multiple systems influence psychological development.

Nonshared environment.   The social and physical factors, distinct for each child in the family, that affect development.

Nonshared environmentality.   A quantitative estimate of the *anonymous influences* of all nonshared environmental factors on a particular measure of development.

Outcome variable.   *See Dependent variable*

Parent-effects model.   A hypothetical sequence beginning with genetic factors that influence a heritable trait in the developing child. This trait, in turn, influences parental processes that then influence the heritable trait. This final influence could be an exaggeration of the trait or it might suppress the trait. The parental influence in this sequence transforms a trait that has little significance for the overall adjustment of the child to a characteristic that plays a notable role in adjustment.

Partialing out.   A statistical procedure for untangling the relationship among at least three variables: "independent variable d," "dependent variable f," and a "third variable, e." The statistical procedure allows the researcher to hold the association of *e* on *f* constant while estimating the association between *d* and *f*. Thus, the confounding effects of *e* on the association between *d* and *f* are "partialed out."

Passive gene-environment correlation.   *See Gene-environment (GE) correlation*

Path coefficient.   The estimated association between two *latent variables*.

Preventive interventions.   *See Selective preventive intervention; Indicated preventive intervention; Universal preventive intervention*

Randomized intervention.   *See Experimental study*

Relationship code.   A term used to express the hypothesis that family relationships are mediating variables in the association between genetic factors and the evolution of many complex behaviors. The hypothesis suggests that different family subsystems and different qualities of relationships within these subsystems mediate specific genetic influences. Hence, family relationships serve as a code by which specific sets of genetic factors correspond to a specific quality of a particular family subsystem at a particular point in development.

Selective preventive intervention.   A preventive intervention that is provided to individuals who show no signs of psychological difficulty but are living in circumstances that add to their future risk of developing such difficulties. For example, children of depressed mothers may show no signs of psychological difficulties soon after a first episode of depression in the parent but are at sufficient risk for developing their own psychological problems. In such cases an indicated preventive intervention should be considered.

Shared environment.   The social and physical factors, shared by siblings in the same family, that affect development.

Shared environmentality. A quantitative estimate of the *anonymous influences* of all shared environmental factors on a particular measure of development.

Single-systems hypothesis. A theory positing that children and adolescents are capable of monitoring social processes in only one subsystem in their family, and that social processes in this single subsystem influence psychological development.

Specific correlation. A *path coefficient* estimating the association between adolescents' experience of their relationships and measures of their psychological adjustment. This coefficient is derived through the use of the *covariance model.*

Specific variable. A variable or process that is specifically measured, in contrast to an *anonymous influence.*

Third variable. A variable that may influence the association between an *independent variable* and a *dependent variable.* Third variables can corrupt a research design if they are unsuspected, unmeasured, and not partialed from estimates of the association between the independent and dependent measures.

Universal preventive intervention. A preventive intervention that is provided to all members of a defined population. Flouridation of the drinking water is a classic example. In the behavioral area, as another example, some universal preventive interventions have been offered in the school setting. These have been aimed at reducing the likelihood of depression or antisocial behavior, but the interventions are not aimed either at children who live in high-risk circumstances, a *selective preventive intervention,* or at children who are showing early signs of psychological difficulty, an *indicated preventive intervention.*

Variable. A characteristic that can be clearly conceptualized and quantitatively assessed. For example, the degree of negativity in mother-child relationships is a variable. So too is the degree of sociability in an adolescent.

References

Abdel-Rahmin, A. R., C. T. Nagoshi, and S. G. Vandenberg. 1990. Twin resemblances in cognitive ability in an Egyptian sample. *Behavior Genetics,* 21(1): 33–43.

Achenbach, T. M., and C. Edelbrock. 1983. *Manual for the Child Behavior Checklist and Revised Child Behavior Folder.* Burlington, Vt.: University of Vermont.

Adelman, M. B., and M. Siemon. 1986. Communicating the relational shift: Separation among adult twins. *American Journal of Psychotherapy,* 40(1): 96–109.

Amato, P. R., and B. Keith. 1991. Parental divorce and the well-being of children: A meta-analysis. *Psychological Bulletin,* 110: 26–46.

Anderson, E., E. M. Hetherington, D. Reiss, and G. Howe. 1994. Parents' nonshared treatment of siblings and the development of social competence during adolescence. *Journal of Family Psychology,* 8: 303–320.

Anderson, K. E., H. Lytton, and D. M. Romney. 1986. Mothers' interaction with normal and conduct-disordered boys: Who affect whom? *Developmental Psychology,* 22: 604–609.

Arnold, J. D. 1985. Adolescent perceptions of family scapegoating: A comparison of parental and sibling involvement. *Journal of Adolescence,* 8(2): 159–165.

Baker, L. A., and D. Daniels. 1990. Nonshared environmental influences and personality differences in adult twins. *Journal of Personality and Social Psychology,* 58: 103–110.

Barkley, R. A. 1989. Hyperactive girls and boys: Stimulant drug effects on mother-child interactions. *Journal of Child Psychology and Psychiatry,* 30: 336–341.

Baumrind, D. 1991. The influence of parenting style on adolescent competence and substance use. *Journal of Early Adolescence,* 11: 56–95.

Bell, D. C., and L. G. Bell. 1983. Parental validation and support in the development of adolescent daughters. In H. D. Grotevant, ed., *Adolescent Development in the Family,* pp. 27–42. San Francisco: Jossey Bass.

Benes, F. M., M. Turtle, Y. Khan, and P. Farol. 1994. Myelination of a key relay zone in the hippocampal formation occurs in the human brain during childhood, adolescence and adulthood. *Archives of General Psychiatry,* 51(6): 477–484.

Bennett, L. A., S. J. Wolin, D. Reiss, and M. A. Teitelbaum. 1987. Couples at risk for transmission of alcoholism: Protective influences. *Family Process,* 26: 111–129.

Bergeman, C. S., R. Plomin, G. E. McClearn, N. L. Pedersen, and L. T. Friber. 1988. Genotype-environment interaction in personality development. *Psychology and Aging,* 3(4): 399–406.

Biederman, J., J. F. Rosenbaum, E. A. Bolduc-Murphy, S. V. Faraone, J. Chaloff, D. R. Hirshfeld, and J. Kagan. 1993. A 3-year follow-up of children with and without behavioral inhibition. *Journal of the American Academy of Child and Adolescent Psychiatry,* 32(4): 814–821.

Biederman, J., J. F. Rosenbaum, D. Hirshfeld, S. V. Faraone, E. A. Bolduc, M. Gerstein, S. R. Meminger, J. Kagan, N. Snidman, and J. S. Reznick. 1990. Psychiatric correlates of behavioral inhibition in young children of parents with and without psychiatric disorders. *Archives of General Psychiatry,* 47(1): 21–26.

Billings, A. G., and R. H. Moos. 1982. Family environments and adaptation: A clinically applicable typology. *American Journal of Family Therapy,* 10(2): 26–38.

Billy, J. O. G., and J. R. Udry. 1985. Patterns of adolescent friendship and effects on sexual behavior. *Social Psychology Quarterly,* 48: 27–41.

Biron, P., J. G. Mongeau, and D. Bertrand. 1977. Family resemblances of body weight/height in 374 homes with adopted children. *Journal of Pediatrics,* 91: 555–558.

Blackston, T. C., R. E. Tarter, C. S. Martin, and H. B. Moss. 1994. Temperament induced father-son family dysfunction: Etiological implications for child behavior problems and subtance abuse. *American Journal of Orthopsychiatry,* 64: 280–292.

Block, J. H., J. Block, and P. F. Gjerde. 1986. The personality of children prior to divorce: A prospective study. *Child Development,* 57: 827–840.

Blos, P. 1962. *On Adolescence: A Psychoanalytic Interpretation.* New York: Free Press.

Booth, A., D. Johnson, and J. N. Edwards. 1983. Measuring marital instability. *Journal of Marriage and the Family,* 45: 387–393.

Bouchard, T. J., Jr., and M. McGue. 1981. Familial studies of intelligence: A review. *Science,* 212: 1055–1059.

Braungart, J. W. 1994. Genetic influence on "environmental measures." In J. C. DeFries, R. Plomin, and D. W. Fulker, eds., *Nature and Nurture during Middle Childhood.* Cambridge, Mass.: Blackwell.

Brody, G., and Z. Stoneman. 1994. Sibling relationships and their association with parental differential treatment. In E. M. Hetherington, D. Reiss, and R. Plomin, eds., *Separate Social Worlds of Siblings: The Impact of Nonshared Environment on Development.* Hillsdale, N.J.: Lawrence Erlbaum.

———. 1996. A risk amelioration model of sibling relationships: Conceptual underpinnings and preliminary findings. In G. H. Brody, ed., *Sibling Relationships: Their Causes and Consequences.* Norwood, N.J.: Ablex.

Brody, G. H., Z. Stoneman, and M. Burke. 1987. Child temperaments, maternal differential behavior, and sibling relationships. *Developmental Psychology,* 23(3): 354–362.

Brooks, J. B. 1981. Social maturity in middle age and its developmental antecedents. In D. H. Eichorn, J. A. Clausen, N. Haan, M. P. Honzik, and P. H. Mussen, eds., *Present and Past in Middle Life,* pp. 244–269. New York: Academic Press.

Brooks-Gunn, J., and E. O. Reiter. 1990. The role of pubertal processes. In S. S. Feldman and G. R. Elliott, eds., *At the Threshold: The Developing Adolescent,* pp. 16–53. Cambridge, Mass.: Harvard University Press.

Brown, B. B. 1990. Peer groups and peer cultures. In S. S. Feldman and G. R. Elliott, eds., *At the Threshold: The Developing Adolescent,* pp. 171–196. Cambridge, Mass.: Harvard University Press.

Browne, M. W., and R. Cudeck. 1993. Alternative ways of assessing model fit. In K. A. Bollen and J. S. Long, eds., *Testing Structural Equation Models,* pp. 136–162. Newbury Park, Calif.: Sage.

Bryant, B. K. 1992. Sibling caretaking: Providing emotional support during middle childhood. In F. Boer and J. Dunn, eds., *Children's Sibling Relationships: Developmental and Clinical Issues,* pp. 55–69. Hillsdale, N.J.: Lawrence Erlbaum.

Buchsbaum, M. S., C. S. Mansour, D. G. Teng, and A. D. Zia. 1992. Adolescent developmental change in the topography of EEG amplitude. *Schizophrenia Research,* 7(2): 101–107.

Buss, A. H., and R. Plomin. 1984. *Temperament: Early Developing Personality Traits.* Hillsdale, N.J.: Lawrence Erlbaum.

Cadoret, R. J., and C. A. Cain. 1981. Genetic-environmental interaction in adoption studies of antisocial behavior. In C. Perris, G. Struwe, and B. Jansson, eds., *Biological Psychiatry,* pp. 97–100. Amsterdam: Elsevier.

Cadoret, R. J., C. A. Cain, and R. R. Crowe. 1983. Evidence for gene-environment interaction in the development of adolescent antisocial behavior. *Behavior Genetics,* 13(3): 301–310.

Cadoret, R. J., E. Troughton, J. Bagford, and G. Woodworth. 1990. Genetic and environmental factors in adoptee antisocial personality. *European Archives of Psychiatry and Neurological Sciences,* 239: 231–240.

Cadoret, R. J., E. Troughton, and T. W. O'Gorman. 1987. Genetic and environmental factors in alcohol abuse and antisocial personality. *Journal of Studies in Alcohol,* 48: 1–8.

Cadoret, R. J., G. Winokur, D. Langbehn, E. Troughton, W. R. Yates, and M. A. Stewart. 1996. Depression spectrum disease, I: The role of gene-environment interaction. *American Journal of Psychiatry,* 153(7): 892–899.

Cadoret, R. J., W. R. Yates, E. Troughton, G. Woodworth, and M. A. Steward. 1995.

Genetic-environmental interaction in the genesis of aggressivity and conduct disorders. *Archives of General Psychiatry,* 52: 916–924.

———. 1996. An adoption study of drug abuse/dependency in females. *Comprehensive Psychiatry,* 37: 88–94.

Capaldi, D. M., and G. R. Patterson. 1994. Interrelated influences of contextual factors on antisocial behavior in childhood and adolescence for males. In D. Fowles, P. Sutker, and S. Goodman, eds., *Progress in Experimental Personality and Psychopathology Research,* pp. 165–198. N.Y.: Springer.

Capron, C., and M. Duyme. 1989. Assessment of effects of socio-economic status on IQ in a full cross-fostering study. *Nature,* 340: 552–554.

Cardon, L. R., D. W. Fulker, J. C. DeFries, and R. Plomin. 1992. Continuity and change in general cognitive ability from 1 to 7 years of age. *Developmental Psychology,* 28(1): 64–73.

Carr, M., and C. Schellenbach. 1993. Reflective monitoring in lonely adolescents. *Adolescence,* 28(111): 737–747.

Chasnoff, I. J., W. J. Burns, S. H. Schnoll, and K. A. Burns. 1985. Cocaine use in pregnancy. *New England Journal of Medicine,* 313: 666–669.

Chasnoff, I. J., D. R. Griffith, C. Freier, and J. Murray. 1992. Cocaine/polydrug use in pregnancy: Two-year follow-up. *Pediatrics,* 89: 284–289.

Chasnoff, I. J., D. R. Griffith, S. MacGregor, K. Dirkes, and K. A. Burns. 1989. Temporal patterns of cocaine use in pregnancy. *Journal of the American Medical Association,* 24(31): 1741–1744.

Chasnoff, I. J., H. J. Landress, and M. E. Barrett. 1993. The prevalence of illicit drug or alcohol use during pregnancy and discrepancies in mandatory reporting in Pinellas County, Florida. *New England Journal of Medicine,* 322: 1202–1206.

Cherlin, A. J., F. F. Furstenburg, L. P. Chase-Lansdale, and K. E. Kiernan. 1991. Longitudinal studies of effects of divorce on children in Great Britain and the United States. *Science,* 252(5011): 1386–1389.

Cherny, S. S., and L. R. Cardon. 1994. General cognitive ability. In J. C. DeFries, R. Plomin, and D. W. Fulker, eds., *Nature and Nurture during Middle Childhood.* Oxford: Blackwell.

Cloninger, C. R., S. Sigvardsson, M. Bowman, and A. von Knorring. 1982. Predisposition to petty criminality in Swedish adoptees II. Cross-fostering analysis of gene-environment interaction. *Archives of General Psychiatry,* 39(11): 1242–1247.

Cohen, D. S. F., W. N. Friedrich, T. M. Jaworski, and D. Copeland. 1994. Pediatric cancer: Predicting sibling adjustment. *Journal of Clinical Psychology,* 50(3): 303–319.

Conger, R. D., K. J. Conger, G. H. Elder, Jr., F. O. Lorenz, R. L. Simons, and L. B. Witbeck. 1992. A family process model of economic hardship and adjustment of early adolescent boys. *Child Development,* 63: 526–541.

————. 1993. Family economic stress and adjustment in early adolescent girls. *Developmental Psychology,* 29(2): 206–219.

Conger, R. D., F. O. Lorenz, G. H. Elder, Jr., J. N. Melby, R. L. Simons, and K. J. Conger. 1991. A process model of family economic pressure and early adolescent alcohol use. *Journal of Early Adolescence,* 11: 430–449.

Conger, R. D., G. R. Patterson, and X. Ge. 1995. It takes two to replicate: A mediational model for the impact of parents' stress on adolescent adjustment. *Child Development,* 66(1): 80–97.

Conger, R. D., and M. R. Rueter. 1996. Siblings, parents and peers: A longitudinal study of social influences in adolescent risk for alcohol use and abuse. In G. H. Brody, ed., *Sibling Relationships: Their Causes and Consequences.* N.Y.: Ablex.

Conger, R. D., M. R. Rueter, and K. J. Conger. 1994. The family context of adolescent vulnerability to alcohol use and abuse. *Sociological Studies of Children,* 6: 55–86.

Covell, K., and R. Abramovitch. 1988. Children's understanding of maternal anger: Age and source of anger differences. *Merrill-Palmer Quarterly,* 34(4): 353–368.

Cowan, C. P., and P. A. Cowan. 1992. *When Partners Become Parents.* N.Y.: Basic Books.

Cudeck, R. 1989. Analysis of correlation matrices using covariance structure models. *Psychological Bulletin,* 105: 317–327.

Cunningham, C. E., and L. S. Siegel. 1987. Peer interactions of normal and attention deficit–disordered boys during free-play, cooperative task and simulated classroom situations. *Journal of Abnormal Child Psychology,* 15: 247–268.

Cunningham, C. E., L. S. Siegel, and D. R. Offord. 1985. A developmental dose response analysis of the effects of methylphenidate on the peer interactions of attention deficit disordered boys. *Journal of Child Psychology and Psychiatry,* 26: 955–971.

Daniels, D. 1986. Differential experiences of children in the same family as predictors of adolescent sibling personality differences. *Journal of Social Psychology and Personality,* 51: 339–346.

Daniels, D., J. Dunn, F. Furstenberg, and R. Plomin. 1985. Environmental differences within the family and adjustment differences within pairs of adolescent siblings. *Child Development,* 56: 764–774.

Daniels, D., and R. Plomin. 1985. Differential experiences of siblings in the same family. *Developmental Psychology,* 21(5): 747–760.

Deal, J., C. Halverson, and K. Wampler. 1994. Sibling similarity as an individual differences variable: Within family measures of shared environment. In E. M. Hetherington, D. Reiss, and R. Plomin, eds., *Separate Social Worlds of Siblings: The Impact of Nonshared Environment on Development,* pp. 205–218. Hillsdale, N.J.: Lawrence Erlbaum.

DeFries, J., R. Plomin, and D. Fulker. 1994. *Nature and Nurture during Middle Childhood.* Oxford: Blackwell.

Derryberry, D., and M. A. Reed. 1996. Regulatory processes and the development of cognitive representations. *Development and Psychopathology,* 8: 215–234.

Dodge, K. A., and J. D. Coie. 1987. Social-information processing factors in reactive and proactive aggression in children's peer groups. *Journal of Personality and Social Psychology,* 53(6): 1146–1158.

Dodge, K. A., and N. R. Crick. 1990. Social information processing bases of aggressive behavior in children. *Personality and Social Psychology Bulletin,* 16(1): 8–22.

Dodge, K. A., and A. M. Tomlin. 1987. Utilization of self-schemas as a mechanism of interpretational bias in aggressive children. *Social Cognition,* 5(3): 280–300.

Dolan, L. J., S. G. Kellam, C. H. Brown, L. Werthamer-Larsoon, G. W. Rebok, L. S. Mayer, J. Laudoff, J. Turkkan, C. Ford, and L. Wheeler. 1993. The short-term impact of two classroom-based preventive interventions on aggressive and shy behavior and poor achievement. *Journal of Applied Developmental Psychology,* 14: 317–345.

Dornbusch, S., P. Ritter, H. Leiderman, D. Roberts, and M. Fraleigh. 1987. The relation of parenting style to adolescent school performance. *Child Development,* 58: 1244–1257.

Dunn, J., and J. Brown. 1994. Affect expression in the family: Children's understanding of emotions and their interactions with others. *Merrill-Palmer Quarterly,* 40(1): 120–137.

Dunn, J., and C. Kendrick. 1982. Siblings and their mothers: Developing relationships within the family. In M. E. Lamb and B. Sutton-Smith, eds., *Sibling Relationships: Their Nature and Significance across the Life Span,* pp. 39–60. Hillsdale, N.J.: Lawrence Erlbaum.

Dunn, J., and S. McGuire. 1993. Young children's nonshared experiences: A summary of studies in Cambridge and Colorado. In E. M. Hetherington, D. Reiss, and R. Plomin, eds., *Nonshared Environments.* Hillsdale, N.J.: Lawrence Erlbaum.

Dunn, J., and P. Munn. 1987. Development of justification in disputes with mother and sibling. *Developmental Psychology,* 23(6): 791–798.

Dunn, J., and R. Plomin. 1986. Determinants of maternal behavior towards three-year-old siblings. *British Journal of Developmental Psychology,* 4: 127–137.

Dunn, J., R. Plomin, and D. Daniels. 1986. Consistency and change in mothers' behavior towards young siblings. *Child Development,* 57: 348–356.

Dunn, J. F., R. Plomin, and M. Nettles. 1985. Consistency of mothers' behavior towards infant siblings. *Developmental Psychology,* 21: 1188–1195.

Dunn, J., C. Stocker, and R. Plomin. 1990. Nonshared experiences within the family: Correlates of behavioral problems in middle childhood. *Development and Psychopathology,* 2: 113–126.

Eaves, L. J., J. L. Silberg, J. M. Meyer, H. H. Maes, E. Simonoff, A. Pickles, M. Rutter, C. A. Reynolds, A. C. Heath, K. R. Truett, M. C. Neale, M. T. Erikson, R. Loeber, and J. K. Hewitt. 1997. Genetics and developmental psychopathology: 2. The main effects of genes and environment on behavioral problems in the Virginia Twin Study of Adolescent Behavioral Development. *Journal of Child Psychology and Psychiatry,* 38(8): 965–980.

Ekman, P. 1994. Strong evidence for universals in facial expressions: A reply to Russell's mistaken critique. *Psychological Bulletin,* 115(2): 268–287.

Ekman, P., and W. V. Friesen. 1971. Constants across cultures in the face and emotion. *Journal of Personality and Social Psychology,* 17(2): 124–129.

Ekman, P., and K. G. Heider. 1988. The universality of a contempt expression: A replication. *Motivation and Emotion,* 12(3): 303–308.

Elder, G. H., Jr. 1974. *Children of the Great Depression: Social Change in Life Experience.* Chicago: University of Chicago Press.

———. 1986. Problem behavior and family relationships: Life course and intergenerational themes. In A. M. Sorensen, F. E. Weinert, and L. R. Sherrod, eds., *Human Development and the Life Course: Multidisciplinary Perspectives,* pp. 293–342. Hillsdale, N.J.: Lawrence Erlbaum.

Emde, R. N., R. Plomin, J. Robinson, J. S. Reznick, J. Campos, R. Corley, J. C. DeFries, D. W. Fulker, J. Kagan, and C. Zahn-Waxler. 1992. Temperament, emotion and cognition at 14 months: The MacArthur Longitudinal Twin Study. *Child Development,* 53: 1437–1455.

Ensminger, M. E., C. H. Brown, and S. G. Kellam. 1982. Sex differences in antecedents of substance use among adolescents. *Journal of Social Issues,* 38(2): 25–42.

Erikson, E. H. 1963. *Childhood and Society.* 2nd ed. N.Y.: Norton.

———. 1968. *Identity, Youth and Crisis.* N.Y.: Norton.

Farhood, L., H. Zurayk, M. Chaya, and F. Saadeh. 1993. The impact of war on the physical and mental health of the family: The Lebanese experience. *Social Science and Medicine,* 36(12): 1555–1567.

Farrington, D. 1988. Studying changes within individuals: The causes of offending. In M. Rutter, ed., *Studies of Psychosocial Risk: The Power of Longitudinal Data,* pp. 158–183. Cambridge, England: Cambridge University Press.

Faust, J., C. G. Baum, and R. Forehand. 1985. An examination of the association between social relationships and depression in early adolescence. *Journal of Applied Developmental Psychology,* 6(4): 291–297.

Fergusson, D. M., and L. J. Horwood. 1996. The role of adolescent peer affiliations in the continuity between childhood behavioral adjustment and juvenile offending. *Journal of Abnormal Child Psychology,* 24(2): 205–221.

Fischbein, S. 1977. Intra-pair similarity in monozygotic and dizygotic twins during adolescence. *Annals of Human Biology,* 4: 417–430.

Fischbein, S., and T. Nordquist. 1978. Profile comparisons of physical growth for monozygotic and dizygotic twin pairs. *Annals of Human Biology,* 5: 321–328.

Fisher, L., D. C. Ransom, H. W. Terry, M. Lipkin, and R. Weiss. 1992. The California family health project: I. Introduction and description of family health. *Family Process,* 31: 231–250.

Fitzpatrick, M. A., and L. D. Ritchie. 1994. Communication schemata within the family: Multiple perspectives on family interaction. *Human Communication Research,* 20(3): 275–301.

Forehand, R., A. M. Thomas, M. Wierson, and G. Brody. 1990. Role of maternal functioning and parenting skills in adolescent functioning following parental divorce. *Journal of Abnormal Psychology,* 99(3): 278–283.

Frazier, P. A., and L. J. Schauben. 1994. Causal attributions and recovery from rape and other stressful life events. *Journal of Social and Clinical Psychology,* 13(1): 1–14.

Freedenfeld, R. N., S. R. Ornduff, and R. M. Kelsey. 1995. Object relations and physical abuse: A TAT analysis. *Journal of Personality Assessment,* 44(3): 552–568.

Furman, W., and D. Buhrmester. 1985. Children's perceptions of the personal relationships in their social networks. *Developmental Psychology,* 21: 1016–1024.

———. 1992. Age and sex differences in perceptions of networks of personal relationships. *Child Development,* 63(1): 103–115.

Furstenburg, F. F., Jr., J. Brooks-Gunn, and S. P. Morgan. 1987. *Adolescent Mothers in Later Life.* N.Y.: Cambridge University Press.

Gauvain, M., and B. Fagot. 1995. Child temperament as a mediator of mother-toddler problem solving. *Social Development,* 4(3): 257–276.

Ge, X., R. D. Conger, R. Cadoret, J. Neiderhiser, W. Yates, E. Troughton, and M. A. Stewart. 1996. The developmental interface between nature and nurture: A mutual influence model of child antisocial behavior and parent behaviors. *Developmental Psychology,* 32(4): 574–589.

Gifford, R., and T. M. Gallagher. 1985. Sociability: Personality, social context, and physical setting. *Journal of Personality and Social Psychology,* 48(4): 1015–1023.

Gjerde, P. F., J. Block, and J. H. Block. 1991. The preschool family context of 18 year olds with depressive symptoms: A prospective study. *Journal of Research on Adolescence,* 1(1): 63–92.

Glaser, R., S. Kennedy, W. P. Lafuse, R. H. Bonneau, C. Speicher, J. Hillhouse, and J. K. Kiecolt-Glaser. 1990. Psychological stress-induced modulation of interleukin 2 receptor gene expression and interleukin 2 production in peripheral blood leukocytes. *Archives of General Psychiatry,* 47: 707–712.

Glyshaw, K., L. H. Cohen, and L. C. Towbes. 1989. Coping strategies and psychological distress: Prospective analyses of early and middle adolescence. *American Journal of Community Psychology,* 17(5): 607–623.

Goffman, E. 1974. *Frame Analysis.* Cambridge, Mass.: Harvard University Press.

Gonzalez, S., P. Steinglass, and D. Reiss. 1989. Putting the illness in its place: Discussion groups for families with chronic medical illnesses. *Family Process,* 28: 69–97.

Goodman, R., and J. Stevenson. 1991. Parental criticism and warmth towards unrecognized monozygotic twins. *Behavior and Brain Sciences,* 14: 394–395.

Gordon, R. S. 1983. An operational classification of disease prevention. *Public Health Reports,* 98: 107–109.

Gorsuch, R. L. 1983. *Factor Analysis,* 2nd ed. Hillsdale, N.J.: Lawrence Erlbaum.

Gottman, J. M. 1979. *Marital Interaction.* N.Y.: Academic Press.

———. 1994. *What Predicts Divorce?* Hillsdale, N.J.: Lawrence Erlbaum.

Gough, H. G. 1966. Appraisal of social maturity by means of the CPI. *Journal of Abnormal Psychology,* 77: 236–241.

Graham, P., and J. Stevenson. 1985. A twin study of genetic influences on behavioral deviance. *Journal of the American Academy of Child Psychiatry,* 24(1): 33–41.

Grant, S. G. N., T. J. O'Dell, K. A. Karl, P. L. Stein, P. Soriano, and E. R. Kandel. 1992. Impaired long-term potentiation, spatial learning, and hippocampal development in *fyn* mutant mice. *Science,* 258: 1903–1910.

Graziano, W. G., C. Leone, L. M. Musser, and G. J. Lautenschlager. 1987. Self monitoring in children: A differential approach to social development. *Developmental Psychology,* 23(4): 571–576.

Grilo, C. M., and M. F. Pogue-Geile. 1991. The nature of environmental influences on weight and obesity: A behavioral genetic analysis. *Psychological Bulletin,* 110: 520–537.

Grotevant, H. D., and C. R. Cooper. 1985. Patterns of interaction in family relationships and the development of identity exploration in adolescence. *Child Development,* 56: 415–428.

———. 1986. Individuation in family relationships. A perspective on individual differences in the development of identity and role-taking skills in adolescence. *Human Development,* 29: 82–100.

Gruenberg, E. M. 1967. The social breakdown syndrome. *American Journal of Psychiatry,* 123(12): 1481–1489.

Grusec, J., and J. J. Goodnow. 1994. Impact of parental discipline methods on the child's internalization of values: A reconceptualization of current points of view. *Developmental Psychology,* 30(1): 4–19.

Gunnar, M. R., F. L. Porter, C. M. Wolf, J. Rigatuso, and M. C. Larson. 1995. Neonatal stress reactivity: Predictions to later emotional temperament. *Child Development,* 66: 1–13.

Gustafson, G. E., and J. A. Green. 1994. Robustness of individual identity in the cries of human infants. *Developmental Psychobiology,* 27(1): 1–9.

Gustafson, G. E., and K. L. Harris. 1990. Women's response to young infants' cries. *Developmental Psychology,* 26(1): 144–152.

Hall, G. S. 1904. *Adolescence: Its Psychology and Its Relations to Physiology, Anthropology, Sociology, Sex, Crime, Religion and Education.* N.Y.: Appleton.

Harter, S. 1982. The perceived competence scale for children. *Child Development,* 53: 87–97.

———. 1988. "The self-perception profile for adolescence." Denver: Unpublished manuscript, University of Denver.

———. 1990a. Processes underlying adolescent self concept formation. In R. Montemayor, G. R. Adams, and T. P. Gullotta, eds., *From Childhood to Adolescence. A Transitional Period?* Newbury Park, Calif.: Sage.

———. 1990b. Self and identity development. In S. S. Feldman and G. R. Elliott, eds., *At the Threshold: The Developing Adolescent,* pp. 352–387. Cambridge, Mass.: Harvard University Press.

Hatch, E. E., and M. B. Bracken. 1986. Effect of marijuana use in pregnancy on fetal growth. *American Journal of Epidemiology,* 124: 986–993.

Heath, A. C., K. Berg, L. J. Eaves, M. H. Sclaas, L. A. Corey, J. Sunder, P. Magnus, and W. E. Nance. 1985. Education policy and the heritability of educational attainment. *Nature,* 314(6013): 734–736.

Heath, A. C., R. Jardine, and N. G. Martin. 1989. Interactive effects of genotype and social environment on alcohol consumption in female twins. *Journal of Studies on Alcohol,* 50(1): 38–48.

Heath, A. C., and N. G. Martin. 1991. Intoxication after an acute dose of alcohol: An assessment of its association with alcohol patterns by using twin data. *Alcoholism: Clinical and Experimental Research,* 15(1): 122–128.

———. 1992. Genetic differences in psychomotor performance decrement after alcohol: A multivariate analysis. *Journal of Studies on Alcohol,* 53(3): 262–271.

Henderson, S. H., E. M. Hetherington, D. Mekos, and D. Reiss. 1996. Stress, parenting, and adolescent psychopathology: A within-family perspective. In E. M. Hetherington and E. A. Blechman, eds., *Advances in Family Research Series,* vol. 5., *Stress, Coping and Resiliency in Children and Families,* pp. 39–66. Hillsdale, N.J.: Lawrence Erlbaum.

Hetherington, E. M. 1991. The role of individual differences and family relationships in children's coping with divorce and remarriage. In P. A. Cowan and E. M. Hetherington, eds., *Family Transitions,* pp. 165–194. Hillsdale, N.J.: Lawrence Erlbaum.

Hetherington, E. M., and W. G. Clingempeel. 1992. Coping with marital transitions: A family systems perspective. *Monographs of the Society for Research in Child Development,* 57(2–3): 1–242.

Hirshberg, L. M. 1990. When infants look to their parents: II. Twelve-month-olds' response to conflicting parental emotional signals. *Child Development,* 61(4): 1187–1191.

Hirshfeld, D. R., J. F. Rosenbaum, J. Biederman, E. A. Bolduc, S. V. Faraone, N. Snidman, J. S. Reznick, and J. Kagan. 1992. Stable behavioral inhibition and its association with anxiety disorder. *Journal of the American Academy of Child and Adolescent Psychiatry,* 31(1): 103–111.

Horowitz, M. J. 1989. Relationship schema formulation: Role-relationship models and intrapsychic conflict. *Psychiatry,* 52(3): 260–274.

Horowitz, M. J., and T. D. Eells. 1993. Case formulations using role-relationship model configurations: A reliability study. *Psychotherapy Research,* 3(1): 57–68.

Horowitz, M. J., T. Eells, J. Singer, and P. Salovey. 1995. Role-relationship models for case formulation. *Archives of General Psychiatry,* 52(8): 625–632.

Hyman, S. E., and E. J. Nestler. 1993. *The Molecular Foundations of Psychiatry.* Washington, D.C.: American Psychiatry Press.

Isberg, R. S., S. T. Hauser, A. M. Jacobson, S. I. Powers, G. Noam, B. Weiss-Perry, and D. Follansbee. 1989. Parental contexts of adolescent self-esteem: A developmental perspective. *Journal of Youth and Adolescence,* 18(1): 1–23.

Jackson, D. 1960. A critique of the literature on the genetics of schizophrenia. In D. Jackson, ed., *The Etiology of Schizophrenia,* pp. 37–90. N.Y.: Basic Books.

Jarrett, R. L. 1995. Growing up poor: The family experiences of socially mobile youth in low-income African-American neighborhoods. *Journal of Adolescent Research,* 10(1): 111–135.

Jenkins, J. M., and M. A. Smith. 1990. Factors protecting children living in disharmonious homes. *Journal of the American Academy of Child and Adolescent Psychiatry,* 29: 60–69.

Jenkins, P. H. 1995. School delinquency and school commitment. *Sociology of Education,* 68(3): 221–239.

Kagan, J., D. Arcus, and N. Snidman. 1993. The idea of temperament: Where do we go from here? In R. Plomin and G. E. McClearn, eds., *Nature, Nurture and Psychology.* Washington, D.C.: American Psychological Association.

Kandel, D. B. 1978. Homophily, selection and socialization in adolescent friendships. *American Journal of Sociology,* 84: 427–436.

Kandel, D. B., and M. Davies. 1986. Adult sequelae of adolescent depressive symptoms. *Archives of General Psychiatry,* 43: 255–262.

Kandel, D. B., M. Davies, D. Karus, and K. Yamaguchi. 1986. The consequences in young adulthood of adolescent drug involvement. *Archives of General Psychiatry,* 43: 746–754.

Kellam, S. G., and G. W. Rebok. 1992. Building developmental and etiological theory through epidemiologically based preventive intervention trials. In J. McCord and R. E. Temblay, eds., *Preventing Antisocial Behavior,* pp. 162–195. N.Y.: Guilford.

Kellam, S. G., G. W. Rebok, N. Ialongo, and L. S. Mayer. 1994. The course and malleability of aggressive behavior from early first grade into middle school: Results of a developmental epidemiologically based preventive trial. *Journal of Child Psychiatry and Psychology,* 35(2): 259–281.

Kellam, S. G., G. W. Rebok, L. S. Mayer, N. Ialongo, and C. Kalodner. 1994. Depressive symptoms over first grade and their response to a developmental epidemiologically based preventive trial aimed at improving achievement. *Development and Psychopathology,* 6: 463–481.

Kelly, D. H. 1975. Tracking and its impact upon self-esteem: A neglected dimension. *Education,* 96(1): 2–9.

Kendler, K. S. 1996. Parenting: A genetic-epidemiologic perspective. *American Journal of Psychiatry,* 153(1): 11–20.

Kendler, K. S., R. C. Kessler, E. E. Walters, C. MacLean, M. C. Neale, A. C. Heath, and L. J. Eaves. 1995. Stressful life events, genetic liability and onset of an episode of major depression in women. *American Journal of Psychiatry,* 152(6): 833–842.

Klinnert, M. D., R. N. Emde, P. Butterfield, and J. J. Campos. 1986. Social referencing: The infant's use of emotional signals from a friendly adult with mother present. *Developmental Psychology,* 22(4): 427–432.

Kovacs, M. 1985. The children's depression inventory (CDI). *Psychopharmacology Bulletin,* 21: 995–998.

Kramer, L., and J. M. Gottman. 1992. Becoming a sibling: "With a little help from my friends." *Developmental Psychology,* 28(4): 685–699.

Lee, C. L., and J. E. Bates. 1995. Mother-child interaction at two years and perceived difficult temperament. *Child Development,* 56: 1314–1325.

Lerner, R. M., J. V. Lerner, and J. Tubman. 1989. Organismic and contextual bases of development in adolescence. In G. R. Adams, R. Montemayor, and T. P. Gullota, eds., *Biology of Adolescent Behavior and Development.* Newbury Park, Calif.: Sage.

Levitt, M. J., N. Guacci-Franco, and J. L. Levitt. 1993. Convoys of social support in childhood and early adolescence: Structure and function. *Developmental Psychology,* 29(5): 811–818.

Loehlin, J. C. 1992a. *Genes and Environment in Personality Development.* Newbury Park, Calif.: Sage.

———. 1992b. *Latent Variable Models: An Introduction to Factor, Path, and Structural Analysis,* 2nd ed. Hillsdale, N.J.: Lawrence Erlbaum.

———. 1996. The Cholesky approach: A cautionary note. *Behavior Genetics,* 26(1): 65–69.

Loehlin, J. C., J. M. Horn, and L. Willerman. 1994. Differential inheritance of mental abilities in the Texas Adoption Project. *Intelligence,* 19: 325–336.

Loehlin, J. C., and R. C. Nichols. 1976. *Heredity, Environment and Personality.* Austin, Tex.: University of Texas Press.

Loevinger, J. 1979. *Ego Development: Concepts and Theories.* San Francisco: Jossey-Bass.

Lyons, M. J., W. R. True, S. A. Eisen, J. Goldberg, J. M. Meyer, S. V. Faraone, L. J. Eaves, and M. T. Tsuang. 1995. Differential heritability of adult and juvenile antisocial traits. *Archives of General Psychiatry,* 52: 906–915.

Lyons-Ruth, K., L. Alpern, and B. Repacholi. 1993. Disorganized infant attachment classification and maternal psychosocial problems as predictors of hostile-aggressive behavior in the preschool classroom. *Child Development,* 64: 575–585.

McCown, W. G., J. L. Johnson, and S. H. Austin. 1985. Patterns of facial affect recognition errors in delinquent adolescent males. *Developmental Psychology,* 21(2): 338–349.

MacGregor, S. N., L. G. Keith, I. J. Chasnoff, M. A. Rosner, G. M. Chisum, P. Shaw, and J. P. Minogue. 1987. Cocaine use during pregnancy: Adverse perinatal outcome. *American Journal of Obstetrics and Gynecology,* 157(3): 686–690.

McGue, M., S. Bacon, and D. T. Lykken. 1993. Personality stability and change in early adulthood: A behavioral genetic analysis. *Developmental Psychology,* 29(1): 96–109.

McGue, M., and T. J. Bouchard, Jr. 1984. Adjustment of twin data for the effects of age and sex. *Behavior Genetics,* 14: 325–343.

McGue, M., T. J. Bouchard, Jr., W. G. Iacono, and D. T. Lykken. 1993. Behavioral genetic of cognitive ability: A life span perspective. In R. Plomin and G. E. McClearn, eds., *Nature, Nurture and Psychology.* Washington, D.C.: American Psychological Assocation.

McGue, M., and D. T. Lykken. 1992. Genetic influence on risk of divorce. *Psychological Science,* 3(6): 368–373.

McGuire, S., and J. Dunn. 1994. Nonshared environment in middle childhood. In J. C. DeFries, R. Plomin, and D. W. Fulker, eds., *Nature and Nurture during Middle Childhood.* Oxford: Blackwell.

McGuire, S., J. M. Neiderhiser, D. Reiss, E. M. Hetherington, and R. Plomin. 1994. Genetic and environmental influences on perceptions of self worth and competence in adolescence: A study of twins, full siblings and step siblings. *Child Development,* 65(3): 785–799.

McHale, S. M., W. T. Bartko, A. C. Crouter, and M. Perry-Jenkins. 1990. Children's housework and psychosocial functioning: The mediating effects of parents' sex role behavior and attitudes. *Child Development,* 61: 1413–1426.

McHale, S. M., and A. C. Crouter. 1996. The family contexts of children's sibling relationships. In G. H. Brody, ed., *Sibling Relationships and Their Consequences,* pp. 173–196. Norwood, N.J.: Ablex.

McHale, S., and T. M. Pawletko. 1992. Differential treatment of siblings in two family contexts. *Child Development,* 63(1): 68–81.

Manke, B., S. McGuire, D. Reiss, E. M. Hetherington, and R. Plomin. 1995. Genetic contributions to adolescents' extrafamilial social interactions: Teachers, best friends, and peers. *Social Development,* 4(3): 238–256.

Marcia, J. E. 1966. Development and validation of ego identity status. *Journal of Personality and Social Psychology,* 3: 551–558.

Marjoribanks, K. 1988. Cognitive and environmental correlates of adolescents' achievement ambitions: Family-group differences. *Alberta Journal of Educational Research,* 34(2): 166–178.

Marsh, H. W. 1990. Influences of internal and external frames of reference on the formation of math and English self-concepts. *Journal of Educational Psychology,* 82(1): 107–116.

Martin, A. O., T. W. Kurezynski, and A. G. Steinberg. 1973. Familial studies of medical and anthropometric variables in a human isolate. *American Journal of Human Genetics,* 25: 581–593.

Martin, N. G. 1975. The inheritance of scholastic abilities in a sample of twins. *Annals of Human Genetics, London,* 39: 219–229.

Mink, I. T., and K. Nihira. 1986. Family life-styles and child behaviors: A study of direction of effects. *Developmental Psychology,* 22(5): 610–616.

Mishler, E. G. 1986. *Research Interviewing.* Cambridge, Mass.: Harvard University Press.

Modell, J., and M. Goodman. 1990. Historical perspectives. In S. S. Feldman and G. R. Elliott, eds., *At the Threshold: The Developing Adolescent,* pp. 93–122. Cambridge, Mass.: Harvard University Press.

Moffitt, T. E. 1993. Adolescence-limited and life-course–persistent antisocial behavior: A developmental taxonomy. *Psychological Review,* 100(4): 674–701.

Moos, R. H., and B. S. Moos. 1981. *Family Environment Scale.* Palo Alto, Calif.: Consulting Psychologists Press.

Mueller, K. E., and W. G. Powers. 1990. Parent-child sexual discussion: Perceived communicator style and subsequent behavior. *Adolescence,* 25(98): 469–482.

Mueller, W. H. 1977. Sibling correlations in growth and adult morphology in rural Columbian populations. *Annals of Human Biology,* 4: 133–142.

Mueller, W. H., and R. M. Malina. 1980. Genetic and environmental influences on growth of Philadelphia black and white schoolchildren. *Annals of Human Biology,* 7: 441–448.

Mulaik, S. A., L. R. James, J. V. Alstine, N. Bennett, S. Lind, and C. D. Stilwell. 1989. Evaluation of goodness-of-fit indices for structural equation models. *Psychological Bulletin,* 105: 430–445.

Neale, M. C., and L. R. Cardon. 1992. *Methodology for Genetic Studies of Twins and Families.* Boston: Kluwer Academic Publishers.

Neiderhiser, J., and S. McGuire. 1994. Competence during middle childhood. In J. C. DeFries, R. Plomin, and D. W. Fulker, eds., *Nature and Nurture during Middle Childhood,* pp. 141–151. Oxford: Blackwell.

Neiderhiser, J., S. McGuire, R. Plomin, E. M. Hetherington, and D. Reiss. (unpublished). Teacher ratings of cognitive and social competence of adolescents: Genetic as well as environmental influences.

Neiderhiser, J. M., T. G. O'Connor, H. M. Chipeur, D. Reiss, E. M. Hetherington, and R. Plomin (unpublished). Adding stepfamilies to genetic designs.

Neiderhiser, J. M., and A. Pike. 1995. Mediation of adolescents' perception on the relationship between parenting and adolescent adjustment: Genetic and environmental influences (abstract). *Behavior Genetics,* 25(3): 281.

Neiderhiser, J., D. Reiss, and E. M. Hetherington. 1996. Genetically informative designs for distinguishing developmental pathways during adolescence: Responsible and antisocial behavior. *Development and Psychopathology,* 8: 779–791.

Neiderhiser, J. M., D. Reiss, E. M. Hetherington, and R. Plomin. 1999. Relationships between parenting and adolescent adjustment over time: Genetic and environmental contributions. *Developmental Psychology,* 35(3): 680–692.

Nichols, R. C. 1978. Heredity and environment: Major findings from twin studies of ability, personality, and interests. *Homo,* 29: 158–173.

Nichols, R. C., and W. C. Bilbro, Jr. 1966. The diagnosis of twin zygosity. *Acta Genetica,* 16: 265–275.

O'Connor, T. G., E. M. Hetherington, D. Reiss, and R. Plomin. 1995. A twin-sibling study of observed parent-adolescent interactions. *Child Development,* 66(3): 812–829.

O'Connor, T. G., S. McGuire, D. Reiss, E. M. Hetherington, and R. Plomin. 1998. Co-occurrence of depressive symptoms and antisocial behavior in adolescence: A common genetic liability. *Journal of Abnormal Psychology,* 107: 27–37.

O'Connor, T. G., J. M. Neiderhiser, D. Reiss, E. M. Hetherington, and R. Plomin. 1998. Genetic contributions to continuity, change, and co-occurrence of antisocial and depressive symptoms in adolescence. *Journal of Child Psychology and Psychiatry and Allied Disciplines,* 39: 323–336.

O'Hara, B. F., K. A. Young, F. L. Watson, H. C. Heller, and T. S. Kilduff. 1993. Immediate early gene expression in brain during sleep deprivation. *Sleep,* 16(1): 1–7.

Olson, D. H., H. McCubbin, H. L. Barnes, A. S. Larsen, M. J. Muxen, and M. A. Wilson. 1989. *Families: What Makes Them Work.* Beverly Hills, Calif.: Sage.

Ordnuff, S. R., R. N. Freedenfeld, R. M. Kelsey, and J. W. Critelli. 1994. Object relations of sexually abused female subjects: A TAT analysis. *Journal of Personality Assessment,* 63(2): 223–238.

Palazolli, M. S., L. Boscolo, G. Cecchin, and G. Prata. 1978. *Paradox and Counterparadox.* N.Y.: Jason Aronson.

Paluszny, M., M. L. Selzer, A. Vinokur, and L. Lewandowski. 1977. Twin relationships and depression. *American Journal of Psychiatry,* 134(9): 988–990.

Patterson, G. 1982. *Coercive Family Process: A Social Learning Approach.* Eugene, Ore.: Castalia.

———. 1985. Family process: Loops, levels and linkages. In *SRCD Study Group on "Interacting Systems in Human Development."* Cornell University, Ithaca, N.Y.

———. 1989. A developmental perspective on antisocial behavior. *American Psychologist,* 44: 329–335.

———. 1993. Orderly change in a stable world: The antisocial trait as a chimera. *Journal of Consulting and Clinical Psychology,* 61(6): 911–919.

Patterson, G. R., and D. M. Capaldi. 1991. Antisocial parents: Unskilled and vulnerable. In P. A. Cowan and M. Hetherington, eds., *Family Transitions,* pp. 195–218. Hillsdale, N.J.: Lawrence Erlbaum.

Patterson, G. R., L. Crosby, and S. Vuchinich. 1992. Predicting risk for early police arrest. *Journal of Quantitative Criminology,* 8: 333–355.

Patterson, G. R., J. B. Reid, and T. J. Dishion. 1992. *A Social Learning Approach,* vol. 4: *Antisocial Boys.* Eugene, Ore.: Castalia.

Perusse, D., M. C. Neale, A. C. Heath, and L. J. Eaves. 1994. Human parental behavior: Evidence for genetic influence and potential implication for gene-culture transmission. *Behavior Genetics,* 24(4): 327–335.

Peterson, J. L., and N. Zill. 1986. Marital disruption, parent-child relationships and behavior problems in children. *Journal of Marriage and the Family,* 48(5): 295–307.

Petrill, S. A., and L. A. Thompson. 1993. The phenotypic and genetic relationships among measures of cognitive ability, temperament and scholastic achievement. *Behavior Genetics,* 23: 511–518.

Pike, A., S. McGuire, E. M. Hetherington, D. Reiss, and R. Plomin. 1996. Family environment and adolescent depression and antisocial behavior: A multivariate genetic analysis. *Developmental Psychology,* 32(4): 590–603.

Pike, A., D. Reiss, E. M. Hetherington, and R. Plomin. 1996. Using MZ differences in the search for nonshared environmental effects. *Journal of Child Psychology and Psychiatry,* 37: 695–704.

Platt, J. E., X. He, D. Tang, J. Slater, and M. Goldstein. 1995. C-fos expression in vivo in human lymphocytes in response to stress. *Progress in Neuropsychopharmacology and Biological Psychiatry,* 19(1): 65–74.

Plomin, R. 1986. *Development, Genetics and Psychology.* Hillsdale, N.J.: Lawrence Erlbaum.

———. 1994. *Genetics and Experience: The Interplay between Nature and Nurture.* Newbury Park, Calif.: Sage.

Plomin, R., H. H. Chipuer, and J. Neiderhiser. 1994. Behavioral genetic evidence for the importance of the nonshared environment. In E. M. Hetherington, D. Reiss, and R. Plomin, eds., *Separate Social Worlds of Siblings: Impact of the Nonshared Environment on Development,* pp. 1–31. Hillsdale, N.J.: Lawrence Erlbaum.

Plomin, R., and D. Daniels. 1987. Why are children in the same family so different from one another? *Behavioral and Brain Sciences,* 10: 1–16.

Plomin, R., J. C. DeFries, and D. W. Fulker. 1988. *Nature and Nurture during Infancy and Early Childhood.* Cambridge, England: Cambridge University Press.

Plomin, R., J. C. DeFries, and J. C. Loehlin. 1977. Genotype-environment interaction and correlation in the analysis of human behavior. *Psychological Bulletin, 84:* 309–322.

Plomin, R., J. C. DeFries, and G. E. McClearn. 1990. *Behavioral Genetics: A Primer,* 2nd ed. N.Y.: W. H. Freeman.

Plomin, R., J. C. DeFries, G. E. McClearn, and M. Rutter. 1997. *Behavioral Genetics.* N.Y.: W. H. Freeman.

Plomin, R., R. N. Emde, J. M. Braungart, J. Campos, R. Corley, D. W. Fulker, J. Kagan, J. S. Reznick, J. Robinson, C. Zahn-Waxler, and J. C. DeFries. 1993. Genetic change and continuity from fourteen to twenty months: The MacArthur Longitudinal Twin Study. *Child Development, 64:* 1354–1376.

Plomin, R., J. C. Loehlin, and J. C. DeFries. 1985. Genetic and environmental components of "environmental" influences. *Developmental Psychology,* 21(3): 391–402.

Plomin, R., G. E. McClearn, N. L. Pedersen, J. R. Nesselroade, and C. S. Bergeman. 1988. Genetic influence on childhood family environment perceived retrospectively from the last half of the life span. *Developmental Psychology,* 24(5): 738–745.

Plomin, R., N. L. Pedersen, P. Lichtenstein, and G. E. McClearn. 1994. Variability and stability in cognitive abilities are largely genetic later in life. *Behavior Genetics,* 24(3): 207–215.

Plomin, R., D. Reiss, E. M. Hetherington, and G. Howe. 1994. Nature and nurture: Genetic influence on measures of family environment. *Developmental Psychology,* 30(1): 32–43.

Radloff, L. S. 1977. The CES-D scale: A self report depression scale for research in the general population. *Applied Psychological Measurement,* 1: 385–401.

Reiss, D. 1967. Individual thinking and family interaction: An introduction to an experimental study of problem solving in families of normals, character disorders and schizophrenics. *Archives of General Psychiatry,* 16: 80–93.

———. 1981. *The Family's Construction of Reality.* Cambridge, Mass.: Harvard University Press.

———. 1982. The working family: A researcher's view of health in the household. *American Journal of Psychiatry,* 139: 1412–1420.

Reiss, D., E. M. Hetherington, R. Plomin, G. Howe, S. Simmens, S. Henderson, T. O'Conner, D. Bussell, E. Anderson, and T. Law. 1995. Genetic questions for environmental studies: Differential parenting of siblings and its association with depression and antisocial behavior in adolescents. *Archives of General Psychiatry,* 52: 925–936.

Reiss, D., and D. Klein. 1987. Paradigm and pathogenesis: A family-centered approach to problems of etiology and treatment of psychiatric disorders. In T. Jacob, ed., *Family Interaction and Psychopathology: Theories, Methods and Findings.* N.Y.: Plenum Publishing.

Reiss, D., and M. E. Oliveri. 1983. The family's construction of social reality and its ties to its kin network: An exploration of causal direction. *Journal of Marriage and the Family,* 45: 81–91.

Reiss, D., M. E. Oliveri, and K. Curd. 1983. Family paradigm and adolescent social behavior. In H. D. Grotevant and C. R. Cooper, eds., *New Directions for Child Development,* 22: 77–92.

Reiss, D., P. Steinglass, and G. W. Howe. 1993. The family's organization around the illness. In R. E. Cole and D. Reiss, eds., *How do families cope with chronic illness?* Hillsdale, N.J.: Lawrence Erlbaum.

Rende, R., R. Plomin, D. Reiss, and E. M. Hetherington. 1993. Genetic and environmental influences on depressive symptomatology in adolescence: Individual differences and extreme scores. *Journal of Child Psychiatry and Psychology,* 34(8): 1387–1398.

Rende, R. D., C. L. Slomkowski, C. Stocker, D. W. Fulker, and R. Plomin. 1992. Genetic and environmental influences on maternal and sibling interaction in middle childhood: A sibling adoption study. *Developmental Psychology,* 28: 484–490.

Rhea, S.-A., and R. P. Corley. 1994. Applied issues. In J. C. DeFries, R. Plomin, and D. W. Fulker, eds., *Nature and Nurture during Middle Childhood,* pp. 295–309. Oxford: Blackwell.

Rose, R. J., and W. B. Ditto. 1983. A developmental-genetic analysis of common fears from early adolescence to early adulthood. *Child Development,* 54: 361–368.

Rowe, D. C. 1981. Environmental and genetic influences on dimensions of perceived parenting: A twin study. *Developmental Psychology,* 17: 203–208.

———. 1983a. A biometrical analysis of perceptions of family environment: A study of twin and singleton sibling kinships. *Child Development,* 54: 416–423.

———. 1983b. Biometrical genetic models of self-reported delinquent behavior: Twin study. *Behavior Genetics,* 13: 473–489.

———. 1994. *The Limits of Family Influence: Genes, Experience and Behavior.* N.Y.: Guilford Press.

Rowe, D. C., S. Calloe, S. G. Harmon-Lasoya, and H. H. Goldsmith. 1992. The transmission of parenting behavior: Rearing or genetic. *Behavior Genetics,* 22(6): 750.

Rutter, M. 1979. *Changing Youth in a Changing Society: Patterns of Adolescent Development and Disorder.* Cambridge, Mass.: Harvard University Press.

Sack, W. H., G. N. Clarke, R. Kinney, and G. Belestos. 1995. The Khmer adolescent project: II. Functional capacities in two generations of Cambodian refugees. *Journal of Nervous and Mental Disease,* 183(3): 177–181.

Scarr, S. 1992. Developmental theories for the 1990s: Development and individual differences. *Child Development,* 63: 1–19.

Scarr, S., and L. Carter-Saltzman. 1979. Twin method: Defense of a critical assumption. *Behavior Genetics,* 9(6): 527–542.

Scarr, S., and K. McCartney. 1983. How people make their own environments: A theory of genotype-environment effects. *Child Development,* 54: 424–435.

Scarr, S., P. I. Webber, R. A. Weinberg, and M. A. Wittig. 1981. Personality resemblance among adolescents and their parents in biologically related and adoptive families. *Journal of Personality and Social Psychology,* 40: 885–898.

Scarr, S., and R. A. Weinberg. 1978. The influence of "family background" on intellectual attainment. *American Sociological Review,* 43: 674–692.

———. 1983. The Minnesota Adoption Studies: Genetic differences and malleability. *Child Development,* 54: 260–267.

Schachar, R., E. Taylor, M. Wieselberg, G. Thorley, and M. Rutter. 1987. Changes in family function and relationships in children who respond to methylphenidate. *Journal of the American Academy of Child and Adolescent Psychiatry,* 26(5): 728–732.

Schaefer, E., and M. Edgerton. 1981. *The Sibling Inventory of Behavior.* Chapel Hill: The University of North Carolina Press.

Schukit, M. A., D. A. Goodwin, and G. Winokur. 1972. A study of alcoholism in half siblings. *American Journal of Psychiatry,* 128(9): 1132–1136.

Segal, N. L., S. M. Wilson, T. J. Bouchard, and D. G. Gitlin. 1995. Comparative grief experiences of bereaved twins and other bereaved relatives. *Personality and Individual Differences,* 18(4): 511–524.

Seligman, M. 1988. Psychotherapy with siblings of disabled children. In M. D. Kahn and K. G. Lewis, eds., *Siblings in Therapy: Life Span and Clinical Issues,* pp. 167–189. N.Y.: W. W. Norton.

Sigafoos, A., C. Feinstein, M. Damond, and D. Reiss. 1988. The measurement of behavioral autonomy in adolescence: A preliminary psychometric study of the autonomous functioning checklist. *Adolescent Psychiatry,* 15: 432–462.

Sigafoos, A., D. Reiss, J. Rich, and E. Douglas. 1985. Pragmatics in the measurement of family functioning: An interpretive framework for methodology. *Family Process,* 24: 189–203.

Silberg, J. L., M. T. Erickson, J. M. Meyer, L. J. Eaves, M. L. Rutter, and J. K. Hewitt. 1994. The application of structural equation modeling to maternal ratings of twins' behavioral and emotional problems. *Journal of Consulting and Clinical Psychology,* 62(2): 510–521.

Sim, H., and S. Vuchinich. 1996. The declining effects of family stressors on antisocial behavior from childhood to adolescence and early adulthood. *Journal of Family Issues,* 17(5): 408–427.

Slavin, R. E. 1987. Ability grouping and student achievement in elementary schools: A best-evidence synthesis. *Review of Educational Research,* 57(3): 293–336.

Snyder, J., T. J. Dishion, and G. R. Patterson. 1986. Determinants and consequences of associating with deviant peers during preadolescence and adolescence. *Journal of Early Adolescence,* 6(1): 29–43.

Sorce, J. F., R. N. Emde, J. J. Campos, and M. D. Klinnert. 1985. Maternal emotional signaling: Its effect on the visual cliff. *Developmental Psychology,* 21(1): 195–200.

Sorensen, A. B., and M. T. Hallinan. 1986. Effects of ability grouping on growth in academic achievement. *American Educational Research Journal,* 23(4): 519–542.

Spanier, G. B. 1976. Measuring dyadic adjustment: New scales for assessing the quality of marriage and similar dyads. *Journal of Marriage and the Family,* 38: 15–28.

Sroufe, J. W. 1991. Assessment of parent-adolescent relationships: Implications for adolescent development. *Journal of Family Psychology,* 5(1): 21–45.

Steiger, J. H. 1990. Structural model evaluation and modification: An interval estimation approach. *Multivariate Behavioral Research,* 25: 173–180.

Stein, A., D. H. Gath, J. Butcher, and A. Bond. 1991. The relationship between postnatal depression and mother-child interaction. *British Journal of Psychiatry,* 158: 46–52.

Stein, D. M., and P. Reichert. 1990. Extreme dieting behaviors in early adolescence. *Journal of Early Adolescence,* 10(2): 108–121.

Steinberg, L. 1987. Impact of puberty on family relations: Effects of pubertal status and pubertal timing. *Developmental Psychology,* 23(3): 451–460.

———. 1989. Pubertal-maturation and parent-adolescent distance: An evolutionary perspective. In G. R. Adams, R. Montemayor, and T. P. Gullotta, eds., *Biology of Adolescent Behavior and Development,* pp. 71–97. Newbury Park, Calif.: Sage.

Steinberg, L., S. Lamborn, S. Dornbusch, and N. Darling. 1992. Impact of parenting practices on adolescent achievement: Authoritative parenting, school involvement, and encouragement to succeed. *Child Development,* 63: 1266–1281.

Stevenson, J., and P. Graham. 1988. Behavioral deviance in 13-year-old twins: An item analysis. *Journal of the American Academy of Child and Adolescent Psychiatry,* 27(6): 791–797.

Stewart, R. B. 1983. Sibling attachment relationships: Child-infant interaction in the strange situation. *Developmental Psychology,* 19(2): 192–199.

Stocker, C., J. Dunn, and R. Plomin. 1989. Sibling relationships: Links with child temperament, maternal behavior, and family structure. *Child Development,* 60: 715–727.

Straus, M. A. 1979. Measuring intrafamily violence and conflict: The conflict tactics (CT) scale. *Journal of Marriage and the Family,* 41: 75–85.

Summers, F., and F. Walsh. 1980. Schizophrenic and sibling: A comparison of parental relationships. *American Journal of Family Therapy,* 8(3): 45–52.

Tambs, K., J. M. Sundet, and K. Berg. 1985. Cotwin closeness in monozygotic and dizygotic twins: A biasing factor in IQ heritability analysis? *Acta Geneticae Medicae et Gemellologiae: Twin Research,* 34(1–2): 33–39.

Tanaka, J. S. 1993. Multifaceted conceptions of fit in structural equation models. In K. A. Bollen and J. S. Long, eds., *Testing Structural Equation Models.* Newbury Park, Calif.: Sage.

Tejerina-Allen, M., B. Wagner, and P. Cohen. 1994. A comparison of across-family and within-family parenting predictors of adolescent psychopathology and suicidal ideation. In E. M. Hetherington, D. Reiss, and R. Plomin, eds., *Separate Social Worlds of Siblings: Impact of the Nonshared Environment on Development*, pp. 143–158. Hillsdale, N.J.: Lawrence Erlbaum.

Teti, D. M., L. A. Bond, and E. D. Gibbs. 1986. Sibling-created experiences: Relationships to birth-spacing and infant cognitive development. *Infant Behavior and Development*, 9(1): 27–42.

Thompson, L. A., D. K. Detterman, and R. Plomin. 1993. Differences in heritability across groups differing in ability, revisited. *Behavior Genetics*, 23: 331–336.

Tienari, P., I. Lahti, A. Sorri, M. Naarla, J. Moring, K.-E. Whalberg, and L. C. Wynne. 1987. The Finnish Adoptive Family Study of Schizophrenia. *Journal of Psychiatric Research*, 21(4): 437–445.

Tienari, P., A. Sorri, and I. Lahti. 1985. Interaction of genetic and psychosocial factors in schizophrenia. *Acta Psychiatric Scandinavica*, 71: 19–30.

Tienari, P., L. C. Wynne, J. Moring, I. Lahti, M. Maarala, A. Sorri, K. E. Wahlberg, O. Saarento, M. Seitamaa, M. Kaleva, and K. Laksy. 1994. The Finnish Adoption Family Study of Schizophrenia: Implications for family research. *British Journal of Psychiatry*, 164: 20–26.

Tremblay, R. E., R. O. Pihl, F. Vitaro, and P. L. Dobkin. 1994. Predicting early onset of male antisocial behavior from preschool behavior. *Archives of General Psychiatry*, 51: 732–739.

Undheim, J. O., and H. Nordvik. 1992. Socio-economic factors and sex differences in an egalitarian educational system: Academic achievement in 16-year-old Norwegian students. *Scandinavian Journal of Educational Research*, 36(2): 87–98.

Vaillant, G. E., and C. O. Vaillant. 1981. Natural history of male psychological health, X: Work as a predictor of positive mental health. *American Journal of Psychiatry*, 138(11): 1433–1440.

van den Boom, D. C. 1994. The influence of temperament and mothering on attachment and exploration: An experimental manipulation of sensitive responsiveness among lower class mothers with irritable infants. *Child Development*, 65: 1457–1477.

van den Boom, D. C., and J. B. Hoeksma. 1994. The effect of infant irritability on mother-infant interaction: A growth curve analysis. *Developmental Psychology*, 30(4): 581–590.

Vernon, P. A., K. L. Jang, H. J. Aitken, and J. M. McCarthy. 1997. Environmental predictors of personality differences: A twin and sibling study. *Journal of Personality and Social Psychology*, 72(1): 177–183.

Volling, B., and J. Belsky. 1992. The contribution of mother-child and father-child relationships to the quality of sibling interaction: A longitudinal study. *Child Development*, 63: 1209–1222.

Vuchinich, S., R. E. Emery, and J. Cassidy. 1988. Family members as third parties in dyadic family conflict: Strategies, alliances and outcomes. *Child Development,* 59(5): 1293–1302.

Wachs, T. D. 1992. *The Nature of Nurture.* Newbury Park, Calif.: Sage.

Walker, D., and C. Leister. 1994. Recognition of facial affect cues by adolescents with emotional and behavioral disorders. *Behavioral Disorders,* 19(4): 269–276.

Walsh, F. W. 1978. Concurrent grandparent death and birth of schizophrenic offspring: An intriguing finding. *Family Process,* 14(4): 457–463.

Werner, E. E., and R. S. Smith. 1982. *Vulnerable but Invincible: A Longitudinal Study of Resilient Children and Youth.* N.Y.: McGraw-Hill.

Westen, D., J. Klepser, S. A. Ruffins, M. Silverman, N. Lifton, and J. Bockamp. 1991. Object relations in childhood and adolescence: The development of working representations. *Journal of Consulting and Clinical Psychology,* 59(3): 400–409.

Whitbeck, L. B., R. L. Simons, R. D. Conger, F. O. Lorenz, S. Huck, and G. H. Elder, Jr. 1991. Family economic hardship, parental support and adolescent self esteem. *Social Psychology Quarterly,* 54: 353–363.

Wierzbicki, M. 1987. Similarity of monozygotic and dizygotic child twins in level and lability of subclinically depressed mood. *American Journal of Orthopsychiatry,* 57: 33–40.

Wiggers, M., and C. F. Vanlishout. 1985. Development of recognition of emotions: Children's reliance on situational and facial expressive cues. *Developmental Psychology,* 21(2): 338–349.

Willerman, L., J. C. Loehlin, and J. M. Horn. 1992. An adoption and a cross-fostering study of the Minnesota Multiphasic Personality Inventory (MMPI) psychopathic deviate scale. *Behavior Genetics,* 22(5): 515–529.

Wishart, J. G. 1986. Siblings as models in early infant learning. *Child Development,* 57(5): 1232–1240.

Woollett, A. 1986. The influence of older siblings on the language environment of young children. *British Journal of Developmental Psychology,* 4(3): 235–245.

Zall, D. S. 1994. The long term effects of childhood bereavement: Impact on roles as mothers. *Omega: Journal of Death and Dying,* 29(3): 219–230.

Zeskind, P. S. 1983. Cross-cultural differences in maternal perceptions of cries of low- and high-risk infants. *Child Development,* 54: 1119–1128.

———. 1987. Adult heart-rate responses to infant cry sounds. *British Journal of Developmental Psychology,* 5: 73–79.

Zill, N. 1985. *Behavior Problems Scale Developed for the 1981 Child Health Supplement to the National Health Interview Survey.* Washington, D.C.: Child Trends, Inc.

Index of Tables and Figures

Tables

2.1 Classification of logical principles of social and genetic analyses of development, 19

5.1 Characteristics of the sample at time 1 and time 2, 111

5.2 Summary of measures of adolescent psychopathology, 114

5.3 Summary of measures of adolescent competence, 116–117

5.4 Summary of measures of family subsystems, 122–125

7.1 Summary of sibling relationships indexed by the specific and cross-correlations of the covariance model, 198

8.1 Genetic and environmental influences on development from earlier to later adolescence, 235

13.1 Sample items from measures of sociability and autonomy, 404–405

A1 Longitudinal phenotypic and sibling correlations for stability in measures of adolescent development, 436

A2 Phenotypic and sibling cross-correlations for stability in family subsystems, 437

A3 Phenotypic and sibling cross-correlations for associations between family relationships and adolescent development at time 1, 438–439

A4 Phenotypic and sibling cross-correlations for associations between family relationships and adolescent development at time 2, 440–441

A5 Phenotypic and sibling cross-correlations for cross-lagged associations between family relationships at time 1 and adolescent development at time 2, 442

A6 Phenotypic and sibling cross-correlations for cross-lagged associations between adolescent development at time 1 and family relationships at time 2, 443

A7 Phenotypic and sibling cross-correlations for interrelationships among measures of adolescent development at time 1, 444

A8 Phenotypic and sibling cross-correlations for interrelationships among family subsystems at time 1, 445

A9 Phenotypic and sibling cross-correlations for interrelationships among family subsystems at time 2, 446–447

B1 Stability in adolescent development from time 1 to time 2: Standardized parameter estimates from bivariate genetic model-fitting, 455

B2 Stability in family environment from time 1 to time 2: Standardized parameter estimates from bivariate genetic model-fitting, 456

B3 Associations between family relationships and adolescent development at time 1: Standardized parameter estimates from bivariate genetic model-fitting, 457–458

B4 Associations between family relationships and adolescent development at time 2: Standardized parameter estimates from bivariate genetic model-fitting, 459–460

B5 Results of genetic cross-lagged analyses examining environment at time 1 and adolescent adjustment at time 2: Fit indexes and standardized maximum-likelihood model-fitting estimates, 461–467

B6 Results of genetic cross-lagged analyses examining adolescent adjustment at time 1 and environment at time 2: Fit indexes and standardized maximum-likelihood model-fitting estimates, 468–474

B7 Specificity in outcome at time 1: Standardized parameter estimates from bivariate genetic model-fitting, 475–476

B8 Specificity in environment at time 1: Standardized parameter estimates from bivariate genetic model-fitting, 477–479

B9 Specificity in environment at time 2: Standardized parameter estimates from bivariate genetic model-fitting, 480

Figures

3.1 Correlations for height in MZ and DZ twins, 47

3.2 Comparison of intraclass correlations across subsamples, 49

3.3 Adoption study correlations in body weight between parents and their four-year-old children, 60

3.4 Correlations of neuroticism measures between pairs of biological siblings and pairs of adopted siblings, 67

4.1 Heritability of scholastic self-perception, 73

4.2 Heritability of general self-worth, 76

4.3 Differential heritability of self-perceived competence, 77

4.4 Cross-correlations between sociability and social competence, 79

4.5 Genetic and environmental components of the association between sociability and social competence and verbal IQ and scholastic competence, 80

4.6 Parent and child reports of conflict and negativity, 83

4.7 Five developmental pathways, 86

4.8 Genetic and environmental components of the association between mother's negativity and adolescent's antisocial behavior and social responsibility, 89

5.1 Comparison of co-residence among full, half-, and blended sibs, 134

5.2 Co-residence and correlations among measures of adolescent adjustment, 136

5.3 Co-residence and correlations among measures of parenting, 137

5.4 Genetic cascade for height, weight, and vocabulary, 140

5.5 Univariate models comparing twins and steps on measures of height, weight, and vocabulary, 141

7.1 Sibling correlations of parent, child, and observer ratings, 170

7.2 Parental ratings of family and peer relationships, 172

7.3 Correlations among ratings of sibs across time, 173

7.4 Self-report of sibling behavior versus observer report, 174

7.5 Patterns of parent-child and marital relationships, 175

7.6 The covariance model, 181

7.7 Simple correlations between mother's negativity and positivity and adolescent's adjustment, 182

7.8 Simple correlations between father's negativity and positivity and adolescent's adjustment, 184

7.9 Simple correlations between mother's knowledge of adolescent's activities and adolescent's adjustment, 185

7.10 Simple correlations between father's knowledge of adolescent's activities and adolescent's adjustment, 186

7.11 Associations between mother's negativity and areas of adolescent development using the covariance model, 187

7.12 Associations between father's negativity and areas of adolescent development using the covariance model, 188

7.13 Associations between mother's positivity and areas of adolescent development using the covariance model, 189

7.14 Associations between father's positivity and areas of adolescent development using the covariance model, 190

7.15 Associations between mother's knowledge of adolescent's activities and adolescent's adjustment using the covariance model, 191

7.16 Associations between father's knowledge of adolescent's activities and adolescent's adjustment using the covariance model, 192

7.17 Associations between mother's attempted control of adolescent and adolescent's adjustment using the covariance model, 193

7.18 Associations between father's attempted control of adolescent and adolescent's adjustment using the covariance model, 194

7.19 Associations between mother's actual control of adolescent and adolescent's adjustment using the covariance model, 195

7.20 Associations between father's actual control of adolescent and adolescent's adjustment using the covariance model, 196

7.21 Simple correlations between sibling's negativity and positivity and adolescent's adjustment, 200

7.22 Associations between sibling's negativity and adolescent's adjustment using the covariance model, 201

7.23 Associations between sibling's positivity and areas of adolescent development using the covariance model, 202

8.1 Heritability of adjustment, 210

8.2 Data contrived to illustrate genetic influences on the stability of scores of antisocial behavior, 213

8.3 Data contrived to illustrate genetic and environmental effects on stability and change, 214

8.4 Balloon diagram showing results of model testing for factors influencing stability and change, 217

8.5 Cross-correlations for estimating the role of genetic and environmental factors in the change and stability of antisocial behavior, 221

8.6 Balloon diagram showing results of model testing for factors influencing stability and change in antisocial behavior, 223

8.7 Cross-correlations for estimating the role of genetic and environmental factors in the change and stability of depressive symptoms, 224

8.8 Balloon diagram showing results of model testing for factors influencing stability and change in depressive symptoms, 226

8.9 Cross-correlations for estimating genetic and environmental factors in the change and stability of cognitive agency, 227

8.10 Balloon diagram showing results of model testing for factors influencing stability and change in cognitive agency, 228

8.11 Balloon diagram showing results of model testing for factors influencing stability and change in sociability, 230

8.12 Balloon diagram showing results of model testing for factors influencing stability and change in autonomy, 231

8.13 Balloon diagram showing results of model testing for factors influencing stability and change in social responsibility, 233

8.14 Balloon diagram showing results of model testing for factors influencing stability and change in global self-worth, 234

9.1 Shared and nonshared components of adjustment measures, 247

9.2 Intraclass correlations for estimating the role of genetic and environmental influences on measures of mother's and father's negativity, 250

9.3   Heritability and environmentality of mothering, 252

9.4   Heritability and environmentality of fathering, 253

9.5   Heritability of sibling relationships and marital conflict measures, 255

9.6   Intraclass correlations among sibling groups on mother-father conflict regarding children, 256

9.7   Balloon diagram showing results of model testing for factors influencing stability and change in mother's conflict and negativity, 260

9.8   The proportion of stability and change in measures of parenting and sibling relationships accounted for by genetic, shared, and nonshared environmental factors, 261

9.9   Coefficients of stability for all measures of family relationships and all measures of adolescent adjustment, 263

10.1   The effects of family relationships and genetic factors on variations in adolescent adjustment, 271

10.2   Schematic contrast between passive model and evocative/active model, 275

10.3   Two possible evocative/active models accounting for overlapping genetic influences on family processes and adolescent adjustment, 278

10.4   Cross-correlations between mother's and father's negativity toward the adolescent and the sibling's antisocial behavior, 281

10.5   Balloon diagram showing components of the association between mother's negativity and adolescent's antisocial behavior, 284

10.6   Analysis of the association between mother's parenting and all seven domains of adolescent adjustment, 286

10.7   Analysis of the association between father's parenting and all seven domains of adolescent adjustment, 291

10.8   Analysis of the association between sibling interaction and all seven domains of adolescent adjustment, 294

10.9   Components of change in the seven domains of adolescent adjustment, 297

10.10   Components of change in the eight domains of family subsystems, 298

10.11   Analysis of the association between mother's parenting and all seven domains of adolescent adjustment at time 2, 300

10.12   Summary of changes in genetic overlap for mothering, 301

10.13   Analysis of the association between father's parenting and all seven domains of adolescent adjustment at time 2, 302

10.14   Summary of changes in genetic overlap for fathering, 303

10.15   Analysis of the association between sibling relationships and all seven domains of adolescent adjustment at time 2, 304

10.16   Summary of changes in genetic and shared environmental overlap for siblings, 305

11.1   Child-initiated family-effects model, 312

11.2   Alternative sequences of effects for evocative models, 313

11.3   Distinctive genetic influences on family relationships, 315

11.4   Genetic and environmental components of the association between parental monitoring and antisocial behavior, 319

11.5   Child-effects model of parental salvage, 320

11.6   Parent-effects model of parental monitoring and antisocial behavior, 321

11.7   Basic phenotypic cross-lagged correlation, 324

11.8   Partialing out confounding relationships from a cross-lagged association, 325

11.9   Model estimating genetic and environmental components of the cross-lagged association from family process to adolescent adjustment, 327

11.10 Model estimating genetic and environmental components of the cross-lagged association from adolescent adjustment to family process, 329

11.11 Partialed cross-lag associations for mother's parenting, 332

11.12 Partialed cross-lag associations for father's parenting, 333

11.13 Partialed cross-lag associations for sibling relationships, 334

11.14 Genetic and environmental components of $f1 \rightarrow a2$ cross-lags at .15 or higher, 335

11.15 Genetic and environmental components of $a1 \rightarrow f2$ cross-lags at .15 or higher, 336

11.16 Summary of eight $f1 \rightarrow a2$ sequences, 337

11.17 Schematic summary of single clear $a1 \rightarrow f2$ sequence, 340

12.1   Genetic and environmental components of the relationship between antisocial behavior and sociability, 348

12.2   Genetic and environmental components of the relationship between antisocial behavior and social responsibility, 350

12.3   Genetic and environmental components of the relationship between all pairs of domains of adjustment where the correlations equaled or exceeded .20, 352

12.4   Comparison of family-relationship variables, 360

12.5   Correlations among measures of family relationships in early adolescence, 364

12.6   Genetic and environmental components of the covariance between mother's and father's negativity at time 1, 367

12.7   Genetic and environmental components of the covariance between mother's and father's positivity at time 1, 369

12.8   Heritability of association for all family measures at time 1, 370

12.9   Heritability of overlap for changes in relationship subsystems from early to late adolescence, 372

12.10 Preventive intervention as a test of the relationship-code hypothesis, 383

13.1   Environmental influences on adolescent adjustment, 387

13.2   Correlates of sociability and autonomy, 390

13.3   The role of shared environment in associations across family subsystems, 391

13.4 Genetic and environmental components of the association between sibling relationships and adolescent adjustment, 397

13.5 Balance of nonshared influences comparing stable and unstable components, 408

13.6 Comparison of common and unique nonshared influences in the associations among measures of adjustment in earlier adolescence where phenotypic correlations equated or exceeded .20, 409

13.7 Comparison of common and unique nonshared influences in the associations among measures of adjustment in later adolescence where phenotypic correlations equated or exceeded .20, 410

A1 Diagram of a bivariate model, 448

A2 Diagram of a genetic cross-lagged model: family process, time 1→adolescent adjustment, time 2, 449

A3 Diagram of a genetic cross-lagged model: adolescent adjustment, time 1→family process, time 2, 450

Ability groupings, 237–238

Abuse, impact on children and adolescents, 161

Active aggression, 200–201

Active genetic influences, 82, 245; in overlapping genetic effects, 277–279

Adjustment variables, 78–81

Adolescence (Hall), 15

Adolescent adjustment: social relationships and, 3, 9–10; genetic influence and, 3–6; cross-sectional genetic analysis of, 70–81; differential heritability in, 75–78; gene-environment correlation and, 84–91; possible genetically based pathways in, 85–88; parenting and, 183–185, 185–197; sibling relationships and, 199–203, 292–295; change and stability in, 212–234; variation in, 270–271; mother's parenting and, 287–290; father's parenting and, 290–292; and changes in genetic and environmental overlap across time, 296–308; family relationships and, 330–340; environmental influences on, 387–389; influence of siblings on, 397–402; decline of parental influences and, 398–399. *See also* Adolescent development

Adolescent adjustment measures, 71–75; overlap with family systems measures, 285–287; analysis for common and specific genetic influences in, 346–352

Adolescent development: parental responsibility for, 2; issues of failure and, 2–3; tasks of, 26; association studies and, 26–27; nonshared environment and, 155–164, 426; parentification and, 203; genetic and environmental influences on, 235–242; difficulties in studying, 317–318; relationship code hypothesis and, 375–384. *See also* Adolescent adjustment

Adolescent development study: interdisciplinary approach to, 1; notions of social relationships and, 3, 9–10; notions of genetic influence and, 3–6; sample size and characteristics, 6, 110–113; notions of measures in, 6–7; data analysis in, 7–8, 126–132, 429–435; logical principles of research and, 8–9, 18–22; conclusions from, 9–10; nonshared environment concept and, 103–105, 109, 147–149; study design, 103–110; measures of psychopathology in, 113–115; measures of competence in, 115–118; measures of family subsystems, 118–121; data collection procedures, 121, 126; step-family design in, 132–143, 481–482; critical reappraisal of, 417–426; explanation of results, 451–454

Adolescent failure, 2–3

Adolescents: changes across time, 207–211; and genetic influences on family relationships and subsystems, 244–246, 376–377; temperament, family relationships and, 377–378; and nongenetic influences

Adolescents *(continued)*
on family relationships, 393–394; decline of parental influences and, 398–399

Adoption, involvement of the birth mother in, 62–63

Adoption studies: types of, 58–60; selective placement and, 60; criticisms of, 60–63; characteristics of adopting families and, 61–62, 93; prenatal effects and, 62; involvement of the birth mother in adoption and, 62–63; on gene-environment interactions, 92–93; preventive intervention design for, 96, 101, 380–384; nonshared environment and, 151

Adoptive families: characteristics of, 61–62, 93; gene-environment interaction studies and, 93, 95–96

Adults, responses to infant crying, 425. *See also* Parents

Aggression: misreading of emotional cues and, 158; between siblings, 199–201

Alcoholism: genetic experiments on, 100; shared environment and, 152–153

Alzheimer's disease, 220

Antisocial behavior: aversive parenting and, 25–26; association studies and, 29; universal preventive interventions and, 42–43; shared environment concept and, 66; mother's negativity and, 89–90, 287; gene-environment interaction studies on, 92–93; measures of, 115; heritability and, 209–211; change and stability in, 220–223; overlapping genetic and environmental influences with parental negativity, 279–280, 281–285; mother's positivity and, 289; sibling relationships and, 293; sibling positivity and, 295, 338; parental monitoring and, 318–323, 339–340; parental positivity and, 338; father's negativity and, 339; sociability and, 347–349; social responsibility and, 349–351

Association studies: strategies within, 23–25; tasks of adolescent development and, 25–62, 26–27; perspectives on family relationships in, 27–29

Assortative affiliation studies, 32, 35–37

At-risk populations, selective preventive interventions and, 42–43

Autonomy: change and stability in, 229, 232; shared environment and, 238–239, 388, 402–406; mother's positivity and, 289; sibling positivity and, 295; changes in overlapping shared environmental influences across time, 304–306; sociability and, 402–406; measures of, 403; positivity and, 403–406

Backing off behavior, 340, 342

Balloon diagrams, 216–218

Behavioral genetics: notions of environment in, 5–6; fundamental characteristics of genes, 50–51; gene expression and, 51

Behavioral genetics research: psychosocial research and, 13–14, 17–18; early influences on, 15–16; equal-environment assumption and, 16–17; logical principles of research, 18–22; social association studies and, 25; approaches to, 44–45; quantitative, 45–46; sample size in, 46; measures of association, 46–49; inferences about environmental and nongenetic influences, 49, 50; fundamental characteristics of genes and, 50–51; twin studies, 52–55; step-family studies, 55–58; adoption designs, 58–63; nongenetic environment factors and, 63–66; environmentality concept and, 64–65; nonshared environment and, 65–66, 152–153; shared environment concept and, 66; analyzing anonymous and specific independent variables, 66–68; measures of adolescent adjustment, 72; longitudinal studies, 96–99; experimental studies, 99–102. *See also* Cross-sectional genetic studies; Genetic experiments

Behavioral pleiotropism, 204, 205

Behavior Events Inventory, 115

Behavior Problems Index, 115

Biological determinism, 17

Birth mothers, involvement in adoption, 62–63
Bivariate models: description of, 432–433; explanation of results from, 451, 452
Blended sibs, 6; levels of genetic relatedness in step-families, 55, 132; co-residence and, 56, 134, 135; nonshared environment and, 150–151
Blos, P., 15

Care-giving behavior, in sibling relationships, 197–198, 202–203
Care-receiving behavior, in sibling relationships, 197–198, 201–202
Cattell, James, 15
Change: considered in longitudinal genetic studies, 97–98; analytic principles for detecting, 212–220; in adolescent adjustment, factors influencing, 220–234, 239–242; in family relationships, factors influencing, 257–265; in genetic and environmental overlap across time, 296–308; difficulties in studying, 317–318
Child Behavior Checklist, 113
Child Depression Inventory, 115
Child effects, 88
Child-effects model, 277, 343; overview of, 311; cross-lagged analysis and, 330
Child-initiated parent-effects model, 277–279, 312–313; cross-lagged analysis on, 341–342. See also Parent-effects model
Children: psychological development, 161–164; effect of siblings on, 162; ratings of parenting and, 170–171. See also Adolescents; Siblings
Chi-square test, 435
Cholesky factoring, 434
Cliques, 35
Closeness: twins and, 254, 255; in families, 406
Cluster analyses, of family relationships, 174–176
Coders, 120–121
Coefficient of change, 216
Coefficient of stability, 218

Coercive relationships, 29–30
Cognitive agency: notions of genetic influence across time and, 98; heritability of, 209; change and stability in, 225, 227; genetic influences on stability in, 235–236; mother's negativity and, 287–288; mother's positivity and, 289; sibling positivity and, 295; parental negativity and, 339
Competence, measures of, 115–118
Concordances, 46
Conflict: genetic analysis of parent and child reports in, 83–84; reports of siblings and parents on, 169–170, 171; marital, possible genetic influences on, 245–246
Conflict, parental: over siblings, genetic and environmental influences on, 255–257; change and stability in, 259–262, 264
Conflict, sibling: forms of, 197–198; simple correlations to adjustment, 199; specific and cross-correlations to adjustment, 199–203
Conflict avoidance variable, 174–175, 176
Control, parental: specific and cross-correlations to adolescent adjustment, 188–194, 196–197; genetic influences and, 252–254, 370–371; change and stability in, 264–265; possible scenarios for genetic effects across time, 371–375; shared environment and, 394–395. See also Monitoring, parental
Controlled trials, 38–39
Cooperative-behavior games, 42–43
Cooperative-learning curricula, 42–43
Co-residency: step-family model and, 55–57, 133–135; analysis of, 481–482
Correlated error, 180–181
Correlational analyses, 430–431
Correlation coefficient, 47
Correlations: described, 46–49; on shared and nonshared influences of parenting, 178–179. See also Cross-correlations; Cross-lagged correlations; Gene-environment correlation; Intraclass correlations; Specific correlations

Covariance model: described, 179–183; applied to comparisons of parenting and adolescent adjustment, 185–197; identifying forms of sibling relationships with, 198–199; applied to comparisons of sibling relationships and adolescent adjustment, 199–203

Cross-correlations: in multivariate genetic analyses, 78–79; in gene-environment correlation, 89; defined, 166, 179; between parenting and adolescent adjustment, 185–197; applied to sibling relationships, 197–100; possible genetic mechanisms in, 204–205; estimating change across time, 212–213; applied to overlapping genetic and environmental influences, 280–285; description of, 430–431. See also Covariance model

Cross-lagged correlations, 431; of parental monitoring and antisocial behavior, 318–323; analytic principles of, 323–330; partialing out confounding relationships, 324–326; genetic and environmental components of, 326–328, 334–340; reversing, 328–330; of parenting, 331; of sibling relationships, 332–333; conclusions from, 341–342; described, 433–435; explanation of results from, 453–454

Cross-sectional genetic studies: overview of, 69–70; gene-environment correlation and, 70, 84–91; gene-environment interactions and, 70, 91–96; analysis of genetic and environmental influences on adolescent adjustment, 70–81; heritability estimates and, 71; measures of adolescent adjustment, 71–75; differential heritability and, 75–78; adjustment variables, 78–81; role of genetic factors in social relationships, 81–84

Cross-sectional studies, logical principles of, 21

Cross-sibling correlations, 430–431

Crowds, 35

Crying, 425

Curvilinear relationships, 178

Darwin, Charles, 15–16

Data: methods of analysis, 7–8, 126–132, 429–432; using effectively, 129–132; visual presentations of, 131–132; preparation, 429

Delinquency, 152–153

Dependent variables, 20, 48

Depression/Depressive symptoms: universal preventive interventions and, 42–43; longitudinal genetic studies on, 99; measures of, 115; genetic and environmental influences on, 211; change and stability in, 224–225; mother's negativity and, 287; mother's positivity and, 289; sibling relationships and, 293; sibling positivity and, 295; changes in overlapping nonshared environmental influences and, 306–307

Deterrence theory, 35

Developmental change, difficulties in studying, 317–318. See also Adolescent development; Longitudinal studies

Developmentally active model of child adjustment, 87–88

Developmentally evocative model of child adjustment, 87, 88

Developmental theory, relationship code hypothesis and, 375–384

Developmental time, encoding of genetic information and, 363

Differential assortative mating, 138–139

Differential heritability, 75–78

Difficult temperament, 28

Disabled children, families with, 418–419

Disjunction, heritability of, 282–283

Divorce: association studies and, 24–25; moderation concept and, 31

Dizygotic twins, 6, 50, 52

Dreams, 412–414

Dream thoughts, 413

Drinking behavior, 91

Dropouts, from studies, 38

Economic conditions, 392–393

Ego development, 26

Emotional cues: adolescent psychological development and, 156–158; framing concept and, 157–158

Encoding. *See* Family encoding

Environmental echo concept, 402

Environmental influences: insights from behavioral genetics on, 5–6; environmentality concept and, 64–65; gene expression and, 65, 219–220; at early and mid-adolescence, 211; analytic principles for detecting across time, 212–219; on change and stability in adolescent adjustment, 220–234; on family relationships, 246–250, 251, 255, 256–257, 258–265, 266–268; on adolescent adjustment, 387–389; critical reappraisal of, 420–421. *See also* Nonshared environment; Shared environment

Environmental influences, overlapping, 273; possible mechanisms for, 279–280; analytic principles for detecting, 280–285; in measures of family systems and adolescent adjustment, 285–287; between mother's parenting and adolescent adjustment, 287–290; between father's parenting and adolescent adjustment, 290–292; between sibling relationships and adolescent adjustment, 292–295; contrasts between parenting and sibling relationships, 295–296; changes from earlier to later adolescence, 296–308. *See also* Nonshared environment, overlapping; Shared environment, overlapping

Environmentality, 5; description of concept, 64–65; of neuroticism, 67; computing, 452

Environmentality of stability, 213–215

Equal-environment assumption, 16–17, 53–55

Erickson, E. H., 15

Error, correlated, 180–181

Evocative genetic influences, 81, 245; in overlapping genetic effects, 277–279

Experimental studies: logical principles of, 19–20; limitations of, 20–21; psychosocial research and, 22; reassembly of families, 37, 38–39; therapeutic interventions, 37, 39–41; overview of, 37–38; preventive interventions, 41–43. *See also* Natural experiments

Extroversion, 92

Families: disruptions, 24–25; reassembly of, 37, 38–39; adoptive, 61–62; unique social history of siblings and, 160–161, 249; constructions of social reality and, 395–396, 418–419; individuation and, 405; relationship of autonomy, sociability, individuation, and closeness in, 405–406; with disabled children, 418–419; relationship code hypothesis and, 424–426

Family-effects model: overview of, 312–313, 343–344, 386; transmission of genetic information and, 313–315; supporting data and, 344, 375; common and specific genetic influences in, 345, 346–352; nature-versus-nurture debate and, 376. *See also* Parent-effects model; Relationship code

Family encoding: overview of, 354–358; notions of developmental time and, 363; analysis of family relationship measures for, 363–371; family relationship development and, 371–375

Family relationships: association studies and, 27–29; perspectives on schizophrenia, 28; mediating and moderating studies in, 29–31; genetic influences and, 81–84, 244–246, 250–258, 259–266, 358–371; nonshared environment and, 152, 159–163, 164, 248–250, 264, 265; framing concept applied to, 159–160; comparing, between siblings, 169–176; analytic principles for detecting genetic and environmental influences in, 244; adolescents' genes and, 244–246; shared environment and, 246–248, 251, 266–268, 390–396; environmental influences on, 246–250, 251, 255, 256–257, 258–265, 266–268; across time, factors influencing, 250–257; stability and change in, 257–265; adoles-

Family relationships *(continued)*
cent adjustment and, 270–271, 330–340; encoding of genetic information and, 354–358, 371–375; genetic specificity and, 358–371; children's temperament and, 377–378; nongenetic influences of adolescents on, 393–394. *See also* Family subsystems; Parent-child relationships

Family subsystems: relationship code hypothesis and, 9, 376–377, 419–420; association studies and, 27–28; adolescent adjustment and, cross-lagged analysis of, 330–340; genetic specificity and, 362–363, 385; shared environment and, 390–396

Family systems measures, 118–121; overlap with measures of adolescent adjustment, 285–287; genetic specificity in, 358–371; comparisons between, 359–362

Father-child relationships, backing off behavior of, 340, 342. *See also* Knowledge of child's activities, father's; Negativity, father's; Parent-child relationships; Parenting, father's; Positivity, father's

Fetal development, 62

First-grade classrooms, universal preventive interventions and, 42–43

Fit indexes, 435, 454

Framing concept: overview of, 157; adolescent psychological development and, 157–158, 163–164; applied to family relationships, 159–160

Fraternal twins, 6, 50, 52

Freud, Sigmund, 15, 412–414

Friendships, 35

Full-sib step-families: co-residency and, 55, 133–134, 135; levels of genetic relatedness in, 55, 132

Galton, Francis, 15–16, 46

Gender, 108–109

Gene-environment correlation: defined, 5; in cross-sectional genetic studies, 70; assessing, 84–91; possible influence on sib-

ling relationships, 204; parental salvage effect as, 205; relationship code hypothesis and, 379

Gene-environment interactions: in cross-sectional genetic studies, 70, 91–96; studies on, 91–93; problems with using, 93–96; influencing heritable conditions through preventive interventions, 96, 100–102; relationship code hypothesis and, 378–379; reappraisal of, 418

Gene expression: significance for behavioral genetics, 51; environmental factors and, 65; change across time and, 99, 219; relationship code hypothesis, 378–379

Gene knockout studies, 99–100

Genes: segregating, 6; independent assortment of, 50; inheritance of, 50; conservative nature of, 50–51; pleiotropism, 204

Genetic cascades, 48, 208

Genetic conditions, possible changes through time, 220

Genetic cross-lagged model, 433–435

Genetic experiments: gene knockout studies, 99–100; on alcoholism, 100; influencing heritable conditions through preventive interventions, 100–102. *See also* Behavioral genetics research; Cross-sectional genetic studies

Genetic influences, 3–6; relationship code hypothesis and, 9, 376–377, 419–420; mediation concept and, 31; genetic cascades and, 48, 208; environment and, 65; on social relationships, 81–84; possible mechanisms of influence on parenting and sibling relationships, 203–205; at early and mid-adolescence, 209–211; analytic principles for detecting across time, 212–220; possible mechanisms for changes across time, 219–220; on change in adolescent adjustment, 220–234, 239–240, 270–271; on stability in adolescent adjustment, 220–234, 235–238; possible mechanisms for stability across time, 236–237; on family relationships, 244–246, 250–258, 259–266; child-effects model and, 311; parent-effects model and, 311, 312–313; synthesis with approaches on environmental factors, 311–317; cross-lagged analysis of, 341; common and specific, 346–352; links to family subsystems, 376–377

Genetic influences, overlapping: overview of, 272–273; changes across time and, 273–274, 307–308; stability and, 274; possible passive mechanisms in, 274–276; possible active/evocative mechanisms in, 277–279; analytic principles for detecting, 280–285; in measures of family systems and adolescent adjustment, 285–287; between mother's parenting and adolescent adjustment, 287–290; between father's parenting and adolescent adjustment, 290–292; between sibling relationships and adolescent adjustment, 292–295; contrasts between parenting and sibling relationships, 295–296; changes from earlier to later adolescence, 296–308

Genetic information: parent-effects model and transmission of, 313–315; encoding in family relationships, 354–358

Genetics: early attitudes of psychosocial research to, 17; fundamental mechanisms in, 50–51; pleiotropism, 204

Genetic specificity: in measures of adolescent adjustment, 346–352; in measures of family relationships, 358–371; family relationship quality and, 362; across family subsystems, 362–363, 385; notions of developmental time and, 363

Girls, identity exploration and, 27

Global Coding Scales, training of coders in, 120–121

Global self-worth: genetic and environmental influences on, 211; change and stability in, 232; mother's negativity and, 288; changes in environmental overlap and, 298; changes in overlapping nonshared environmental influences and, 306, 307; measures of, 307

Good-behavior games, 42–43

Half-sibs, 6; levels of genetic relatedness in step-families, 55, 132; co-residency and, 56, 134, 135

Hall, G. Stanley, 15

Harter Perceived Competence Scale, 116

Hereditary Genius (Galton), 16

Heritability, 5; study of, 44–45; of neuroticism, 67; of scholastic self-confidence, 73–74; model-testing, 74; in parent and child reports of conflict and negativity, 83–84; comparing results from twin and step-family models, 141–142; of the measures of adolescent adjustment, 209; of antisocial behavior, 209–211; of stability, 213–215; of association, 282, 283; of disjunction, 282–283, 358; of change, 297; computing, 452

Heritability estimates: significance to behavioral genetics research, 71; methods in, 74

Hetherington, Mavis, 120

Homophily, 36

Hospitalization, social breakdown syndrome and, 30

Identical twins: biology of, 50, 52–53; frequency of occurrence, 52; twin transfusion syndrome and, 53; nonshared environment studies and, 150, 154–155

Identity exploration: in adolescent development, 26; association studies on, 27

Identity formation, 71–72

Idiosyncratic experiences, 414–415

Incarceration: social breakdown syndrome and, 30; temporal sequencing studies on, 35; as maintenance factor in antisocial behavior, 40

Independent gene assortment, 50

Independent variables, 20, 24, 48

Indicated preventive interventions, 42

Individuation, 405, 406

Infants: effect of siblings on, 162; crying and, 425

Inner conflict, 423

Intelligence, 98. *See also* Cognitive agency

Interaction, statistical, 177–178

Interpretation of Dreams (Freud), 413

Interviewers: data collection by, 121, 126; training of, 121

Intraclass correlation coefficient, 121

Intraclass correlations, 47, 430

IQ, shared environment studies and, 153

Jackson, Don, 16

James, William, 15

Keying, 157

Knockout studies, 99–100

Knowledge of child's activities, father's: adolescent adjustment and, 185, 292; antisocial behavior and, 339–340. *See also* Monitoring, parental

Knowledge of child's activities, mother's: adolescent adjustment and, 183–185, 289–290; antisocial behavior and, 339. *See also* Monitoring, parental

Latent variables, 216

Life-changing events, 241–242, 249

Loehlin, John, 104

Logical principles of research, 8–9, 18–22

Longitudinal studies: logical principles of, 21–22; overview of, 31–32; assortative affiliation studies, 32, 35–37; on adolescent psychopathology, 32–33; temporal sequencing studies, 32–35; third variable concept and, 33–34; developmental precursors and, 34; success of, 34; natural experiments and, 34–35; dropouts from, 38; genetic, 96–99; difficulties in, 317–318

Marital conflict, 245–246

Marital status, effects on drinking behavior, 91

Maternal health, 62

Maturational hypothesis, 398

Measurement errors, data on nonshared environment and, 411–412

Measures: principles of, 6–7; of psychopathology, 113–115; of antisocial behavior, 115; of competence, 115–118; of family subsystems, 118–121; of sociability, 388, 402–403; of autonomy, 403; of positivity, 404–405. *See also* Adolescent adjustment measures; Family systems measures

Mediating studies, 29–31

Mendel, Gregor, 15–16, 20, 50

Model-fitting analyses: bivariate models, 432–433; genetic cross-lagged model, 433–435; fit indexes, 435

Model-testing heritabilities, 74

Moderating studies, 29–31

Monitoring, parental: simple correlations to adolescent adjustment, 184–185; specific and cross-correlations to adolescent adjustment, 188–190; adolescent adjustment and, 289–290, 292; antisocial behavior and, 318–323, 339; possible scenarios for genetic effects across time, 371–375. *See also* Control, parental

Monozygotic twins, 6; biology of, 50; nonshared environment and, 150, 154–155. *See also* Identical twins

Mother-child relationships, 83–84; nonshared environment studies and, 153–154. *See also* Knowledge of child's activities, mother's; Negativity, mother's; Parent-child relationships; Positivity, mother's

Multiple-system nonshared environments, 159–163, 164, 198

Multivariate genetic analyses: overview of, 78; cross-correlations in, 78–79

Mutual coercion, 29–30

Natural experiments: in longitudinal studies, 34–35; twin studies, 52–55; step-family studies, 55–58

Nature-versus-nurture debate, 376

Negativity, father's: adolescent adjustment and, 90, 91, 186–187, 292; changes in overlapping nonshared environmental influences and, 306–307; antisocial behavior and, 339; cognitive agency and,

339; social responsibility and, 340; correlation with mother's negativity, 366–368

Negativity, in families: genetic analysis of parent and child reports in, 83–84; reports of siblings and parents on, 169–170, 171; shared environmental influence on, 251

Negativity, mother's: antisocial behavior and, 89–90, 287; social responsibility and, 89–90, 287–288; adolescent adjustment and, 185–187, 287–288; cognitive agency and, 339; correlation with father's negativity, 366–368

Negativity, parental: antisocial behavior and, 89–90, 279–280, 281–285; social responsibility and, 89–90, 340; simple correlations to adolescent adjustment, 183; cross-correlations to adolescent adjustment, 185–187, 192; genetic influences on, 251–252; change and stability in, 259–262, 264; changes in overlapping nonshared environmental influences and, 306; cognitive agency and, 339; correlation with parental positivity, 363; correlations between, 366–368

Negativity, sibling: forms of, 197–198; simple correlations to adjustment, 199; specific and cross-correlations to adjustment, 199–203; genetic influence on, 254; shared environmental influence on, 255; change and stability in, 265; adolescent adjustment and, 293; inverse correlation with sibling positivity, 366

Neiderhiser, Jenae, 63

Neuroticism: behavioral genetics on, 67–68; nonshared environment and, 68

Nonfamily variables, and nonshared environment, 411–412

Nonshared environment: description of, 5, 65–66; twin transfusion syndrome and, 53; neuroticism and, 68; estimating in cross-sectional genetic studies, 74–75; John Loehlin and, 104; Robert Plomin and, 104; comparing results from twin

and step-family models, 141–143; evidence for, 150–155; family relationships and, 152, 159–163, 164, 248–250, 264, 265; adolescent adjustment and, 155–164, 211, 240–242; single-system hypothesis, 156–158, 163–164; multiple-system hypothesis, 159–163, 164; notions of the unique history of siblings, 160–161, 249; bi-directionality effects in, 164; guidelines for assessing, 164–167; direct effects in, 165; analytic principles for detecting across time, 212–219; antisocial behavior and, 223; cognitive agency and, 225, 227; depressive symptoms and, 225; global self-worth and, 232; social responsibility and, 232; possible explanations for instability in, 241–242; life-changing events as, 241–242, 249; parental conflict/negativity and, 264; conclusions regarding, 406–407; characteristics of, 407–411; nonfamily variables and, 409–411; measurement errors and, 411–412; idiosyncratic experiences and, 414–415; critical reappraisal of, 421–422; psychoanalytic perspective on, 422–423; reappraisal of research methods and, 423–424; psychological development and, 426; computational estimation, 452

Nonshared environment, overlapping: possible mechanisms for, 280; in associations between parenting and adolescent adjustment, 296; significance of changes across time, 306–307

Nonshared environmentality of association, 283

Nonshared environment thesis: development of, 103–105, 147–149; evidence for, 150–155; adolescent development and, 155–164; guidelines for assessment, 164–167; genetic factors and, 206–207; problems with, 309–310; synthesis with approaches on genetic factors, 315–316

Outcome variables, 20
Owness concept, 135–138

Pairwise concordance, 46

Parental influences, 398–399

Parental protection, 320–323

Parental salvage, 205, 320, 321–323, 340

Parent-child relationships, 83–84; mediating and moderating studies in, 29–31; multiple-system nonshared environment hypothesis and, 159–161, 162–163; social referencing concept, 163; single-system nonshared environment hypothesis and, 163–164; effect on sibling relationships, 267, 401–402. *See also* Control, parental; Family relationships; Monitoring, parental; Negativity, parental; Positivity, parental

Parent effects, 88

Parent-effects model, 277–279; overview of, 310–311, 312–313; cross-lagged analysis and, 330, 341–342. *See also* Family-effects model

Parentification, 203

Parenting: association studies and, 25–26; differential, in step-families, 57–58, 135–138; comparisons between siblings, 169–173; detecting shared and nonshared influences of, 177–183; simple correlations to adolescent adjustment, 183–185; specific and cross-correlations to adolescent adjustment, 185–197; possible mechanisms of genetic influence on, 203–205; adolescent adjustment and, 287–292; overlapping influences and, 296; cross-lagged analysis of, 331

Parenting, father's: adolescent adjustment and, 290–292; changes in genetic and environmental overlap across time, 299, 301; cross-lagged analysis of, 331

Parenting, mother's: adolescent adjustment and, 287–290; changes in genetic and environmental overlap across time, 299; cross-lagged analysis of, 331

Parents: responsibility for adolescent development and, 2; ratings of parenting and, 170–171; genetic influences in family subsystems and, 376–377; responses to infant crying, 425

Partialing, 324–326

Passive aggression, 200

Passive-effects model, 343

Passive genetic influences, 82, 245; in overlapping genetic effects, 274–276

Passive model of child adjustment, 85–86

Path coefficients, 216, 218

Path diagrams, 216–218

Peer relationships, 400–401; types of, 35; characteristics of, 35–36; homophily and, 36; assortative affiliation studies and, 36–37

Perceived self-confidence: defined, 71–72; heritability of, 73–74; differential heritability in, 75–78; cross-correlation with sociability, 78–79; verbal IQ and, 80–81

Perceptual bias: as nonshared environmental influence, 150; as shared environmental influence, 248

Personality traits, 153

Pharmacological interventions, 38–39

Phenylketonuria, 17, 236

Placebo effect, 22

Pleiotropism, 204

Plomin, Robert, 104

Positivity: relationship with autonomy, sociability, and shared environment, 403–406; measures of, 404–405

Positivity, father's: adolescent adjustment and, 292; antisocial behavior and, 338; correlation with mother's positivity, 366, 368; shared environment and, 392–393

Positivity, mother's: adolescent adjustment and, 183–184; 288–290; sociability and, 336, 338; antisocial behavior and, 338; correlation with father's positivity, 366, 368; shared environment and, 392–393

Positivity, parental: reports of siblings and parents on, 171, 173–174, 176; simple correlations to adolescent adjustment, 183–184; cross-correlations to adolescent adjustment, 187, 195–196; change and stability in, 264; sociability and, 336,

338; antisocial behavior and, 338; correlation with parental negativity, 363; correlations between, 366, 368; shared environment and, 392–393
Positivity, sibling: genetic influence on, 254; shared environmental influence on, 255; change and stability in, 265; adolescent adjustment and, 293–295; antisocial behavior and, 338; inverse correlation with sibling negativity, 366
Prenatal effects, 62
Preventive interventions: universal, 41, 42–43; selective, 41–42; to influence heritable conditions, 96, 101, 380–384
Primary frames, 157, 158, 163–164
Probandwise concordance, 46
Psychoanalysis: early work on adolescents, 15; perspectives on nonshared environment, 411–416, 422–423
Psychology: early adolescence studies, 15; of representation, 411–416, 422–423
Psychopathology: social breakdown syndrome, 30; temporal sequencing studies and, 32–33; therapeutic interventions and, 39–41; etiologic factors, 40; maintenance factors, 40; treatment milieu and, 40; gene-environment interaction studies on, 92–93; measures of, 113–115; studies on heritability and nonshared environment in, 153; studies on shared environment in, 153–154
Psychosocial research: reconciling with behavioral genetics research, 13–14, 17–18; split with behavioral genetics research, 16; biological determinism and, 17; early attitudes toward genetics, 17; logical principles of research, 18–22; measures of association, 48–49; inferences about genetic influences and, 49–50; measures of adolescent adjustment, 71–72; benefits of genetic studies to, 203. See also Social analyses studies

Quantitative genetics, 44–45. See also Behavioral genetics research

Reassembly of families, 37, 38–39
Receiver-oriented sibling systems, 198, 199–203
Reciprocity, peer groups and, 401
Reciprocity-oriented sibling systems, 198–199, 267–268, 400
Reiss, David, 63
Relationship code, 10; fundamental propositions of, 353–354; encoding of genetic information and, 354–358; genetic effects in parental control/monitoring, 371–375; developmental theory and, 375–384; on links between genetic influences and family subsystems, 376–377; child temperament studies and, 377–378; gene-environment interactions and, 378–379; gene-environment correlations and, 379; evidence needed to support, 379–384; proposed prevention research design, 380–384; summary of, 384; critical reappraisal of, 424–426. See also Family-effects model
Representation, psychology of, 411–416, 422–423
Research: logical principles of, 19–22; methods, critical reappraisal of, 423–424
Ritalin, 38
RMSEA, 435

Schizophrenia: family systems perspective on, 28; social breakdown syndrome and, 30; gene-environment interaction studies on, 93; notions of unique individual history and, 160–161, 249; notions of nonshared environmental influence and, 249
Scholastic self-confidence, 73–74
Schools, ability groupings and, 237–238
Segregating genes, 6
Selective placement, 60, 95–96
Selective preventive interventions, 42–43
Self-acceptance, 26
Self-confidence. See Perceived self-confidence
Self-esteem, 26

Self-representation, 411–416, 422–423

Self-worth, global. *See* Global self-worth

Sequestrational hypothesis, 399–400

Shared environment: description of, 5–6, 66; estimating in cross-sectional genetic studies, 74–75; comparing results from twin and step-family models, 141–143; alcoholism and, 152–153; delinquency and, 152–153; IQ studies and, 153; effects on adolescent adjustment, 211, 238–239; antisocial behavior and, 223; sociability and, 227, 229, 238–239, 388, 389, 402–406; change in autonomy and, 229, 232; family relationships and, 246–248, 251, 266–268; influence on sibling positivity and negativity, 255, 265; parental conflict/negativity and, 256–257, 264; sibling relationships and, 266, 267–268, 397–402; association with autonomy and sociability, 388, 389, 402–406; significance of, 389–390; coordination of family subsystems and, 390–396; critical reappraisal of, 420–421; computational estimation, 452

Shared environment, overlapping: possible mechanisms for, 279–280; in associations between parenting and adolescent adjustment, 296; changes across time, 303–306

Shared environmentality of association, 283

Sibling relationships: siblings' ratings of, 173–174; care-giving and care-receiving behavior, 197–198, 201–203; adolescent adjustment and, 197–199, 292–295, 397–402; forms of, 197–199; specific and cross-correlations to adjustment, 197–199; receiver- and reciprocity-oriented, 198–203, 400; simple correlations to adjustment, 199; possible mechanisms of genetic influence on, 203–205; shared environment and, 266, 267–268, 397–402; parent-child relationships and, 267, 401–402; overlapping influences and, 296; changes in genetic and environmental

overlap across time, 301, 303; changes in overlapping shared environmental influences across time, 303–304; cross-lagged analysis of, 332–333; maturational hypothesis and, 398; sequestrational hypothesis and, 399–400; asymmetries in, 400; peer relationships and, 400–401

Siblings: neuroticism and, 67–68; acquisition of unique social history and, 160–161, 249; comparisons on aspects of family environment, 169–176; shared environmental influences and, 246–248, 266–268; nonshared environmental influences and, 248–250; parental conflict over, 255–257; influence on adolescent adjustment, 397–402; maturational hypothesis and, 398

Sibling studies, 52–58, 106–107; nonshared environment factors and, 65–66. *See also* Adolescent development study; Step-family model; Twin studies

Simple active model of child adjustment, 88

Simple evocative model of child adjustment, 86–87, 88

Single-system nonshared environments, 156–158, 163–164, 198

Sociability: perceived self-confidence and, 78–79; change and stability in, 227, 229; shared environment and, 227, 229, 238–239, 388, 389, 402–406; positivity and, 289, 295, 336, 338, 403–406; changes in overlapping shared environmental influences across time, 304–306; antisocial behavior and, 347–349; measures of, 388, 402–403; autonomy and, 402–406

Social analyses studies: overview of, 22–23; association studies, 23–29; mediating and moderating studies, 29–31. *See also* Psychosocial research

Social breakdown syndrome, 30

Social constructions: by families, 395–396; in families with disabled children, 418–419

Social environment: comparing between siblings, 169–176; parent-effects model and, 311

Social reality, family constructions of, 395–396, 418–419

Social referencing, 163

Social relationships: significance to adolescent development, 3, 9–10; role of genetic factors in, 81–84; gene-environment correlation and, 84–91. *See also* Family relationships

Social responsibility: mother's negativity and, 89–90, 287–288; change and stability in, 232; mother's positivity and, 289; sibling relationships and, 293; sibling positivity and, 295; father's negativity and, 340; antisocial behavior and, analysis of genetic influences on, 349–351

Specific correlations: described, 166, 179, 180; applied to sibling relationships, 197–100; possible genetic mechanisms in, 204. *See also* Covariance model

Stability: considered in longitudinal genetic studies, 97–98; analytic principles for detecting, 212–220; concept of, 213; environmentality and heritability of, 213–215; coefficient of, 218; in adolescent adjustment, 220–234, 235–239; in family relationships, factors influencing, 257–265; significant presence of, 318; of family relationship measures, 371

Stability coefficient, 218

Statistical interaction, 31

Step-families: instability of, 107; characteristics of, 133

Step-family model, 106–107; levels of sibling genetic relatedness in, 55, 132; co-residency and, 55–57, 133–135, 481–482; differential parenting behavior and, 57–58, 135–138; groupings of siblings in, 110; characteristics of step-families and, 133; differential assortative mating and, 138–139; compared to twin studies, 139–143; nonshared environment concept and, 150–151

Substance abuse, 62

Support, in families: reports of siblings and parents on, 171, 173–174, 176. *See also* entries *at* Positivity

Support, sibling: forms of, 197–198; simple correlations to adjustment, 199; covariance model analysis of, 201–203

Teen parenthood, 24, 25

Temperament, children's, 377–378

Temporal sequencing studies: on adolescent psychopathology, 32–33; third variable concept and, 33–34; developmental precursors and, 34; natural experiments and, 34–35

Tests of fit, 330

Therapeutic interventions: overview of, 37, 39; factors limiting, 39–41; third variable problem in, 41

Third variable problem, 33–34; in therapeutic interventions, 41; in assessing gene-environment correlation, 84–88

Threshold effects, 178

Titchener, Edward, 15

Transmission of genetic information, 313–315

Trauma, 161

Triangular decomposition, 434

Twins/Twinning, 6; biology of, 50, 52–53; frequency of, 52; twin transfusion syndrome and, 53; shared environment concept and, 66; closeness and, 254, 255; determining zygosity of, 110

Twin studies: equal-environment assumption and, 16, 53–55; advantages of, 53; limitations of, 53–55; gene-environment interaction studies and, 91–92; preventive interventions and, 101; step-family model compared to, 139–143; nonshared environment and, 150, 151, 154–155

Univariate results, 452

Universal preventive interventions, 41, 42–43

Variables: dependent and independent, 20, 48; latent, 216

Variance, 208
Verbal IQ, 80–81

Warmth, in families: reports of siblings and parents on, 171, 173–174, 176. *See also entries at* Positivity

Warmth, sibling: forms of, 197–198; simple correlations to adjustment, 199; covariance model analysis of, 201–203
Women: identity exploration and, 27; drinking behavior, 91
World views, of families, 395–396